ADVANCES IN
GEOPHYSICS

VOLUME 20

Contributors to This Volume

E. L. KOSCHMIEDER
BARRY SALTZMAN
SURENDRA N. TIWARI
J. VAN MIEGHEM

Advances in
GEOPHYSICS

Edited by
BARRY SALTZMAN

Department of Geology and Geophysics
Yale University
New Haven, Connecticut

VOLUME 20

1978

Academic Press • New York San Francisco London
A Subsidiary of Harcourt Brace Jovanovich, Publishers

Copyright © 1978, by Academic Press, Inc.
ALL RIGHTS RESERVED.
NO PART OF THIS PUBLICATION MAY BE REPRODUCED OR
TRANSMITTED IN ANY FORM OR BY ANY MEANS, ELECTRONIC
OR MECHANICAL, INCLUDING PHOTOCOPY, RECORDING, OR ANY
INFORMATION STORAGE AND RETRIEVAL SYSTEM, WITHOUT
PERMISSION IN WRITING FROM THE PUBLISHER.

ACADEMIC PRESS, INC.
111 Fifth Avenue, New York, New York 10003

United Kingdom Edition published by
ACADEMIC PRESS, INC. (LONDON) LTD.
24/28 Oval Road, London NW1 7DX

LIBRARY OF CONGRESS CATALOG CARD NUMBER: 52-12266

ISBN 0-12-018820-1

PRINTED IN THE UNITED STATES OF AMERICA

CONTENTS

LIST OF CONTRIBUTORS	vii
PREFACE	ix

Models for Infrared Atmospheric Radiation
SURENDRA N. TIWARI

1. Introduction	1
2. Absorption by Spectral Lines	3
3. Band Absorption	23
4. Evaluation of Transmittance and Integrated Absorptance of Selected Infrared Bands	59
5. Upwelling Atmospheric Radiation	69
6. Concluding Remarks	78
List of Symbols	79
References	80

Scale Analysis of Large Atmospheric Motion Systems In All Latitudes
J. VAN MIEGHEM

1. Introduction, Generalities, and Statement of the Problem	87
2. Scale Parameters	91
3. Characterization of an Atmospheric Large-Scale Motion System	97
4. Scale Analysis of the Continuity Equation	100
5. Scale Analysis of the Equation of Vertical Motion	102
6. Scale Analysis of the Equations of Horizontal Motion	103
7. Scale Analysis of the Equation of Adiabatic Transformations	106
8. Order of Magnitude of the Divergence Δ of the Horizontal Wind	109
9. Scale Analysis of the Vorticity Equation	113
10. Scale Analysis of the Equation of the Horizontal Wind Divergence	116
11. Discussion of the Basic Scale Equations	117
12. Scale Analysis of the Equation of Balance of Kinetic Energy	128
References	130

Symmetric Circulations of Planetary Atmospheres
E. L. KOSCHMIEDER

1. Introduction	131
2. Convection on a Nonuniformly Heated Sphere	134

3. Azimuthal Disturbances.. 163
4. Nonlinearity... 170
5. Summary and Conclusions..................................... 175
List of Symbols... 176
References... 177

A Survey of Statistical–Dynamical Models of the Terrestrial Climate

Barry Saltzman

1. Introduction... 184
2. Fundamental Equations Governing the Terrestrial Climatic System   185
3. Climatic Averaging... 193
4. Climate Models—An Overview................................. 206
5. The Vertically Integrated, Thermodynamic Models............... 225
6. Momentum Models 250
7. Modeling the Evolution of Climate............................. 267
Appendix. A Simple Sea Ice–Ocean Temperature
Oscillator Model 281
List of Symbols 290
References 295

Subject Index.. 305

LIST OF CONTRIBUTORS

Numbers in parentheses indicate the pages on which the author's contributions begin.

E. L. KOSCHMIEDER, *Atmospheric Science Group, College of Engineering, University of Texas at Austin, Austin, Texas 78712* (131)

BARRY SALTZMAN, *Department of Geology and Geophysics, Yale University, New Haven, Connecticut 06520* (183)

SURENDRA N. TIWARI, *School of Engineering, Old Dominion University, Newport News, Virginia 23606* (1)

J. VAN MIEGHEM, *Avenue des Armures, 34, B-1190 Brussels, Belgium* (87)

PREFACE

The rapid growth of geophysical research over the past 25 years has led to an increasing need for syntheses, summaries, and interpretive reductions of the profusion of emerging results. Four examples of such reviews in the area of atmospheric science appear in this volume, covering aspects of the theory of radiation, planetary convection, scaling of dynamic processes, and climate modeling. It is our hope that future volumes will emphasize other areas of interest, in keeping with the role of *Advances in Geophysics* as a forum for reviews of recent progress in understanding all of the geophysical domains ranging from the cores of the celestial bodies to their outer exospheres. It is a pleasure, as the new editor of this serial publication, to invite contributions from scientists in all of the many geophysical disciplines that are embraced by this vast realm. Although such contributions would naturally be mainly of a review nature, this need not preclude the inclusion of substantial new material that can lend a creative and critical aspect to the reduction process and convey a sense of movement in new directions.

BARRY SALTZMAN

MODELS FOR INFRARED ATMOSPHERIC RADIATION

SURENDRA N. TIWARI

School of Engineering
Old Dominion University, Newport News, Virginia

1. Introduction 1
2. Absorption by Spectral Lines 3
   2.1 Radiative Transmittance by Spectral Lines 8
   2.2 Absorption of an Isolated Spectral Line in an Infinite Spectral Interval 11
   2.3 Absorption of a Spectral Line in a Finite Spectral Interval 17
   2.4 Absorption of an Overlapping Line in a Finite Spectral Interval 21
3. Band Absorption 23
   3.1 Limiting Forms of the Total Band Absorptance 24
   3.2 Narrow Band Models 28
   3.3 Wide Band Models 43
   3.4 Band Absorptance Correlations 51
   3.5 Comparison of Wide Band Absorptance Results 54
   3.6 Band Emissivity (Total Emissivity) 57
4. Evaluation of Transmittance and Integrated Absorptance of Selected Infrared Bands   59
   4.1 Transmittance Models and Computational Procedures 60
   4.2 Transmittance of Selected Infrared Bands 62
   4.3 Integrated Absorptance of Selected Infrared Bands 65
5. Upwelling Atmospheric Radiation 69
   5.1 Procedure for Calculating the Upwelling Radiance 71
   5.2 Results of Model Calculations 72
6. Concluding Remarks 78
   List of Symbols 79
   References 80

1. INTRODUCTION

The study of radiative transmission in real (nonhomogeneous) atmospheres requires a detailed knowledge of the atmospheric constituents that absorb and emit significantly in the spectral range of interest. One of the important quantities required for calculating the atmospheric transmittance is the absorption coefficient of the atmospheric constituents. An accurate model for the spectral absorption coefficient is of vital importance in the correct formulation of the radiative flux equations that are employed for the reduction of data obtained from either direct or remote measurements.

A systematic representation of the absorption by a gas, in the infrared, requires the identification of the major infrared bands and evaluation of the line parameters of these bands. The line parameters depend upon the temperature, pressure, and concentration of the absorbing molecules, and, in general, these quantities vary continuously along a nonhomogeneous path through the atmosphere. Even though it is quite difficult to reproduce the real nonhomogeneous atmosphere in the laboratory, considerable efforts have been expended in obtaining the absorption coefficients of important atmospheric constituents. With the availability of high-resolution spectrometers, it is now possible to determine the line positions, intensities, and half-widths of spectral lines quite accurately (McClatchey et al., 1972, 1973; Selby and McClatchey, 1975; Ludwig et al., 1973; Science Application Incorporated, 1973). As a result, absorption by the strong infrared bands of gases like CO, CO_2, N_2O, H_2O, CH_4, NH_3, and O_3 are now known quite well.

In theoretical calculations of transmittance (or absorptance) of a band, a convenient line or band model, for the variation of the spectral absorption coefficient, is used. High spectral resolution measurements make it necessary to employ line-by-line models for transmittance calculations. If, however, the integrated signals are measured over a relatively wide spectral interval, then one could employ an appropriate band model. The line models usually employed in the study of atmospheric radiation are Lorentz, Doppler, and combined Lorentz–Doppler (Voigt) line profiles. A complete formulation (and comparison) of the transmittance (and absorptance) by these lines, in an infinite and a finite spectral interval, is given in Tiwari (1973), Tiwari and Batki (1974), and Tiwari and Reichle (1974). The band models available in the literature are the narrow band models (such as Elsasser, statistical, random-Elsasser, and quasi-random) and the wide band models (such as coffin, modified box, exponential, and axial). The expressions for wide band absorptance are obtained from the general formulations of the narrow band models. In radiative transfer analyses, the use of band models results in a considerable reduction in computational time. Essential information on various narrow band models is available in Elsasser (1942), Plass (1958, 1960), Wyatt et al. (1962), Goody (1964a), Kunde (1967), Gupta and Tiwari (1975), and Tiwari and Batki (1975) and on wide band models in Tiwari (1976), Tiwari and Batki (1975), Tien (1968), Cess and Tiwari (1972), and Edwards (1976). The most appropriate model for atmospheric application is the quasi-random narrow band model which is discussed in detail in Wyatt et al. (1962), Kunde (1967), and Gupta and Tiwari (1975).

The earth's surface, with its temperature in the vicinity of 300°K, emits like a black body from the near to the far-infrared region of the spectrum. The emission in the infrared range (between 2 and 20 μm) is particularly important because most of the minor atmospheric constituents (i.e., CO_2, N_2O, H_2O, CO, CH_4, NH_3, etc.) absorb and emit this spectral region. The upwelling infrared radiation from the earth's atmosphere, therefore, consists of the modulated surface radiation and the radiation from the atmosphere. This radiation carries the spectral signature of all the minor atmospheric constituents, among which gases such as CO, CH_4, and NH_3 are called the atmospheric pollutants. Ludwig et al. (1973) have explored the possibilities of measuring the amount of atmospheric pollutants through remote sensing. An important method of measuring the pollutant concentration by remote sensing is the passive mode (also called the nadir experiment) in which the earth-oriented detector receives the upwelling atmospheric radiation. The near-infrared region is particularly suitable for passive mode measurements simply because the radiation in this region is practically free from the scattering effects. Radiation in the visible and ultraviolet regions is severely affected by the scattering processes which make meaningful passive mode measurements impossible.

The purpose of this study is to review various line and band models and to present analysis procedure for calculating the atmospheric transmittance and upwelling radiance. Various expressions for absorption by different line and band models are presented in Sections 2 and 3. Theoretical formulations of atmospheric transmittance are given in Section 4, where homogeneous path transmittances are calculated for selected infrared bands. The basic equations for calculating the upwelling atmospheric radiance are presented in Section 5, where model calculations are made to study the effects of different interfering molecules, water vapor profiles, ground temperatures, and ground emittances on the upwelling radiance.

2. Absorption by Spectral Lines

In order to describe the infrared absorption characteristics of a radiating molecule it is necessary to consider the variation of the spectral absorption coefficient for a single line. In general, for a single line centered at the wave number ω_j, this is expressed as

(2.1) $$\kappa_{\omega j} = S_j f_j(\omega, \gamma_j)$$

where S_j is the intensity of the jth spectral line and is given by

(2.2) $$S_j = \int_{-\infty}^{\infty} \kappa_{\omega j} \, d(\omega - \omega_j)$$

The line intensity may be described in terms of the molecular number density and Einstein coefficients, i.e., it depends upon the transition probabilities between the initial and final states and upon the populations of these states. For a perfect gas it may be shown that S_j is a function solely of temperature. The quantity $f_j(\omega, \gamma_j)$ is the line shape factor for the jth spectral line. It is a function of the wave number ω and the line half-width γ_j and is normalized on $\omega - \omega_j$ such that

$$(2.3) \qquad \int_{-\infty}^{\infty} f_j(\omega - \omega_j)\, d(\omega - \omega_j) = 1$$

Several approximate line profiles have been described in the literature. The most commonly used profiles are rectangular, triangular, Lorentz, Doppler, or Voigt (combined Lorentz and Doppler) profiles. The study of line shapes and line broadening is an active research field. For various reviews on the subject, reference should be made to Goody (1964a), Mitchell and Zemansky (1934), Penner (1959), Baranger (1962), Allen (1963), Griem (1964), Cooper (1966), Jefferies (1968), Kondratyev (1969), and Armstrong and Nicholls (1972). Lorentz, Doppler, and Voigt profiles are of special interest in the atmospheric studies, and these are discussed in some detail here.

The line profile usually employed for studies of infrared radiative transfer in the earth's atmosphere is the Lorentz pressure-broadened line shape for which the shape factor is such that the expression for absorption coefficient is found to be

$$(2.4) \qquad \kappa_{\omega j} = S_j f_j(\omega, \gamma_L) = S_j \gamma_L / \{\pi[(\omega - \omega_j)^2 + \gamma_L^2]\}$$

where γ_L is the Lorentz line half-width. From simple kinetic theory it may be shown that γ_L varies with pressure and temperature according to the relation

$$(2.5) \qquad \gamma_L = \gamma_{L0}(P/P_0)^m(T_0/T)^n$$

where γ_{L0} is the line half-width corresponding to a reference temperature T_0 and a pressure P_0. The values of m and n depend, in general, on the collision parameters and on the nature of the molecules. A discussion on the variation of γ_L with P and T is given in Penner (1959), Yamamoto et al. (1969), Ely and McCubbin (1970), and Tubbs and Williams (1972). The value of $m = 1$ and $n = 0.5$ is usually employed for most atmospheric studies.

The maximum absorption coefficient occurs at $\omega = \omega_j$ and is given by the expression

$$(2.6) \qquad (\kappa_{\omega j})_{\omega = \omega_j} = S_j / \pi \gamma_L$$

The variation of κ_ω over a specific wave number range containing n independent lines is given by

$$(2.7) \qquad \kappa_\omega = \sum_{j=1}^{n} \kappa_{\omega j}$$

For Lorentzian line profiles, Eq. (2.7) can be expressed as

$$(2.8) \qquad \kappa_\omega = \sum_j \kappa_{\omega j} = \sum_j S_j \gamma_{Lj} / \{\pi[(\omega - \omega_j)^2 + \gamma_{Lj}^2]\}$$

Note that for the Lorentz line profile, γ_L varies linearly with the pressure. Thus, in a spectral interval containing many lines, the discrete line structure will be smeared out at sufficiently high pressure.

For Doppler-broadened lines, the absorption coefficient (Mitchell and Zemansky, 1934; Penner, 1959) is given by the relation

(2.9a) $$\kappa_{\omega j} = S_j f_j(\omega, \gamma_D)$$

where

(2.9b) $$f_j(\omega, \gamma_D) = (1/\gamma_D)(\ln 2/\pi)^{1/2} \exp[-(\omega - \omega_j)^2(\ln 2/\gamma_D^2)]$$

$$\gamma_D = (\omega_j/c)(2kT \ln 2/m)^{1/2}$$

In this equation γ_D represents the Doppler half-width, c is the speed of light, κ is the Boltzmann constant, and m is the molecular mass. Doppler broadening is associated with the thermal motion of molecules. From Eqs. (2.9) it is clear that the Doppler width depends not only on temperature but also on molecular mass and the location of the line center. For certain atmospheric conditions, therefore, the Doppler and Lorentz widths may become equally important for a particular molecule radiating at a specific frequency. For comparable intensities and half-widths, however, the Doppler line has more absorption near the center and less in the wings than the Lorentz line (Fig. 1).

For radiative transfer analyses involving gases at low pressures (upper atmospheric conditions), it becomes imperative to incorporate the combined influence of the Lorentz and the Doppler broadening. The shape factor for

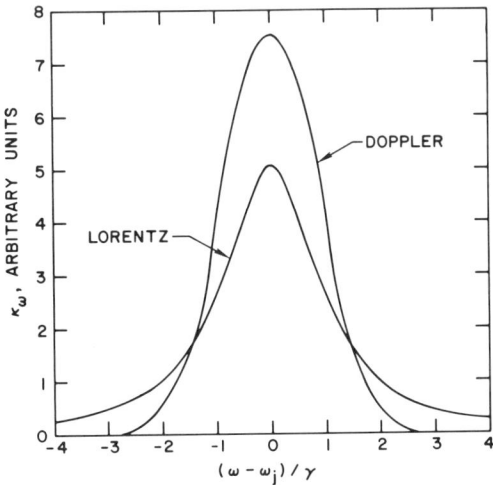

FIG. 1. Comparison of Lorentz and Doppler line profiles having equal line intensities.

the combined profile (Mitchell and Zemansky, 1934; Penner, 1959) is given by

(2.10) $$\kappa_{\omega j} = S_j f_j(a, v) = \kappa_0 K(a, v)$$

where

$$f_j(a, v) = (1/\gamma_D)(\ln 2/\pi)^{1/2} K(a, v)$$

$$K(a, v) = (a/\pi) \int_{-\infty}^{\infty} \{\exp(-t^2)/[a^2 + (v - t)^2]\} \, dt$$

$$\kappa_0 = (S_j/\gamma_D)(\ln 2/\pi)^{1/2} = (S_j/\sqrt{\pi})(a/\gamma_L)$$

$$v = [(\omega - \omega_j)/\gamma_D](\ln 2)^{1/2} = (\omega - \omega_j)(a/\gamma_L)$$

$$a = [(\gamma_c + \gamma_N)/\gamma_D](\ln 2)^{1/2}$$

The function $K(a, v)$ defined in Eq. (2.10) is called the Voigt function, γ_c is the collisional broadened half-width, and γ_N is the natural line half-width. Usually γ_N can be neglected in comparison to γ_c. At very low pressures (upper atmospheric conditions), however, the contribution from γ_N becomes significant. The combined effect of γ_c and γ_N is frequently represented by γ_L, and the parameter a is interpreted as the ratio of the Lorentz and Doppler half-widths. As the pressure becomes small, γ_L approaches zero and Eq. (2.10) reduces to the absorption coefficient for the Doppler broadened lines which may now be expressed as

(2.11) $$\kappa_{\omega j} = \kappa_0 \exp(-v^2)$$

On the other hand, for large pressures, the quantity γ_L/γ_D becomes large and Eq. (2.10) reduces to the Lorentzian case, Eq. (2.4). In other words, the Voigt profile assumes the Lorentzian shape in the limit of large v and reduces to the Doppler profile for small a. The Voigt function is referred to as the reduced absorption coefficient, with κ_0 representing its dimensional constant. The origin and properties of the Voigt function and methods of computing it are reviewed in some detail in Armstrong and Nicholls (1972) where it is shown that $K(a, v)$ is the real part of the complex error function.

Several alternate and approximate forms of the Voigt function, and a number of tabulations of these forms, are available in Mitchell and Zemansky (1934), Penner (1959), Armstrong and Nicholls (1972), Reiche (1913), Born (1933), Van de Hulst and Reesinck (1947), Harris (1948), Penner and Kavanagh (1953), Plass and Fivel (1953), Posener (1959), Fried and Conte (1961), Hummer (1964a,b, 1965), Finn and Mugglestone (1965), Young (1965a,b), and Armstrong (1967). With the aid of computer programs developed by Hummer (1964a,b, 1965), Young (1965a,b), Armstrong (1967), Chiarella and Reichel (1968), and Gautschi (1969, 1970), it is now possible to calculate the Voigt function (for an extended range of parameters) to an accuracy of better than six significant figures.

A simple closed form approximation to the Voigt profile that is valid over a useful range of parameters is given, in terms of the present nomenclature (Van de Hulst and Reesinck, 1947; Posener, 1959; Whiting, 1968), by

(2.12a) $\quad \kappa_{\omega j}/\kappa_{\omega 0} = [1 - (\gamma_L/\gamma_V)] \exp\{-11.088[(\omega - \omega_j)/\gamma_V]^2\}$
$\quad\quad\quad\quad + (\gamma_L/\gamma_V)/\{1 + 16[(\omega - \omega_j)/\gamma_V]^2\}$

where Voigt half-width is expressed in terms of γ_L and γ_D as

(2.12b) $\quad\quad\quad \gamma_V = (\gamma_L/2) + [(\gamma_L^2/4) + \gamma_D^2]^{1/2}$

and

(2.12c) $\quad\quad \kappa_{\omega 0} = S_j/\{\gamma_V[1.065 + 0.447(\gamma_L/\gamma_V) + 0.058(\gamma_L/\gamma_V)^2]\}$

This form is very convenient for numerical computation, and it matches the Voigt profile within 5% under worst conditions. Generally the error is within 3%, with maximum errors occurring near zero pressures. A somewhat better approximation for the Voigt function is suggested by Kielkopf (1973).

The radiative transmittance at a single wave number is given by the relation

(2.13) $\quad\quad\quad\quad \tau_{\omega j} = \exp\left(-\int_0^X \kappa_{\omega j}\, dX\right)$

where $X = \int_0^\ell \rho_a\, d\ell$ is the mass of the absorbing gas per unit area, $\kappa_{\omega j}$ is the mass absorption coefficient for the jth spectral line, ℓ is the length measured along the direction of the path which makes an angle θ with the vertical, and ρ_a is the density of the absorbing gas. For a homogeneous path, Eq. (2.13) becomes

(2.14) $\quad\quad\quad\quad\quad \tau_{\omega j} = \exp(-\kappa_{\omega j} X)$

The total absorption of a single line, in an infinite spectral interval, is given by

(2.15) $\quad\quad\quad\quad A_j = \int_{-\infty}^{\infty} (1 - \tau_{\omega j})\, d(\omega - \omega_j)$

where ω_j represents the wave number at the line center of the jth spectral line. For a homogeneous path, this can be expressed as

(2.16) $\quad\quad\quad A_j = \int_{-\infty}^{\infty} [1 - \exp(-\kappa_{\omega j} X)]\, d(\omega - \omega_j)$

For small values of the quantity $\kappa_{\omega j} X$, Eq. (2.16) reduces to an important limiting form which is independent of any spectral model used for $\kappa_{\omega j}$. This is the conventional optically thin (or linear) limit in radiative transfer. This limit is obtained by expanding the exponential in Eq. (2.16) and retaining only the first two terms in the series such that

$$(2.17) \qquad A_j = X \int_{-\infty}^{\infty} \kappa_{\omega j} \, d(\omega - \omega_j) = X S_j$$

Another limit, which does depend upon the particular model employed for $\kappa_{\omega j}$, is the square-root limit or the strong line approximation for which the total absorption occurs in the vicinity of the line center. To find expressions for absorptance in this limit, it is required that $(\kappa_{\omega j} x) \gg 1$ for $\omega = \omega_j$.

The average absorption $\bar{A}$ of a single line, which is a member of a group of lines, is given by

$$(2.18) \qquad \bar{A}_j = \frac{1}{d} \int_{-\infty}^{\infty} [1 - \exp(-\kappa_{\omega j} X)] \, d(\omega - \omega_j)$$

where d is the average spacing between lines. This is related to the so-called *equivalent width* of a line $W_j(X)$ by $W_j(X) = \bar{A}_j d$, where the expression for $W_j(X)$ is exactly the same as given by Eq. (2.16). Thus,

$$(2.19) \qquad W_j(X) = A_j(X) = \int_{-\infty}^{\infty} A_j(\kappa_{\omega j} X) \, d(\omega - \omega_j) = \bar{A}_j d$$

The equivalent width is interpreted as the width of a rectangular line (whose center is totally absorbed) having the same absorption area as that of the actual line.

The mean transmittance of a single line, in a finite wave number interval $D = 2\delta$, may be expressed as

$$(2.20) \qquad \bar{\tau}_{j, D} = \frac{1}{2\delta} \int_{-\delta}^{+\delta} \exp(-\kappa_{\omega j} X) \, d(\omega - \omega_j)$$

where δ is the wave number interval from the center of the line. The mean absorption over this interval, therefore, becomes

$$(2.21) \qquad \bar{A}_{j, D} = 1 - \bar{\tau}_{j, D} = \frac{1}{\delta} \int_{0}^{\delta} [1 - \exp(-\kappa_{\omega j} X)] \, d(\omega - \omega_j)$$

Note that $\bar{\tau}_{j, D}$ and $\bar{A}_{j, D}$ are in nondimensional form.

2.1. Radiative Transmittance by Spectral Lines

For a homogeneous atmosphere, the radiative transmittance of a line with Lorentz profile is obtained by combining Eqs. (2.4) and (2.14) as

$$(2.22) \qquad \tau_{\omega j} = \tau_L(\omega) = \exp[-2x/(y^2 + 1)] = \tau_L(x, y)$$

where

$$x = S_j X/2\pi\gamma_L \qquad y = (\omega - \omega_j)/\gamma_L$$

It should be noted that, for large y (i.e., away from the line center), Eq. (2.22) approaches to unity for all x while for small x it approaches to unity for all y.

The transmittance of a Doppler-broadened line is obtained by combining Eqs. (2.11) and (2.14) as

(2.23a) $$\tau_{\omega j} = \tau_D(\omega) = \exp[-x_D \exp(-v^2)] = \tau_D(v, x_D)$$

where

$$x_D = \kappa_0 X = [(S_j/\gamma_D)(\ln 2/\pi)^{1/2}]X$$

represents the optical path at the line center. For large v (i.e., away from the line center), the transmittance approaches a value of unity, while in the vicinity of the line center it may be expressed by

(2.23b) $$\tau_D(\omega) = \exp(-x_D)$$

As in the case of the Lorentz line profile, $\tau_D(\omega)$ also approaches a value of unity in the linear limit.

The transmittance of a combined Lorentz-Doppler (Voigt) line profile is obtained by combining Eqs. (2.10) and (2.14) as

(2.24) $$\tau_{\omega j} = \tau_V(\omega) = \exp(-x_D K(a, v)) = \tau_V(a, v, x_D)$$

Note that the transmittance of a Voigt line profile also approaches unity in the linear limit. It can be shown that Eq. (2.24) reduces to the Lorentzian case for large a and to the Doppler case in the limit of small a.

The transmittance by the Lorentz line profile, Eq. (2.22), can be expressed in terms of the quantities x_D, a, and v as

(2.25) $$\tau_L(\omega) = \exp\{-x_D[(a/\sqrt{\pi})/(v^2 + a^2)]\} = \tau_L(a, v, x_D)$$

For transmittance at the line center, this can be written as

(2.26) $$\tau_L(a, x_D) = \exp(-x_D/a\sqrt{\pi})$$

Equation (2.25) is a convenient expression for comparing the results with the transmittance of the Doppler and Voigt line profiles.

The transmittance of the three line profiles (Lorentz, Doppler, and Voigt) at the line center are illustrated in Fig. 2 for various values of the parameter a. As would be expected, for $a = 0 \to 0.1$, the Voigt line transmittance is analogous to that given by the Doppler line profile. For values of $a \geq 5$, the transmittance by Lorentz and Voigt lines is identical for all path lengths.

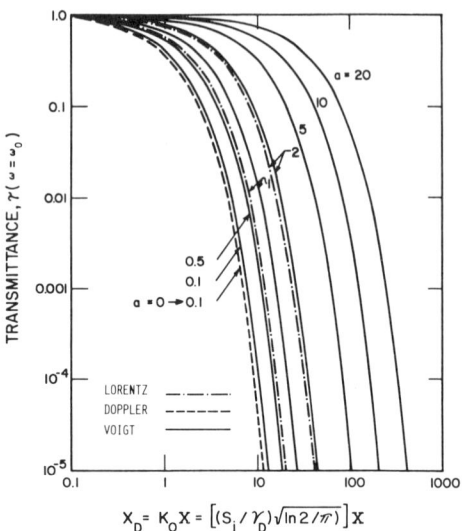

Fig. 2. Transmittance of Lorentz, Doppler, and Voigt lines evaluated at the line center.

Comparisons of the transmittance by the three line profiles are also shown in Figs. 3 and 4 for $x_D = 10$ and for values of a equal to 0.1 and 1, respectively. It should again be noted that at $a = 0.1$, the transmittance of a Voigt line can be approximated quite accurately by the transmittance of a Doppler line. At $a = 1$, however, the transmittance by the Lorentz line provides a better approximation for the Voigt line transmittance.

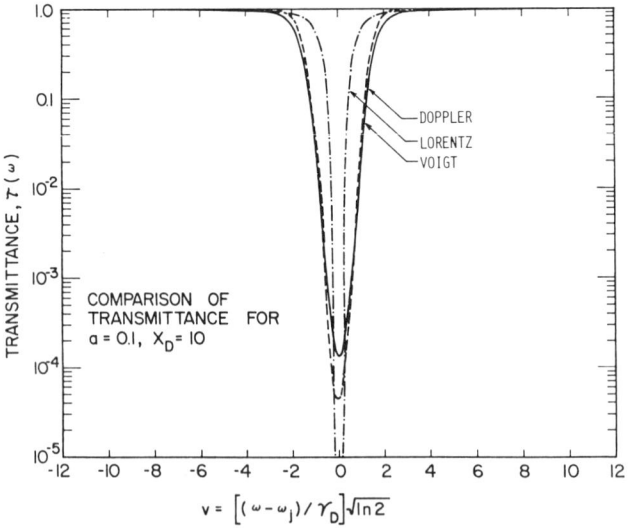

Fig. 3. Comparison of transmittance for $a = 0.1$ and $x_D = 10$.

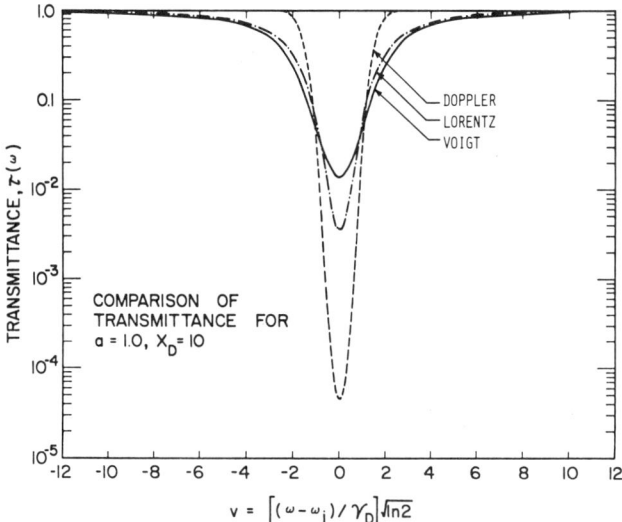

FIG. 4. Comparison of transmittance for $a = 1.0$ and $x_D = 10$.

2.2. Absorption of an Isolated Spectral Line in an Infinite Spectral Interval

For homogeneous atmospheric paths, the absorption of an isolated spectral line of Lorentz shape (in an infinite spectral interval) A_L may be obtained by combining Eqs. (2.4) and (2.16) as

$$(2.27) \quad A_L = 2\gamma_L \int_0^\infty \{1 - \exp[-2x/(y^2 + 1)]\}\, dy$$

where x and y are defined in Eq. (2.22). An exact solution of this equation is found in terms of Ladenberg-Reiche function $L(x)$ (Elsasser, 1942; Goody, 1964a) as

$$(2.28a) \quad A_L/2\pi\gamma_L = L(x) = x\,\exp(-x)(I_0(x) + I_1(x))$$

where I_0 and I_1 are the Bessel functions of imaginary arguments. By multiplying Eq. (2.4) by the mass of the absorbing gas X it can be shown that the quantity x represents one-half the optical path at the line center. In the linear limit (small x), Eq. (2.16) reduces to

$$(2.28b) \quad A_L = 2\pi\gamma_L x = S_j X$$

while in the square-root limit (large x), it becomes

$$(2.28c) \quad A_L = 2\gamma_L(2\pi x)^{1/2} = 2(S_j \gamma_L X)^{1/2}$$

For further discussions on absorption by the Lorentz lines, one should refer to Tiwari (1973), Tiwari and Batki (1974), Tiwari and Reichle (1974), Plass (1958), Goody (1964a), and Penner (1959). For these lines, absorption calculations for nonisothermal paths have been carried out by Simmons (1967), Yamamoto and Aida (1970), and Cogley (1970).

Several approximate solutions of Eq. (2.27) are suggested in the literature. These are:

(2.29a) $\quad A_L/2\pi\gamma_L = (2x/\pi)^{1/2}\{1 - \exp[-(\pi x/2)^{1/2}]\}$ (Tien, 1966, 1968)

(2.29b) $\quad A_L/2\pi\gamma_L = \{(2x/\pi)[1 - \exp(-\pi x/2)]\}^{1/2}$
(Varanasi, cited in Tien, 1966)

(2.29c) $\quad A_L/2\pi\gamma_L = 1/[1 + (\pi x/2)]^{1/2}$ (Rodgers, cited in Tien, 1968)

(2.29d) $\quad A_L/2\pi\gamma_L = x/[1 + (\pi x/2)]^{1/2}$ (Goldman, 1968)

(2.29e) $\quad A_L/2\pi\gamma_L = x/[1 + (\pi x/2)]^{1/2\alpha}$ (Goldman, 1968)

The value of α in Eq. (2.29e) range between $1 < \alpha < 3/2$. For $\alpha = 1$, Eq. (2.29e) reduces to Eq. (2.29d). A value of $\alpha = 5/4$ is recommended by Goldman [58] for better approximation. Each equation in Eqs. (2.29) reduces to the correct asymptotic limits. Over the entire range of x, however, the approximations listed in Eqs. (2.29) agree with the exact solution, Eq. (2.28a), by a varying degree of accuracy. As discussed by Goldman, a maximum error of about 17 % occurs in using Eq. (2.29a) while less than 8 % errors are encountered in using Eqs. (2.29b) and (2.29d). Equation (2.29c) agrees with the exact solution within 3 %, and for a value of $\alpha = 5/4$, Eq. (2.29e) gives accurate results within 1 % over the entire range of x. In calculating the total absorptance over a band pass, by employing the line-by-line model, a tremendous amount of computational time is saved if an appropriate form of Eq. (2.29) is used.

The absorption of a Doppler broadened line A_D is obtained by combining Eqs. (2.15) and (2.23a) as

(2.30a) $\quad A_D^* = A_D/A_n = \int_0^\infty \{1 - \exp[-x_D \exp(-v^2)]\}\, dv$

where

$$A_n = 2\gamma_D/(\ln 2)^{1/2}$$

As $x_D \to 0$ (linear limit), Eq. (2.30a) reduces to

(2.30b) $\quad A_D^* = (\sqrt{\pi}/2)x_D \quad$ or $\quad A_D = S_j X$

and in the limit of large x_D it yields

(2.30c) $\quad A_D^* = (\ln x_D)^{1/2}$

Equation (2.30b) is identical to the result for the Lorentz profile in the linear limit (Eq. 2.28b). Since the linear limit can be obtained from Eq. (2.15), independent of line shape, the absorption by any line profile should reduce to the same expression in this limit. From Eq. (2.30c), it should be noted that in the limit of large x_D the absorption increases very slowly with the amount of absorbing gas. For further discussions on absorption by the Doppler lines, reference should be made to Tiwari (1973), Tiwari and Reichle (1974), Tiwari and Batki (1974), Plass (1958), Goody (1964a), Penner (1959), and Yamada (1967).

For conditions where both the Doppler and the Lorentz broadenings are important, the total absorption A_V is obtained by combining Eqs. (2.15) and (2.24) as

$$(2.31) \quad A_V^*(x_D, a) = A_V/A_n = \int_0^\infty [1 - \exp(-x_D K(a, v))] \, dv \equiv A_{VE}^*(x_D, a)$$

It can easily be shown that Eq. (2.31) reduces to the Lorentzian case for large values of a and to the Doppler case in the limit of small a.

By employing the numerical procedure described by Young (1965a, b), Eq. (2.31) was solved for a range of parameter a, and the results are illustrated in Fig. 5. Similar results were also obtained by Jansson and Korb (1968), who employed a somewhat modified version of Armstrong's (1967) computer program. As would be expected, for $a = 0$, the results correspond to the case of a pure Doppler profile, while for $a > 0$, the results correspond to the case of a pure Doppler profile and for $a > 10$, they correspond to the Lorentzian shape. Kyle (1968) has given results for Voigt line profiles in an isothermal atmosphere with an exponentially decreasing pressure.

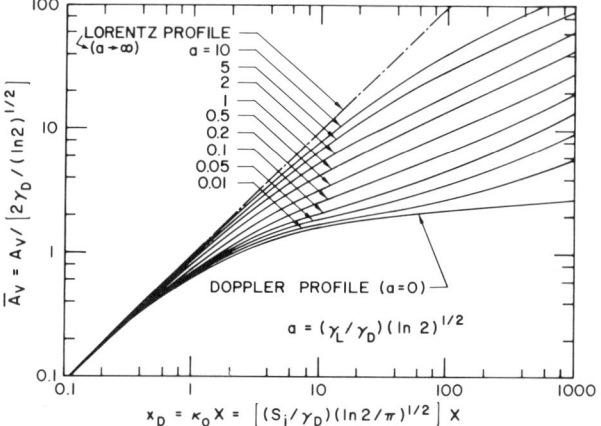

FIG. 5. Absorption of a single line of Voigt shape.

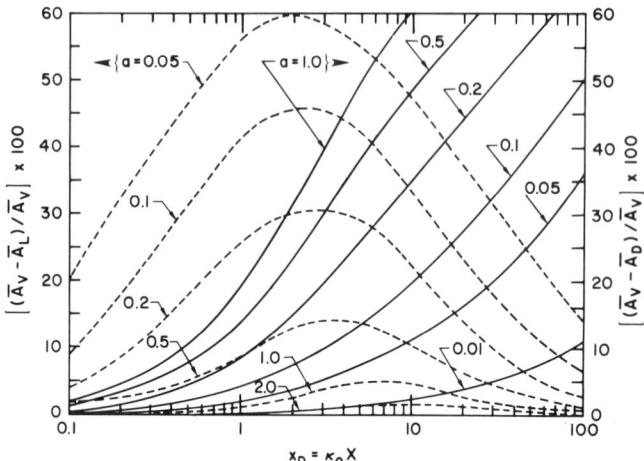

FIG. 6. Errors in the absorption by using the Lorentz and Doppler line shapes instead of the Voigt line profile.

The errors, in the integrated line absorption, encountered in using the Lorentz shape to approximate the Voigt profile, are illustrated in Fig. 6, for a range of the parameter a. The nondimensional form of the integrated absorption for the Lorentz profile A_L^* is obtained by dividing Eq. (2.28) by A_n and by noting that $x = x_D/(2a\sqrt{\pi})$. As would be expected, maximum errors occur for intermediate path lengths and for lower values of a. The errors, however, are not significant in the two limiting cases of small and large path lengths. As discussed earlier, for small path lengths (linear limit), the line absorption is independent of the line shapes. In the limit of large path lengths, however, the central portion of the line becomes opaque, and absorption occurs only in the wing regions. Since the Voigt profile is essentially Lorentzian in the line wings, the error encountered in the large path length limit becomes insignificant.

The errors resulting from using the Doppler line profile to approximate the absorption by the Voigt line profile are illustrated in Fig. 6 by the solid lines. Since $a = 0$ corresponds to the case of pure Doppler absorption, the errors are expected to be higher for larger values of a. Maximum errors, in this case, are found in the large path length limit. This is because, in this limit, the absorption occurs essentially in the line wings, and the Voigt profile in the wing regions is a Lorentzian rather than a Doppler shape.

It should be emphasized that for cases of intermediate path lengths and for moderate values of a ($0.1 < a < 10$) the use of the Voigt profile becomes almost essential. This situation corresponds to the radiative transmittance in the earth's troposphere and lower stratosphere. Consequently, for radiative

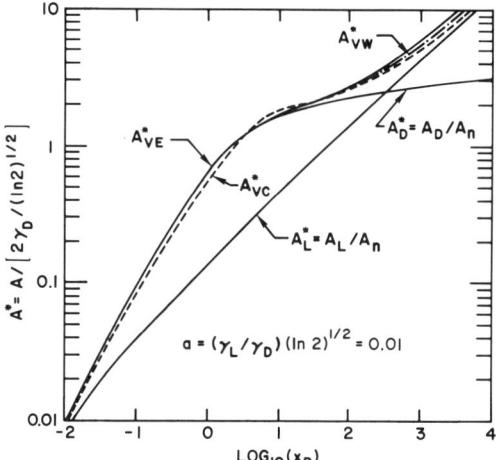

FIG. 7. Comparison of line absorption for $a = 0.01$.

modeling of the lower atmosphere, consideration must be given to the application of combined line profiles.

In Figs. 7 and 8, comparisons of results for all three line profiles (Lorentz, Doppler, and Voigt) are made for $a = 0.01$ and 0.1. Comparison of results for other values of a is available in Tiwari (1973), Tiwari and Reichle (1974), and Yamada (1968). Figures 7 and 8 clearly illustrate the range of validity of the absorption by the three line profiles. The curves for A_L^* and A_D^* intersect

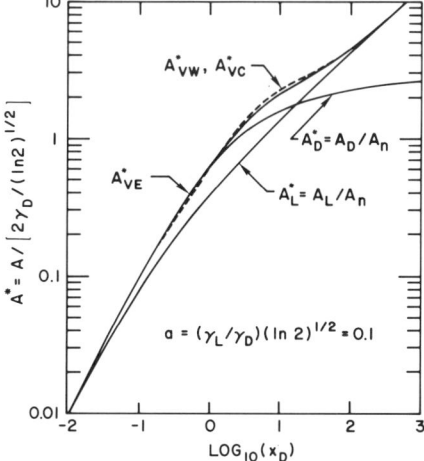

FIG. 8. Comparison of line absorption for $a = 0.1$.

at some point, the location of which increases with decreasing a. For $a > 2$, A_L^* and A_V^* (or A_{VE}^*) curves become identical for all path lengths.

In atmospheric problems involving conduction, convection, and radiation, one ends up with an integro-differential form of the energy equation. If use of the Voigt line profile is made in formulating the expression for absorptance, then the divergence of the radiative flux in the energy equation will involve an equation with triple integrals even for the simple case of energy transfer between plane-parallel atmospheres. The number of integrals appearing in the governing equations remain the same even if one is only concerned with the problem of purely radiative equilibrium. In order to reduce the mathematical complexities and save computational time, it often becomes necessary to express the absorptance of the Voigt line profile by fairly accurate approximate forms. A few such forms have been suggested in the literature.

Curtis [unpublished observations, cited in Gille and Ellingson (1968) and Rodgers and Williams (1974)] has suggested a simple form in terms of the absorptance by Lorentz and Doppler lines as

$$A_V \simeq A_L + A_D - A_L A_D / A_S \tag{2.32}$$

where $A_S = S_j X$ is the absorptance in the linear limit and is independent of any line shape. This is a convenient form for mathematical operations (especially differentiation), and it reduces to the correct limiting forms. According to Gille and Ellingson (1968), however, this form overestimates the absorption [as compared with the exact solution of Eq. (2.31)], especially when $A_L \simeq A_D$, with errors up to 50%. Rodgers and Williams (1974) have suggested a modified form of the Curtis approximation as

$$A_V = [A_L^2 + A_D^2 - (A_L A_D / A_S)^2]^{1/2} \tag{2.33}$$

This also reduces to the correct limiting forms, is easily differentiable, and is accurate within 8%. In final calculation of A_V, however, approximate forms of A_L and A_D (expressed in terms of series coefficients) are employed which makes this form somewhat cumbersome. If the approximate relation suggested by Whiting, Eq. (2.12), is employed, then the absorption of a Voigt line (Tiwari, 1973) can be expressed as

$$A_{VW}^* = A_{VW}/A_n = \int_0^\infty [1 - \exp(-x_D W(a, v))] \, dv \tag{2.34}$$

where

$$W(a, v)/[(\ln 2/\sqrt{\pi})F(a)] = (1 - \gamma_{LV}) \exp\{-11.088[(v/a)\gamma_{LV}^2]\}$$
$$+ \gamma_{LV}/\{1 + 16[(v/a)\gamma_{LV}]^2\}$$
$$F(a) = [a(1.065/\gamma_{LV} + 0.447 + 0.058\gamma_{LV})]^{-1}$$
$$\gamma_{LV} = \gamma_L/\gamma_V = \{0.5 + [0.25 + \ln 2/a^2)]^{1/2}\}^{-1}$$

The results of this equation are found to be accurate within 5% (Tiwari, 1973). Use of this equation has been made in radiative transfer models for detection of atmospheric pollutants (Ludwig et al., 1973; SAI, 1973). The drawbacks of this form are that it involves integration over one variable and its use in the final form of the energy equation requires a considerably long computational time. Another accurate approximation to the absorption by a Voigt line profile can be obtained by employing the expression for Voigt function suggested by Kielkopf (1973). However, as with the Whiting's approximation, this form will also involve integration over one variable (while calculating the total line absorption) and consequently would require long computational time.

While Eqs. (2.33) and (2.34) provide quite accurate approximations to the integrated absorption of a Voigt line profile, a relatively simple (and equally accurate) form can be obtained by employing the approximate formulations for A_L [as given by Goldman (1968)] and A_D (cf. Tiwari, 1973; Penner, 1959) into Eq. (2.33) to yield

(2.35)
$$A^*_{VC} = (\sqrt{\pi}/2)\{F(L)[1 - \exp(-\sqrt{x_D}) + x_D^2 \exp(-\sqrt{x})]\}^{1/2} \quad 0.1 \leq x_D \leq 20$$
$$A^*_{VC} = [(\pi/4)F(L) + \ln x_D(1 - F(L)/x_D^2)]^{1/2} \quad x_D \geq 20$$
where
$$F(L) = x_D^2/[1 + (x_D\sqrt{\pi}/4a)^\alpha]^{1/\alpha} \quad \alpha = 5/4$$

This equation possesses the mathematical simplicity offered by Eqs. (2.32) and (2.33) and reduces to the correct limiting forms.

The solutions of Eq. (2.35) are compared with the exact solution A^*_{VE} in Figs. 7 and 8 for values of the parameter a equal to 0.01 and 0.1. The solutions of Whiting's approximation are also illustrated in these figures. It should be noted that while the agreement between the results is not very good for $a = 0.01$, it is excellent for $a = 0.1$. Depending on the nature of a particular problem, use of the correlation, as given by Eq. (2.35), in radiative flux equations could result in significant saving of computational time.

2.3. Absorption of a Spectral Line in a Finite Spectral Interval

Expressions for absorption by nonoverlapping spectral lines in finite spectral intervals are important in obtaining meaningful formulations for narrow and wide band models. For brevity, attention is directed only to the Lorentzian line profiles in the subsequent formulations. The procedure, however, can easily be extended to obtain formulations for other line profiles.

The average transmission of a Lorentz line in a finite spectral interval may be obtained by combining Eqs. (2.4) and (2.20) as

$$(2.36) \quad \bar{\tau}_{L,\delta} = \frac{1}{\delta} \int_0^\delta \exp\{-(S_j X \gamma_L)/\pi[(\omega - \omega_j)^2 + \gamma_L^2]\} \, d(\omega - \omega_j)$$

By defining nondimensional quantities

$$\eta = \gamma_L/\delta \qquad \xi = (\omega - \omega_j)/\delta$$

Eq. (2.36) can be transformed into

(2.37) $$\bar{\tau}_{L,\delta} = \int_0^1 \exp\{-2x/[(\xi/\eta)^2 + 1]\}\, d\xi$$

The relation for the mean absorption over the interval can, in turn, be expressed by

(2.38) $$\bar{A}_{L\delta} = 1 - \bar{\tau}_{L,\delta} = \int_0^1 (1 - \exp\{-2x/[1 + (\xi/\eta)^2]\})\, d\xi$$

It is important to note that the parameter η in Eqs. (2.37) and (2.38) represents the ratio of the line half-width and the wave number interval from the line center. If required, the limiting forms of Eq. (2.38) can easily be obtained.

Upon introducing a new variable

(2.39a) $$\omega - \omega_j = \gamma_L(1 + \cos t)^{1/2}/(1 + 2\eta^2 - \cos t)^{1/2}$$

and a new dimensionless path length

(2.39b) $$\bar{x} = S_j X/2\pi\gamma_L(1 + \eta^2)$$

Eq. (2.36) is transformed to give

(2.40a) $$\bar{\tau}_{L,\delta} = \eta(1 + \eta^2)\exp[-\bar{x}(1 + 2\eta^2)]\int_0^\pi \exp(\bar{x}\cos t)f(t)\, dt$$

where

(2.40b) $$f(t) = \sin t/[(1 + \cos t)^{1/2}(1 + 2\eta^2 - \cos t)^{3/2}]$$

Several limiting and approximate forms of this equation have been obtained by Yamamoto and Aida (1967). As in the previous case, the expression for the mean absorption over the interval can be expressed by $\bar{A}_{L\delta} = 1 - \bar{\tau}_{L,\delta}$.

The total absorption (not the average absorption) of a single line over the wave number interval $D = 2\delta$ can be obtained from the expression

(2.41) $$A_{L,D} = 2D \int_0^{1/2} (1 - \exp\{-2x/[1 + (\bar{\xi}/\bar{\eta})^2]\})\, d\bar{\xi}$$

where $\eta = \gamma_L/D$, and $\bar{\xi} = (\omega - \omega_j)/D$. Equations (2.38), (2.40), and (2.41) are useful in calculating the absorption over a band pass of nonoverlapping spectral lines.

If there are n nonoverlapping lines in a finite wave number interval $\Delta\omega$

(with an average spacing between the lines of d), then the absorption of a single line over the interval $D = \Delta\omega = nd$ can be expressed by

$$(2.42) \qquad A_{L,D} = \int_{-D/2}^{D/2} [1 - \exp(-\kappa_{\omega j} X)] \, d(\omega - \omega_j)$$

For a line with Lorentz shape, this becomes

$$(2.43) \qquad A_{L,D} = (d/\pi) \int_0^{n\pi} (1 - \exp\{-2x/[1 + (z/\beta)^2]\}) \, dz$$

where $\beta = 2\pi\gamma_L/d$ and $z = 2\pi(\omega - \omega_j)/d$. It can easily be shown that, in the linear limit, Eq. (2.43) reduces to

$$(2.44) \qquad A_{L,D} = (2d\beta x/\pi) \tan^{-1}(n\pi/\beta) = 4\gamma_L x \tan^{-1}(n\pi/\beta)$$

while for large x it becomes

$$(2.45) \qquad A_{L,D} = \zeta\sqrt{\pi}(1 - \text{erf}(\zeta)) + 1 - \exp(-\zeta^2)$$

where

$$\zeta^2 = 2x(\beta/n\pi)^2 \qquad \text{erf}(\zeta) = (2/\sqrt{\pi}) \int_0^\zeta \exp(-t^2) \, dt$$

The quantity β, which expresses the ratio of line width to line spacing, is called the line structure parameter. Since $\gamma_L \sim P$, the limit of large pressure corresponds to $\beta \to \infty$. Expressions for the average absorption can be obtained by dividing Eqs. (2.43) through (2.45) by the wave number interval nd.

The average absorption of a single line, in the spectral interval equal to the average spacing between the lines, can be given by

$$(2.46) \qquad \bar{A}_{L,d} = (1/\pi) \int_0^\pi (1 - \exp\{-2x/[1 + (z/\beta)^2]\}) \, dz$$

The limiting forms of Eq. (2.46) are obtained from Eqs. (2.44) and (2.45) simply by replacing n with unity.

Since the solution of a single line of Lorentz shape (in an infinite spectral interval) is known from Eq. (2.28a), one can obtain an alternate form for the average absorption over the wave number interval $\Delta\omega = d$ as

$$(2.47) \qquad \bar{A}_{L,d} = (1/d)A_{L,\infty} - (1/\pi) \int_\pi^\infty (1 - \exp\{-2x/[1 + (z/\beta)^2]\}) \, dz$$

where

$$A_{L,\infty}/d = \beta x \exp(-x)(I_0(x) + I_1(x))$$

The absorption model expressed by Eq. (2.47) is called the Schnaidt's model and is often employed for calculating the total absorptance of a band of overlapping lines. The effect of the overlap is accounted for by cutting off each line at displacements $\pm (d/2)$ from its center. Under normal atmospheric conditions, the line spacing greatly exceeds the line half-width (i.e., $z/\beta \gg 1$), and Eq. (2.47) can be approximated by

(2.48) $\qquad \bar{A}_{L,d} = (1/d)A_{L,\infty} + 1 - \exp(-u^2) - u\sqrt{\pi}\,\text{erf}(u)$

where
$$u^2 = 2x(\beta/\pi)^2$$

All expressions for absorptance in this section are written in nondimensional form. It should be noted that the expressions for absorption over a finite spectral interval do not reduce to the correct linear limit, Eq. (2.28b).

The solutions of Eq. (2.46), along with the limiting solutions, are illustrated in Fig. 9 for various values of the line structure parameter β. As

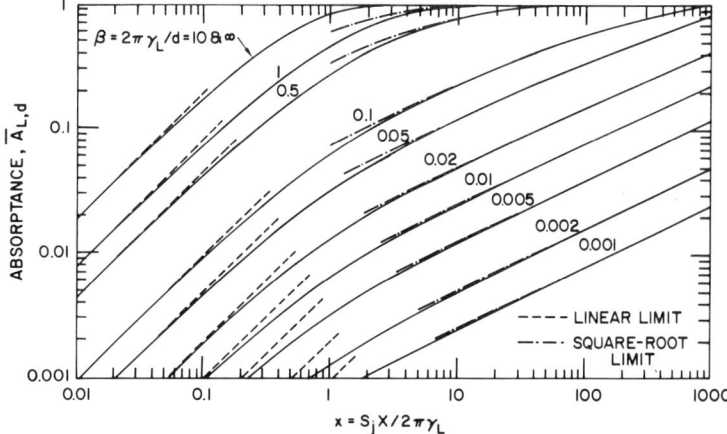

FIG. 9. Absorption of a single line of Lorentz profile in the spectral interval equal to the average spacing between the lines.

discussed earlier, the line structure parameter, which expresses the ratio of line half-width to line spacing, is essentially a pressure parameter. It is obvious from Fig. 9 that the total absorption increases with the path length. This is physically realistic because at a fixed pressure, if the path length is increased, then more molecules become available to interact with the radiative processes of emission and absorption. For any particular path length, however, the total absorption increases as the pressure is increased (i.e., $\bar{A}_{L,d}$ increases with β). This increase in absorption with β continues until the large β limit is achieved. At this point, lines become saturated (completely pressure broadened), and a further increase in pressure does not contribute to the absorption.

2.4. Absorption of an Overlapping Line in a Finite Spectral Interval

Spectral lines of an infrared band, in general, have different line intensities, half-widths, and spacing. At moderately high pressures, these lines overlap (especially in the wing regions). While calculating the transmittance of a single line in a finite spectral interval $D = 2\delta$, one must account for the contribution of the overlapping lines which occur outside the interval δ. The absorption coefficient of a Lorentz line, in such cases, may be expressed by

$$(2.49) \quad \kappa_{\omega j} = S_j \gamma_{Lj}/\pi[(\omega - \delta_j)^2 + \gamma_{Lj}^2] \quad j = 0, 1, 2, \ldots$$

where $j = 0$ refers to the main line under consideration and δ_js are the wave number locations (of center of jth lines) measured from the center of the main line. The expression for the spectral transmittance, therefore, becomes

$$(2.50) \quad \tau_{\omega j} = \exp\left[-\left(\kappa_{\omega 0} + \sum_{j=1}^{n} \kappa_{\omega j}\right)X\right]$$

and the average absorption over the interval $D = 2\delta$ is given by

$$(2.51) \quad \bar{A}_{L,\delta} = \frac{1}{2\delta}\int_{-\delta}^{+\delta}\left(1 - \tau_{\omega 0}\exp\left\{-\sum_{j=1}^{n}S_j\gamma_{Lj}X/\pi[(\omega - \delta_j)^2 + \gamma_{Lj}^2]\right\}\right)d(\omega - \omega_0)$$

Note that $\omega_0 = \delta_0$ denotes the center of the main line.

By introducing quantities

$$n_j = \gamma_{Lj}/\delta \quad \xi = (\omega - \omega_0)/\delta \quad x_j = S_j X/2\pi\gamma_{Lj}$$

Eq. (2.51) can be written as

$$(2.52\text{a}) \quad \bar{A}_{L,\delta} = (1/2)\int_{-1}^{1}\left(1 - \exp\left(-\sum_{j=0}^{n}2x_j/\{1 + [(\xi/n_j) + (\omega_0 - \delta_j)/\gamma_{Lj}]^2\}\right)\right)d\xi$$

For a particular atmospheric condition, evaluation of this equation requires the knowledge of δ_j, γ_{Lj}, and S_j for each interfering line. If in the interval δ there are no contributions from the lines whose centers lie outside the interval, then Eq. (2.52a) directly reduces to Eq. (2.38).

By defining a new dimensionless quantity

$$\varepsilon_j = (\omega_0 - \delta_j)/\gamma_{Lj}$$

Eq. (2.52a) can be written as

$$(2.52\text{b}) \quad \bar{A}_{L,\delta} = A_{L,\delta}(x_j, n_j, \varepsilon_j)$$
$$= (1/2)\int_{-1}^{1}\left(1 - \exp\left(\sum_{j=0}^{n} - 2x_j/\{1 + [(\xi/n_j) + \varepsilon_j]^2\}\right)\right)d\xi$$

Since for $\varepsilon_j = 0$ the mean absorption of a line is independent of the influence of neighboring lines, then ε_j may be treated as an overlapping parameter.

Expressions similar to Eq. (2.52) can easily be written for the total absorption of a line over any spectral interval. The absorption of an overlapping line, in a finite spectral interval equal to the mean line spacing d, is given by

$$(2.53) \quad A_{L,d} = (d/2\pi) \int_{-\pi}^{\pi} \left(1 - \exp\left(\sum_{j=0}^{n} -2x_j/\{1 + [(z/\beta_j) + \varepsilon_j]^2\}\right)\right) dz$$

where $\beta_j = 2\pi \gamma_{Lj}/d$ and $z = 2\pi(\omega - \omega_0)/d$.

For a fixed value of β_j (i.e., same value of β for all lines), the effect of overlapping can be studied simply by assigning different values to ε_j and completing the indicated summation in the exponential. For a fixed average spacing between the lines, it should be obvious from the definitions of the parameters ε_j and β_j that while ε_j varies inversely with pressure, β_j is a direct function of pressure such that the product $\varepsilon_j \beta_j$ becomes a constant. This, of course, will not be the case if one considers the variation in spacing between the lines. Since the quantity $\varepsilon_j \beta_j$ solely depends on the locations of the main and interfering lines and on the average spacing between the lines, then more realistic parameters for this problem will be β_j and $\varepsilon_j \beta_j$ rather than ε_j and β_j. In terms of these new parameters, Eq. (2.53) can be expressed as

$$(2.54) \quad A_{L,d} = A_{L,d}(x_j, \beta_j, \zeta_j)$$

$$= (d/2\pi) \int_{-\pi}^{\pi} \left(1 - \exp\left(\sum_{j=0}^{n} -2x_j/\{1 + [(z + \zeta_j)/\beta_j]^2\}\right)\right) dz$$

where

$$\zeta_j = \varepsilon_j \beta_j = 2\pi(\omega_0 - \delta_j)/d$$

For a fixed average spacing between the lines, a large ζ_j value physically means that the line centers of the interfering lines are at larger wave number from the main line center, and therefore their influence on the absorption of the main line will be smaller.

If all lines in a spectral interval are equally intense and have the same half-widths, then Eq. (2.54) can be written as

$$(2.55) \quad A_{L,d} = (d/2\pi) \int_{-\pi}^{\pi} \left(1 - \exp\left(-2x \sum_{j=0}^{n} \{1 + [(z + \zeta_j)/\beta_j]^2\}^{-1}\right)\right)$$

If there is significant interference due only to one neighboring line, then Eq. (2.55) is written, for average absorption, as

$$(2.56) \quad \bar{A}_{L,d} = \bar{A}_{L,d}(x, \beta, \zeta_1)$$

$$= (1/2\pi) \int_{-\pi}^{\pi} (1 - \exp(-2x/\{1 + [(z + \zeta_1)/\beta]^2\})) \, dz$$

This is a convenient form to study the effect of overlapping in a spectral interval equal to the mean line spacing. It should be noted that for $\zeta_1 = 0$, Eq. (2.56) directly reduces to Eq. (2.46). It should further be noted that, in general, the integrands in Eqs. (2.52) through (2.56) are not symmetric because lines on either side of the main line may have different line intensities and half-widths. Therefore, one cannot simply integrate these equations between the limit of zero to π and multiply the results by 2. This, however, is possible for the special case considered in Eq. (2.56).

The results of Eq. (2.56) are illustrated in Fig. 10 for $\beta = 0.1$. For a fixed β (i.e., at a constant pressure), the absorption of the main line is influenced

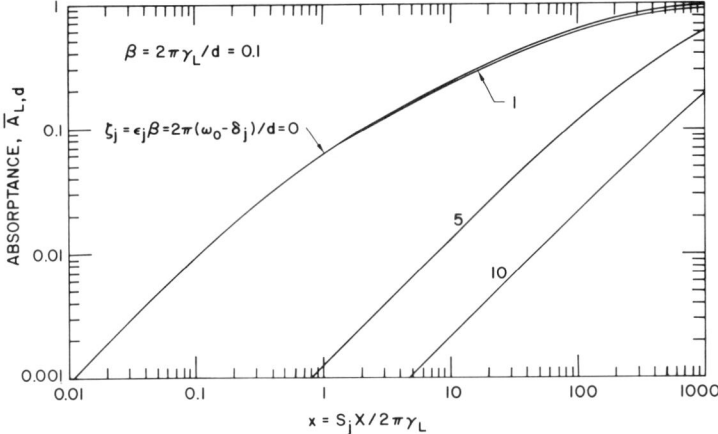

FIG. 10. Effect of interference of a neighboring line on the absorption of the main line for $\beta = 0.1$.

solely due to the wave number location of the interfering lines. The influence is stronger if the neighboring line centers are closer to the main line center. Thus, absorption will be higher for smaller values of the parameter ζ_1. This is clearly evident from the results presented in Fig. 10. Other results of absorption (and transmission) by a spectral line, in various spectral intervals, are available in Tiwari (1973), Tiwari and Reichle (1974), and Tiwari and Batki (1974).

3. Band Absorption

The absorptions by atmospheric gases, under certain atmospheric conditions, result in overlapping spectral lines. Thus, the total absorption within certain frequency intervals cannot accurately be represented simply by summing the absorption by individual lines. This is because the absorption in a region of overlapped lines is always less than the absorption calculated by considering the contributions of individual nonoverlapping lines.

The total absorption of a band of overlapping lines strongly depends upon the line intensity, the line half-width, and the spacing between the lines. In a particular band, the absorption coefficient varies very rapidly with the frequency, and therefore it becomes a very difficult and time-consuming task to evaluate the total band absorptance by numerical integration over the actual band contour. Consequently, several approximate band models have been proposed which represent absorption from an actual band with reasonable accuracy. These are discussed in some detail in this section.

The spectral absorptance of a narrow band (consisting of a sufficiently large number of spectral lines) may be expressed by

$$\alpha_\omega = 1 - \tau_\omega = 1 - \exp\left(-\int_0^X \kappa_\omega \, dX\right) \tag{3.1}$$

where κ_ω is the mass absorption coefficient for the band. The physical interpretation of α_ω is that it is the fraction of energy which is absorbed when a beam of radiant energy passes through a slab of gas of thickness l. For a homogeneous path, the total absorptance of a narrow band is given by

$$A = \int_{\Delta\omega} \alpha_\omega \, d\omega = \int_{\Delta\omega} [1 - \exp(-\kappa_\omega X)] \, d\omega \tag{3.2}$$

where limits of integration are considered over the narrow band pass.

The total band absorptance of a wide band may, in turn, be expressed by

$$A = \int_{-\infty}^{\infty} [1 - \exp(-\kappa_\omega X)] \, d(\omega - \omega_0) \tag{3.3}$$

where the limits of integration are over the entire band pass and ω_0 is the wave number at the center of the wide band.

In subsequent sections, various band models are considered to illustrate the basic features of the total band absorptance. There are, however, specific situations where absorption of a band may be represented quite accurately by one of the limiting forms of the total band absorptance discussed in the next subsection.

3.1. Limiting Forms of the Total Band Absorptance

Various limiting forms of the total band absorptance are: nonoverlapping line approximation, linear approximation, weak line approximation, strong line approximation, square-root limit, limit of large pressure, and the large path length limit. Physically realistic expressions for band absorptance in these limits depend upon the specific band model employed. Certain characteristics of band absorption, however, can be discussed in general terms independent of any band model.

3.1.1. Nonoverlapping Line Approximation. For this approximation, it is assumed that the spectral lines in a given interval $\Delta\omega$ do not overlap appreciably. The total absorptance due to n lines, in the interval, is obtained by summing the contributions of individual lines, i.e., $A = \sum_{j=1}^{n} A_j$, where A_j is given by Eq. (2.16).

As long as the criteria of nonoverlapping lines are satisfied, this approximation is valid regardless of the value of the absorption at the line centers. Since there is no influence of overlapping in this approximation, the absorption is independent of whether the spacing between the lines is regular or random. The absorption may, however, depend slightly on the distribution of line intensities in the band. The expressions of band absorptance for this approximation, therefore, depend on the particular band model employed.

For this approximation to be valid, it is required that the integrand in Eq. (2.16) approach zero for $\omega - \omega_j = 0(d)$, where d is the line spacing. For lines with Lorentz profile, the nonoverlapping limit will be satisfied if

(3.4) $$\kappa_{\omega j} X = (S_j X/\pi)[\gamma_j/(\gamma_j^2 + d^2)] \ll 1$$

and if, in addition, $\gamma_j \ll d$ (Cess and Tiwari, 1972). The conditions for achieving the nonoverlapping limit can, therefore, be stated as

(3.5a) $$(\gamma_j/d) \ll 1 \quad (S_j X \gamma_j/\pi d^2) \ll 1$$

Alternately, this can be expressed by

(3.5b) $$\beta \ll 1 \quad \beta^2 x \ll 1$$

The nonoverlapping approximation is especially useful in extrapolating the absorption to small values of pressure and path lengths.

3.1.2. Linear Approximation (Linear Limit). One of the important limiting forms of the band absorptance, which is completely independent of the band model, is the linear limit (or linear approximation). This approximation is valid when the total absorption due to all of the spectral lines is small at every frequency within the interval (i.e., $\kappa_\omega X \ll 1$ at all wave numbers). As pointed out earlier, this is the conventional optically thin limit in radiative transfer. Upon expanding the exponential in Eq. (3.3) and retaining only the first two terms in the series, the expression for band absorptance in the linear limit is found to be

(3.6a) $$A = X \int_{-\infty}^{\infty} \kappa_\omega \, d(\omega - \omega_0) = XS$$

where S is the band intensity. The total band absorptance in this limit increases linearly with the path length and the line intensity but is independent of the rotational line structure. Since all of the absorption from spectral lines (whose centers are within the wave number interval considered) occurs within $\Delta\omega$, this approximation is valid only when there is no overlapping of the spectral lines. In essence, therefore, the linear approximation is valid when this is an appropriate approximation for a single line and when in addition there is no appreciable overlapping of lines in the spectral range of interest. Equation (3.6a), therefore, is a consequence of summing the absorption from individual lines, i.e.,

$$(3.6b) \qquad A = \sum_j A_j = X \sum_j \int_{\Delta\omega j} \kappa_{\omega j}\, d(\omega - \omega_j) = X \sum_j S_j = SX$$

The linear approximation fails if the spectral lines begin to overlap strongly.

3.1.3. Weak Line Approximation. The weak line approximation is valid when the absorption is sufficiently small at all frequencies in the band (i.e., the absorption is small for each line in the wave number interval considered). Thus, the effect of different spectral lines is additive even if there is strong overlapping of lines. Note the difference between this and the linear approximation which is valid only for the case of nonoverlapping lines.

When weak line approximation is valid, the amount of radiation absorbed by each line is always a small fraction of the incident flux (Plass, 1958, 1960). If, however, many such weak lines overlap in a given spectral interval, then almost all of the radiation can be absorbed in that interval. The absorption coefficient is no longer a rapidly varying function of the wave number (i.e., it is almost constant), and therefore the particular arrangement of the spectral lines in the band (regular or random) does not influence the band absorptance significantly.

Physically, the weak line approximation is valid when the path length is small and the pressure is sufficiently large such that the absorption at the line centers is small. As will be shown later, the mean absorption in this case is a function of a single variable $\beta x = S_j X/d$, and is given by the relation

$$(3.7) \qquad \bar{A} = 1 - \exp(-S_j X/d) = 1 - \exp(-\beta x)$$

It should be noted that the linear limit, Eq. (3.6), is a special case of the weak line approximation, Eq. (3.7).

3.1.4. Strong Line Approximation. The strong line approximation is valid when the incident radiation is completely absorbed near the centers of the strongest spectral lines in the band and when these lines are primarily responsible for the total band absorption. In this approximation, therefore, the centers of the main absorbing lines are opaque, and further radiative contributions are primarily from the wing regions. The physical conditions for this approximation to be valid occur either at large optical path lengths or for low pressure values. The approximation is valid for overlapping as well as nonoverlapping lines in the band. The expressions for band absorption, however, do depend on the particular arrangement of the spectral lines (regular or random) in the band.

The strong line approximation requires that $\kappa_\omega X \gg 1$ for $\omega = \omega_j$. For spectral lines of Lorentz shape, the strong line condition is obtained from Eq. (2.4) as

(3.8) $$S_j X / \pi \gamma_j \gg 1$$

As will be shown later, the band absorptance in this case is a function of a single variable $\beta^2 x = 2\pi \gamma S_j X / d^2$.

The strong line approximation should not be confused with the square-root approximation (discussed in the next subsection) which is valid only when the spectral lines do not overlap. The strong line approximation results are useful for extrapolating the absorption to small pressure and large path length values.

3.1.5. Square-Root Limit (Strong Nonoverlapping Line Limit). For line radiation, the square-root limit (square-root law, square-root absorption, or square-root approximation) is achieved at relatively large optical path lengths. In this limit the central portion of the line is completely absorbed, and the mean absorptance of the line is proportional to the square root of the optical path length (Goody, 1964a).

For band radiation, the square-root approximation is valid when this is a suitable approximation for a single line and when in addition, there is no overlapping of the spectral lines. Consequently, the square-root limit is a special case of the strong line approximation. This is also referred to as the strong nonoverlapping line limit. As the name implies, the limit requires that two separate conditions (strong lines and nonoverlapping lines) be satisfied. These conditions are described by Eqs. (3.8) and (3.5), respectively.

3.1.6. Limit of Large Pressure (Large β Limit). Pressure plays a dual role in gaseous radiation. It appears in the path length as well as in the line structure parameter β. Its appearance in the pressure path length py (or $p\ell$) is due simply to the fact that absorption is dependent upon the number of molecules which are present along a line of sight. Pressure enters in the line structure parameter because the line half-widths are directly dependent on pressure. For sufficiently high pressures, the discrete line structure is smeared out, and in this limit (large pressure limit) pressure enters solely through the pressure path length py (or through ρy). The absorption coefficient in this limit is given by the expression $\kappa_\omega = S_j/d$. The average absorption in this case, therefore, is a function of a single variable $S_j X/d$ and is given by the Eq. (3.7). As such, the weak line approximation may be considered as a special case of the large β limit.

3.1.7. Large Path Length Limit (Logarithmic Limit). If, in addition to high pressure (large pressure limit), the pressure path length is sufficiently large, then the total band absorptance reaches a logarithmic asymptote [i.e., $A \sim \ell n$ (path length)]. In this limit the central portion of the band becomes opaque, and radiation transfer within the gas takes place solely in the wing regions of the band. As such, the strong line approximation may be considered as a special case of the large path length limit. The large path length limit is an appropriate limit for a wide rather than a narrow band model.

3.2. Narrow Band Models

The absorption within a narrow frequency interval of a vibration-rotation band can be represented quite accurately by so-called "narrow band models." Four commonly used narrow band models are Elsasser, statistical, random Elsasser, and quasi-random. The application of any model to a particular case depends upon the nature of the absorbing-emitting molecule. For example, one model may provide an excellent agreement with experimental results for linear molecules, but it may fail completely for asymmetric and spherical top molecules.

3.2.1. Elsasser (Regular) Band Model. The absorption of some vibration-rotation band, in a sufficiently narrow frequency range, may be represented quite accurately by the regular Elsasser band model (Elsasser, 1942), which consists of equally spaced Lorentz lines of equal half-width and intensity. This is an appropriate model to use for most diatomic gases and some linear triatomic molecules such as CO_2 and N_2O (if the path length is not too large).

In an Elsasser band, the absorption coefficient is a periodic function (with the period of the line spacing), and is given by Eq. (2.8) which is rewritten in a slightly different form as

$$(3.9) \qquad \kappa_\omega = \sum_{n=-\infty}^{\infty} \{S_j \gamma_L / \pi[(\omega - nd)^2 + \gamma_L^2]\}$$

where ω is the distance from the center of any line and d is the distance between adjacent lines. Elsasser [9] showed that Eq. (3.9) can be expressed in an alternate form as

$$(3.10) \qquad \kappa_\omega = (S_j/d) \sinh \beta /(\cosh \beta - \cos z)$$

where β and z are same as defined in Eq. (2.43).

The average absorptance of the periodic line pattern over the line spacing d is obtained by combining Eqs. (3.2) and (3.10) as

$$(3.11) \quad \bar{A}_N(x, \beta) = 1 - (1/2\pi) \int_{-\pi}^{\pi} \exp[-\beta x \sinh \beta/(\cosh \beta - \cos z)] \, dz$$

where A_n represents the absorptance of a narrow Elsasser band, and the quantity x is defined in Eq. (2.27). Elsasser [9] expressed this equation in the form

$$(3.12) \quad \bar{A}_N(x, \beta) = \sinh \beta \int_0^y I_0(t)[\exp(-t \cos \beta)] \, dt$$

where $y = S_j X/(d \sinh \beta) = \beta x/\sinh \beta$, and $I_0(t)$ is the modified Bessel function of order zero. Solutions of Eq. (3.12) are tabulated in Kaplan (1953) for $\beta = 0.0001$ to 1.0, and for $y = 0.02$ to $1.5(10)^5$. Equation (3.12) can be written in terms of the tabulated Schwarz function $I_e(k, v)$ as

$$(3.13) \quad \bar{A}_N(x, \beta) = \tanh \beta(I_e(\text{sech } \beta, \beta x \coth \beta))$$

where

$$I_e(k, v) = \int_0^v I_0(kt) \exp(-t) \, dt$$

Although the numerical solution of Eq. (3.11) can be easily obtained, several attempts have been made (Kaplan, 1953, 1954; Craig, 1963; Seitz and Lundholm, 1964; Zachor, 1967) to find the exact solution of this equation by employing either form (3.12) or (3.13). For exact solutions particularly applicable to the problems of atmospheric radiation, reference should be made to Kaplan (1953) and Zachor (1967).

Numerical solutions of Eq. (3.11) are illustrated in Figs. 11 through 14 in four different ways. Figure 11 is simply a plot of band absorption versus path length for different β values and illustrates that absorption increases with pressure and path lengths. The results illustrated in Figs. 12 through 14 are convenient for comparison with the approximate solutions discussed below. Several useful forms of Eqs. (3.11) through (3.13) can be obtained in the various limits and some of these are discussed here.

For large β (i.e., high pressure), $\sinh \beta \to \cosh \beta \to \infty$ such that $\sinh \beta/(\cosh \beta - \cos z) \to 1$ and Eq. (3.11) reduces to

$$(3.14) \quad \bar{A}_N(x, \beta) = 1 - \exp(-\beta x)$$

In this limit, the mean absorption is a function of a single variable $\beta s = S_j X/d$, the lines overlap, and there is no fine structure (i.e., absorption is independent of the line structure parameter β). As would be expected,

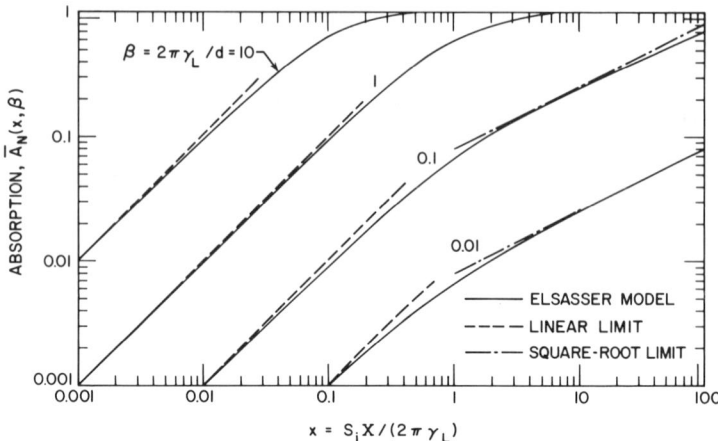

FIG. 11. Absorption by a narrow Elsasser band.

Eq. (3.14) is the same as Beer's law (i.e., the absorption of a band is given by a simple exponential law and the transmittance is independent of pressure).

An important form of the band absorptance is obtained in the limit of weak line approximation which is valid when the path length is small and the pressure is sufficiently high. For small x, therefore, Eq. (3.14) represents the weak line approximation. Note that Eq. (3.14) is exactly the same as Eq. (3.7).

The results of Eqs. (3.11) and (3.14), as functions of βx, are illustrated in Fig. 12. The uppermost curve is the solution of Eq. (3.14). Note that the absorption can never be greater than that given by Eq. (3.14). For $x < 0.2$

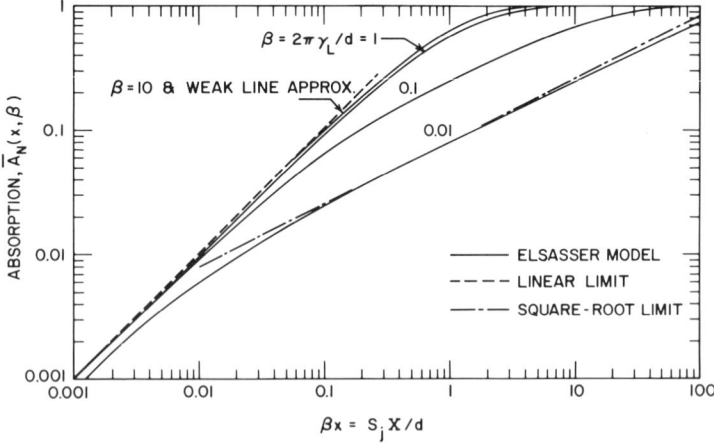

FIG. 12. Absorption as a function of βx for the Elsasser model.

and $\beta < 0.1$, the results of Eq. (3.14) are within 10 % of the exact solution of Eq. (3.11). For a particular β value, better accuracies are obtained at lower x values. For $\beta \geq 1$, the results of Eq. (3.14) are in good agreement with the exact solution for all x values (Plass, 1960).

For $x \ll 1$, Eq. (3.14) reduces to the linear limit as given by Eq. (3.6). This is illustrated in Fig. 12 by the broken line. As noted earlier, this limit is independent of the spectral model for band absorptance.

Another important form of Eq. (3.11) is obtained when the strong line approximation is justified. With reference to Eq. (3.8), the condition for the strong line approximation is that $x \gg 1$, and following Goody (1964a), Eq. (3.3) may be reduced to

(3.15a) $$\bar{A}_N(x, \beta) = \mathrm{erf}[(\tfrac{1}{2}\beta^2 x)^{1/2}] \qquad x \gg 1$$

where

(3.15b) $$\mathrm{erf}(t) = (2/\sqrt{\pi}) \int_0^t \exp(-t^2)\, dt$$

As mentioned earlier, the mean absorption, in this case, is a function of a single variable $\beta^2 x$. The expansion of Eqs. (3.15) in powers of $\beta^2 x$ results in

(3.16) $$\bar{A}_N(x, \beta) = \beta(2x/\pi)^{1/2}[1 - (\beta^2 x/6) + \cdots].$$

Note that the first term in this equation is identical with the strong line approximation for a single line (provided $\Delta\omega = d$).

The results of Eqs. (3.11) and (3.15), as functions of $\beta^2 x$, are illustrated in Fig. 13. The uppermost curve is the solution of Eq. (3.15). Note that the

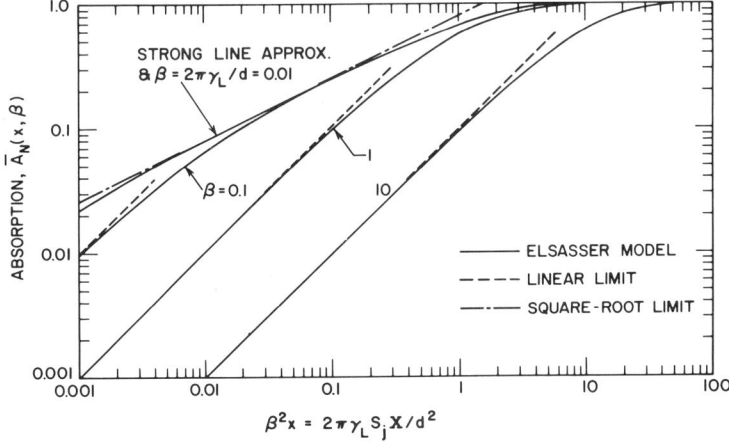

FIG. 13. Absorption as a function of $\beta^2 x$ for the Elsasser model.

absorption can never be greater than this limiting solution. For $x > 1.6$, the results of Eq. (3.15) are within 10% of the results of Eq. (3.11) for all β values (Plass, 1960). For $\beta > 1$, the results compare well even for smaller x values. As discussed by Plass (1960), the influence of Doppler broadening on the absorption can be determined from the results illustrated in the form of Fig. 13.

Another important form for the band absorptance is obtained in the limit of nonoverlapping lines. The conditions for achieving this approximation are given by Eq. (3.5). The mean absorption of an Elsasser band, in this case, may be expressed by

$$\bar{A}_N(x, \beta) = nA_j/\Delta\omega = A_j/d = \beta L(x) \qquad \beta \ll 1 \qquad \beta^2 x \ll 1 \tag{3.17a}$$

and

$$\bar{A}_N/\beta = L(x) = xe^{-x}(I_0(x) + I_1(x)) \tag{3.17b}$$

Thus, in this limit the expression for mean absorption of the Elsasser band is the same as that of a single Lorentz line.

The results of Eq. (3.11) and Eqs. (3.17) are illustrated in Fig. 14. The uppermost curve represents the nonoverlapping line approximation. For lower x values, this curve has a slope of unity indicating the region where the weak line approximation is valid. For large x values, the curve has a slope of one-half which is the region where the strong line approximation is valid. For $\beta < 1$, the nonoverlapping line approximation accurately represents the absorption in the transitional region near $x = 1$.

A final form for the narrow band absorptance is obtained in the limit of strong nonoverlapping lines (the square-root limit). For the Elsasser band,

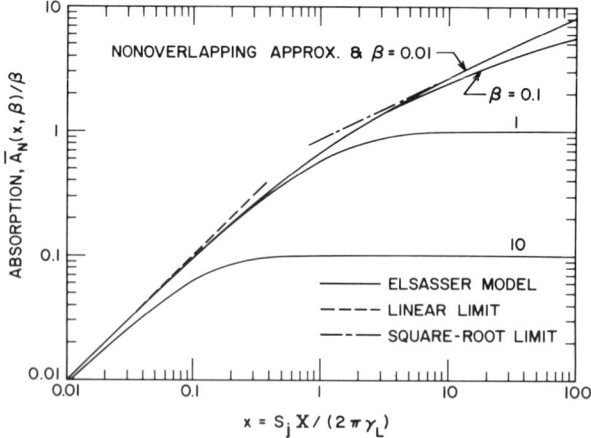

FIG. 14. Absorption divided by β as a function of x for the Elsasser model.

the expression for absorptance in this limit can be obtained from either Eq. (3.16) or Eqs. (3.17). While this will be given by the first term of Eq. (3.16), it will be the limiting form of Eqs. (3.17) for large x [as given by Eq. (2.28c)]. For this limit, therefore, one can write

$$(3.18) \qquad \bar{A}_N(x, \beta) = \beta(2x/\pi)^{1/2} \qquad x \gg 1 \qquad \beta \ll 1 \qquad \beta^2 x \ll 1$$

The results of Eq. (3.18) are illustrated in Figs. 11 through 14, and they clearly reveal the fact that the square-root limit is the limiting solution for low pressure and large path lengths.

The Elsasser band model was originally developed for the pressure-broadened Lorentz line profiles. Recently, the model has been extended to the Doppler line shape by Kyle (1967), Golden (1967, 1968), and Young (1975a,b), and to the Voigt profile by Golden (1969).

3.2.2. Statistical (Mayer-Goody) Model. The statistical band model is based upon the assumption that, in a given wave number interval, the spectral lines are spaced randomly, and the intensity of these lines can be specified by some distribution function (Plass, 1958; Goody, 1964a). The model is useful in describing the absorption characteristics of CO_2 (especially at larger path lengths), H_2O, and other relatively complex polyatomic molecules.

The model considers a spectral interval of width $\Delta\omega = D = nd$ containing n total lines with each individual line having intensity S_j and wave number ω_j occurring within $\Delta\omega$. Let $N(\omega_j) d\omega_j$ represent the probability that the jth line has an intensity between S_j and $S_j + dS_j$. If ω_js and S_js are mutually independent, then the probability of finding the set of n lines with the distribution $\omega_1 \cdots \omega_n, S_1 \cdots S_n$ is

$$(3.19) \qquad \sum_{j=1}^{n} N(\omega_j) P(S_j) \, d\omega_j \, dS_j$$

The average transmittance at wave number ω is obtained by averaging the transmittance at ω over the probability distribution of the set of n lines as

$$(3.20) \quad \bar{\tau}(\omega) = \int_{\Delta\omega} \cdots \int_{\Delta\omega} \int_0^\infty \cdots \int_0^\infty \sum_{j=1}^{n} N(\omega_j) \exp(-S_j X f_j(\omega, \omega_j)) \, d\omega_j \, dS_j$$

By assuming that the line positions are distributed at random in $\Delta\omega$ and that all the lines in $\Delta\omega$ have the same intensity, Eq. (3.20) can be expressed as

$$(3.21) \qquad \bar{\tau}(\omega) = \left[(1/D) \int_{\Delta\omega} \exp(-S_j X f_j(\omega, \omega_j)) \, d\omega_j \right]^n$$

The absorptance over the wave number interval $\Delta\omega = D = nd$ may, in general (Plass, 1958; Goody, 1964a), be expressed by

(3.22a) $$\bar{A}_N = 1 - [1 - (\tilde{A}_{j,D}/D)]^n$$

where

(3.22b) $$\tilde{A}_{j,D}(S_0, X, p) = \int_0^\infty A_{j,D}(S_j, X, p) P(S_j, S_0)\, dS_j$$

In these equations, $A_{j,D}$ is the absorptance of a single line over the entire wave number interval D, $P(S_j, S_0)$ is the normalized probability of finding a spectral line with the intensity S_j and $S_j + dS_j$, and S_0 is the parametric mean line intensity that occurs in the intensity distribution function.

For a large number of lines in the interval D, the term nd may be considered to approach infinity even if the mean spacing d is held constant. By employing one definition of the exponential, it may be shown that, for a large number of lines, Eqs. (3.22) become

(3.23) $$\bar{A}_N = 1 - \exp(-\tilde{A}_j/d) \qquad n \gg 10$$

where $\tilde{A}_j$ without the subscript D represents the absorptance of a single line for an infinite spectral interval. Physically, this assumes that there is no appreciable absorption by the single line outside the interval D. If this is not the case, then a more accurate expression for the absorption of a single line in the finite spectral interval (as given in Sections 2.3 and 2.4) should be employed.

Since the lines are assumed to be distributed randomly in the statistical model, the absorption by this model is always less than that by the Elsasser model. The advantage of the statistical model is that it can easily be applied to any line shape. Depending upon the variation in the intensity of individual spectral lines in a particular band, the statistical model is usually divided into two subclasses: uniform-statistical model (equally intense lines), and general statistical model (exponential distribution of line intensity).

3.2.2.1. *Uniform Statistical Model—Equally Intense Lines.* If in a narrow spectral interval all spectral lines are assumed to be equally intense such that $P(S_j) = \delta(S_j - S_0)$, then in Eq. (3.22)

(3.24) $$\tilde{A}_{j,D}(S_0, X, p) = A_{j,D}(S_0, X, p)$$

and one can now write

(3.25) $$\bar{A}_N = 1 - [1 - (A_{j,D}/D)]^n$$

An appropriate relation for the absorption by a specific line profile (such as Lorentz, Doppler, or Voigt), over a specified spectral interval, should be used in Eq. (3.25). At this point, it should be pointed out that Eq. (3.25) could have been obtained directly by combining Eqs. (3.1), (3.2), and (3.21).

In a special case, if all spectral lines are equally intense, the use of the Lorentz line shape is justified, if the absorption in a narrow band is due to a large number of spectral lines (with an average spacing d), and if there is no absorption by a single line outside the interval D, then Eq. (3.24) can be expressed as

$$\tilde{A}_j/d = A_j/d = \beta L(x) = \beta x e^{-x}(I_0(x) + I_1(x)) \tag{3.26}$$

The expression for the narrow band absorptance can be obtained now by combining Eqs. (3.23) and (3.26) as

$$\bar{A}_N(x, \beta) = 1 - \exp[-\beta x e^{-x}(I_0(x) + I_1(x))] \qquad n \gg 10 \tag{3.27}$$

By employing the approximate form of the Ladenberg-Reiche function [see Eqs. (2.27), (2.28a), and (2.29e)], Eq. (3.27) can be expressed in an alternate form as

$$\bar{A}_N(x, \beta) = 1 - \exp\{-\beta x/[1 + (\pi x/2)^\alpha]^{1/2\alpha}\} \tag{3.28}$$

With $\alpha = 5/4$, results of Eq. (3.28) are within 1 % of the results of Eq. (3.27). In atmospheric radiance calculations, use of Eq. (3.28) results in considerable savings of computational time.

The weak line approximation for this model can be obtained by substituting the appropriate value for $(A_{j,D}/D)$ in Eq. (3.25). For small x, the absorption by a Lorentz line in an infinite spectral interval is given by Eq. (2.28b), and therefore one can write

$$A_j/D = (2\pi\gamma_L x)/d = \beta x \tag{3.29a}$$

$$A_j/D = A_j/(nd) = \beta x/n \tag{3.29b}$$

Upon substituting Eq. (3.29) into Eqs. (3.25) and (3.26), the expressions for the weak line approximation are obtained as

$$\bar{A}_N(x, \beta) = 1 - [1 - (\beta x/n)]^n \tag{3.30a}$$

or

$$\bar{A}_N(x, \beta) = 1 - \exp(-\beta x) \tag{3.30b}$$

Note that Eq. (3.30b) could have been obtained directly from Eq. (3.27) by using the linear form of $L(x) = x$ for small x. Equation (3.30b) is exactly the same as Eqs. (3.7) and (3.14). This, however, should be expected because the particular arrangement of the spectral lines in the band does not influence the absorption when the weak line approximation is valid.

The strong line approximation for this model can be obtained from the asymptotic form of $(A_{j,D}/D)$ in Eq. (3.25) for large x. For large x, the absorption by a Lorentz line in an infinite spectral interval is given by Eq. (2.28c) such that

(3.31) $$A_j/d = [2\gamma_L(2\pi x)^{1/2}]/d = \beta(2x/\pi)^{1/2}$$

By substituting this in Eqs. (3.25) and (3.26), the expressions for the strong line approximation are found to be

(3.32a) $$\bar{A}_N(x, \beta) = 1 - [1 - (2\beta^2 x/\pi)^{1/2}]^n$$

or

(3.32b) $$\bar{A}_N(x, \beta) = 1 - \exp[-(2\beta^2 x/\pi)^{1/2}] \quad n \gg 10$$

Note that Eq. (3.32b) could have been obtained directly from Eq. (3.27) by using the asymptotic form of $L(x) = (2x/\pi)^{1/2}$ for large x. Furthermore, in calculating the results for the strong line approximation from Eq. (3.32), an appropriately defined average value of the line intensity should be used for finding the value of x. Otherwise, the term in the square bracket of (3.32a) should be evaluated for each of the n spectral lines and the results multiplied together.

The nonoverlapping approximation for the statistical model, in general, can be obtained from Eq. (3.23) by expanding the exponential and only retaining the first term, such that

(3.33) $$\bar{A}_N = \tilde{A}_j/d = \int_0^\infty (A_j/d) P(S_j, S_0) \, dS_j$$

where again, A_j represents the absorption of a single line in an infinite spectral interval. For spectral lines of Lorentz shape, Eq. (3.33) can be expressed as

(3.34) $$\bar{A}_N/\beta = \int_0^\infty L(x) P(S_j, S_0) \, dS_j$$

where the right-hand side is only a function of x.

For the present case of equally intense lines, Eq. (3.34) yields

(3.35) $$\bar{A}_N/\beta = L(x) = xe^{-x}(I_0(x) + I_1(x))$$

Note that this expression could have been obtained directly from Eq. (3.27) by expanding the exponential and retaining only the first term. Furthermore, it should be noted that Eq. (3.35) is exactly the same as the nonoverlapping approximation for the Elsasser model [Eq. (3.17)]. This, however, should be expected because it is the intensity distribution function (and not the regular or random spacing of spectral lines) that influences the absorption in this approximation.

The square-root limit (strong nonoverlapping line approximation) of Eq. (3.27) can be obtained by expanding the exponential in Eq. (3.32b) and retaining only the first term (because for this limit $\beta^2 x \ll 1$). This will be exactly the same expression as given by Eq. (3.18) or Eq. (3.31).

Various results (general and limiting) or the uniform statistical model are discussed in detail in Tiwari and Batki (1975).

3.2.2.2. General Statistical Model—Exponential Distribution of Line Intensities. If an exponential distribution of line intensities is assumed (i.e., in a band, the probability of finding a spectral line of intensity S_j in a given intensity range decreases exponentially), then

$$P(S_j, S_0) = [\exp(-S_j/S_0)]/S_0 \tag{3.36}$$

For a particular line shape, the line absorptance in Eq. (3.22b) is calculated by using this line intensity distribution.

The absorption by the general statistical band model consisting of Lorentz lines is obtained by combining Eqs. (3.36), (3.22), and (3.23) as [10]

$$\bar{A}_N(x_0, \beta) = 1 - \{1 - [\beta x_0/n(1 + 2x_0)^{1/2}]\}^n \tag{3.37a}$$

and

$$\bar{A}_N(x_0, \beta) = 1 - \exp[-\beta x_0/(1 + 2x_0)^{1/2}] \qquad n \geqslant 10 \tag{3.37b}$$

where

$$x_0 = S_0 X/2\pi\gamma_L$$

It should be emphasized here that the expression for absorption by a single line in an infinite spectral interval as given by Eq. (2.28a) was used in obtaining Eq. (3.37). Expressions similar to Eq. (3.37) can be obtained for lines in a band having Doppler or Voigt line profiles (Plass, 1958; Kyle, 1967; Golden, 1967, 1968, 1969; Young, 1975a, b). The results of Eq. (3.37b) are illustrated in Figs. 15 through 17 along with the limiting solutions which are discussed below.

The weak line approximation for the general statistical model is also given by Eqs. (3.7), (3.14), or (3.30), but x is replaced by x_0. The results of this approximation are shown in Fig. 15 along with the solution of Eq. (3.37b). The weak line approximation is always valid within 10% for $x_0 < 0.1$.

The strong-line approximation for this case is given by Eq. (3.32), where again x is replaced by x_0. The results are shown in Fig. 16 along with the solution of Eq. (3.37b). For $x_0 > 2.4$, the strong line approximation results are within 10% of the exact solution for all β values. For $\beta \geq 1$, the results compare well even for smaller x values (Plass, 1960).

The nonoverlapping line approximation for this case is obtained by combining Eqs. (3.34) and (3.36) as

$$\bar{A}_N/\beta = x_0(1 + 2x_0)^{-1/2} \tag{3.38}$$

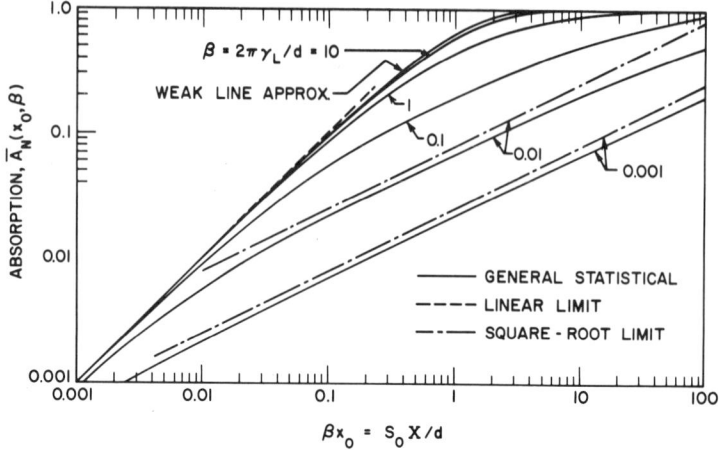

FIG. 15. Absorption as a function of βx for the general statistical model.

The results of Eqs. (3.37b) and (3.38) are illustrated in Fig. 17 where the uppermost curve is the solution of Eq. (3.38).

The square-root limit for this case is given by Eq. (3.18) or Eq. (3.31), where again x is replaced by x_0. The results are illustrated in Figs. 15 to 17. It should be noted that the region of validity of the square-root limit as illustrated in Figs. 15 to 17 is different than those for the Elsasser model (Figs. 12 to 14).

For the sake of brevity, it is often desirable to express the results of

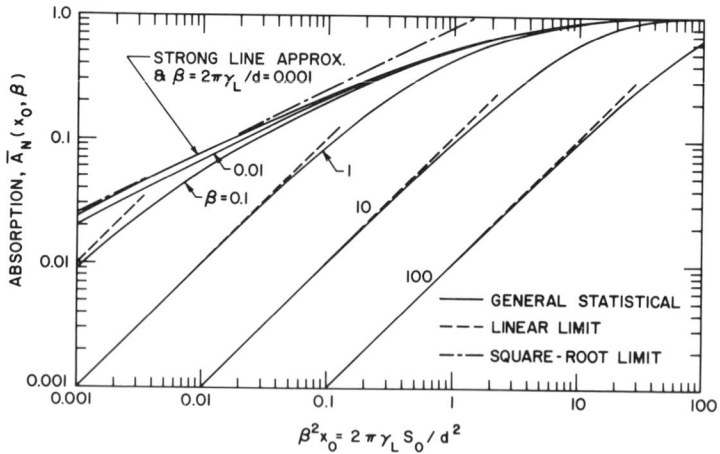

FIG. 16. Absorption as a function of $\beta^2 x$ for the general statistical model.

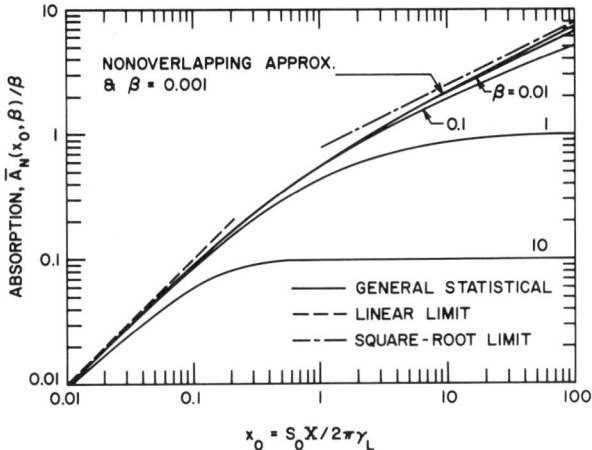

FIG. 17. Absorption divided by β as a function of x for the general model.

statistical models (uniform and general) on a single graph. According to Plass (1958, 1960), this is possible if one employs the relation $x = (\pi/4)x_0 = S_0 X/8\gamma_L$ in plotting the results.

From the comparison of results presented in Figs. 12 and 15, it is noted that the absorption curve for $\beta = 1$ is much closer to the weak line approximation for the Elsasser model than for the statistical model. This is because the absorption saturates at a smaller value of βx for the Elsasser model.

A comparison of results illustrated in Figs. 14 and 17 indicates that the spectral lines begin to overlap at considerably larger path lengths for the Elsasser model than for the statistical model. Thus, the nonoverlapping line approximation has a considerably larger region of validity for the Elsasser model than for the statistical model.

Another line intensity distribution sometimes employed in the statistical band formulation is the inverse-first-power model which is given by Goody (1964a).

(3.39) $$P(S_j) = kS_j^{-1}$$

where k is a normalization constant. For proper normalization, the above distribution must be cut off at some upper limit S'_j on S_j. Depending upon its application, a lower cutoff limit for S_j might also be essential in determining the correct value of k. For Lorentz lines, the expression for $(\tilde{A}_j/d)$ in Eq. (3.23) is found to be

(3.40) $$\tilde{A}_j/d = \beta[e^{-x}I_0(x) + 2xe^{-x}(I_0(x) + I_1(x)) - 1]$$

Substitution of this into Eq. (3.23) will result in another expression for absorption by a narrow statistical model. Different solutions of this model can also be obtained in the manner discussed for the previous two cases.

Malkmus (1967, 1968) and Rodgers (1968) have suggested yet another distribution of line intensity in a band. This constitutes a superposition of the exponential and the inverse-first-power distribution of line intensity. For detailed discussion of this model, reference should be made to Malkmus (1967, 1968) and Rodgers (1968).

3.2.3. Random Elsasser Band Model.

In actual vibration-rotation bands, lines are arranged neither completely at random nor at regular intervals. Within a band, there may be a number of strong lines in a certain narrow spectral region, whereas in other regions only very weak lines may be present. There also may occur a superposition of equally intense lines in another spectral region of the band. In cases like this, a more accurate representation of band absorption is provided by the random Elsasser model, which assumes the random superposition of several different Elsasser bands. Each of the superposed bands may have different line intensities, half-widths, and spacing. As many different Elsasser bands as necessary may be superimposed in this model. Thus, all the weak spectral lines that contribute to the absorption for the path lengths and pressure considered can be included in the absorption calculations. As the number of superposed Elsasser bands becomes large, the absorption by the entire band approaches that given by the statistical band model. For N randomly superposed Elsasser bands (Plass, 1958; Goody, 1964a; Kaplan, 1954), the absorption is given by the relation

$$(3.41) \qquad \bar{A}_N(x, \beta) = 1 - \prod_{i=1}^{N}[1 - \tilde{A}_{E,i}(x_i, \beta_i)/\delta_i]$$

where $\tilde{A}_{E,i}(x_i, \beta_i)$ is the absorptance of the ith Elsasser band and is given by

$$(3.42) \qquad \tilde{A}_{E,i}/\delta_i = \int_0^\infty \tilde{A}_{E,i}(x_i, \beta_i) P_E(S_i, S_{0i}) \, dS_i$$

With known intensity relations between individual sub-bands (e.g., two Elsasser bands with one having five times more intensity than the other, etc.), the general result for the absorption from a random superposition of N Elsasser bands can be obtained from the preceding equations.

As a special case, if exponential distribution of intensity for narrow Elsasser bands is assumed, then

$$(3.43) \qquad P_E(S_i, S_{0i}) = (1/S_{0i}) \exp(-S_i/S_{0i})$$

and Eq. (3.42) becomes

$$(3.44) \qquad \tilde{A}_{E,i}/\delta_i = (\beta_i x_{0i} \sinh \beta_i)/[(\beta_i x_{0i} \sinh \beta_i + \cosh \beta_i)^2 - 1]^{1/2}$$

A combination of Eqs. (3.41) and (3.44) yields the expression for absorptance by a modified random Elsasser band model as

$$(3.45) \quad \bar{A}_N(x_0, \beta) = (\beta x_0 \sinh \beta)/[(\beta x_0 \sinh \beta + \cosh \beta)^2 - 1]^{1/2}$$

The weak line approximation for the random Elsasser model, in general, can easily be found to be

$$(3.46) \quad \bar{A}_N(x, \beta) = 1 - \prod_{i=1}^{N} [\exp(-\beta_i x_i)]$$

The weak line approximation corresponding to Eq. (3.45) will be the same as for the general statistical model. The strong line approximation for the random Elsasser model, in general, is given by

$$(3.47) \quad \bar{A}_N(x, \beta) = 1 - \prod_{i=1}^{N} \{1 - \mathrm{erf}[(\tfrac{1}{2}\beta_i^2 x_i)^{1/2}]\}$$

For the modified random Elsasser model the strong line approximation is given by Eq. (3.32b) where x is replaced by x_0. The nonoverlapping line approximation and the square-root limit for the modified random Elsasser model are given by the analogous expressions for the general statistical model. The results of Eq. (3.45), along with the limiting solutions, are discussed in detail in Tiwari and Batki (1975).

3.2.4 Quasi-Random Band Model. The quasi-random model is probably the best model to represent the absorption of a vibration-rotation band quite accurately. It assumes neither a regular nor a random spacing of the spectral lines. The essential feature of this model is to divide the wider frequency interval of the actual band into much narrower subintervals. In each of these subintervals, the spectral lines are assumed to have random spacing. In this manner, the model accounts for the actual intensity distribution of strong as well as weak spectral lines. The absorption of the narrow spectral interval is calculated from the relation of a single line absorption over a finite interval. The absorption of each of the n lines in the narrow interval is calculated separately, and the results are combined by assuming a random position for the lines within the interval. The total band absorptance is calculated by averaging the results from the smaller intervals.

The fundamental features of the quasi-random band model are discussed, in detail, in Wyatt *et al.* (1962), Kunde (1967), and Gupta and Tiwari (1975). The procedure for calculating the atmospheric transmittance by employing the quasi-random band model is discussed, in detail, in Gupta and Tiwari (1975), where a listing of the computer program is also provided. Since, in the final analysis, the quasi-random band model requires detailed spectro-

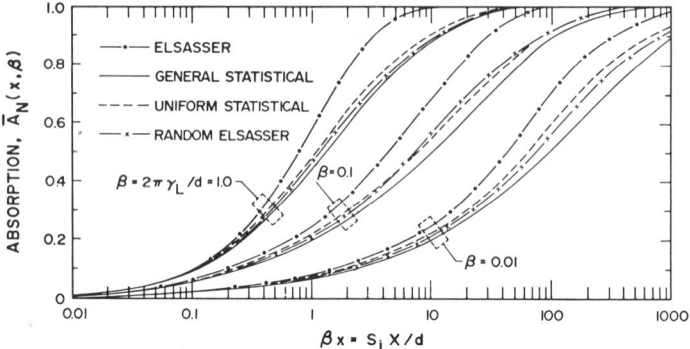

Fig. 18. Comparison of absorption by narrow band models.

scopic information on the specific molecule under consideration, it has not been included in the present general parametric study.

The absorption by four narrow band models (Elsasser, general statistical, uniform statistical, and modified random Elsasser) are compared in Fig. 18 for three different values of β. Since the absorption at small path lengths is a function solely of the total intensity of absorbing lines, the results by all models agree well in this region. At larger path lengths, however, the Elsasser theory predicts higher absorption than the general or uniform statistical model. As pointed out earlier, this is because there is always more overlapping of the spectral lines in the statistical models than in the regular Elsasser band.

The narrow band model formulations discussed thus far represent absorptions of gases under homogeneous conditions. They can be extended to nonisothermal nonhomogeneous optical paths by employing various scaling approximations available in the literature. The original Curtis-Godson approximation (Curtis, 1952; Godson, 1953) for computing atmospheric transmission along an inhomogeneous path has been updated, and several improved approximations are now available in the literature (Goody, 1964a,b; Young, 1975a,b; Walshaw and Rodgers, 1963; Armstrong, 1968; Yamamoto *et al.*, 1972; Lindquist and Simmons, 1972; Weinreb and Nevendorffer, 1973; Aida, 1975; McMillin and Fleming, 1976). In infrared signature work, it becomes essential to use an absorption model for a single highly inhomogeneous optical path (Young, 1975a). For many atmospheric applications, however, the atmosphere can be divided into an appropriate number of homogeneous layers, and the models discussed in this report can be employed directly (Ludwig *et al.*, 1973; Kunde, 1967; Gupta and Tiwari, 1975; Weinreb and Nevendorffer, 1973; McMillin and Fleming, 1976). In most cases this gives better results than any scaling technique employed to treat the entire atmosphere as a single layer.

3.3. Wide Band Models

Aside from the narrow band models discussed in the previous section, there are also available in the literature the so-called wide band models, which provide correlations that are valid over the entire band pass (Tiwari and Batki, 1975; Tiwari, 1976; Tien, 1968; Cess and Tiwari, 1972; Edwards, 1976; Edwards and Menard, 1964; Edwards and Balakrishnon, 1972; Edwards et al., 1967; Felske and Tien, 1974). Besides possessing the conventional linear and square-root limits, these models also possess another asymptotic limit, which is called the logarithmic limit. Even though the use of these models may be restricted for atmospheric applications, they do provide quick and accurate information regarding transmittance of gases at moderately high temperatures. As such, these models are quite useful in many engineering applications (for some recent applications, see Cess and Tiwari, 1972; Echigo et al., 1971; Tiwari and Cess, 1971; Donovan and Breif, 1975; Martin and Hwang, 1975; Edwards and Balakrishnon, 1973; Wassel and Edwards, 1974, 1976; Wassel et al., 1975).

3.3.1. Box (or Coffin) Model. The simplest of the wide band models is the box model which was introduced first by Penner (1959). For this model, it is assumed that the absorption coefficient κ_ω is constant over an effective band width $\Delta\omega$. The expression for the total absorptance by this model is given as

$$(3.48) \quad A = \int_{\Delta\omega} [1 - \exp(-\kappa_\omega X)] \, d\omega = (\Delta\omega)_e [1 - e^{-\bar{\kappa} X}]$$

where $(\Delta\omega)_e$ is the effective band width and $\bar{\kappa}$ is the mean absorption coefficient for the interval $(\Delta\omega)_e$. At relatively high temperatures, the spectral intervals between fundamental bands of infrared active gases are filled by the combination and overtone bands. Under these conditions, modified forms of the box model become useful in radiative transfer calculations (Gilles and Vincenti, 1970). Further discussion and application of the box model is available in Tiwari and Batki (1975), Penner (1959), and Cess et al. (1967).

3.3.2. Exponential Wide Band Model. Edwards et al. (1967; Edwards and Menard, 1964) have considered various wide band models (rigid rotator, nonrigid rotator, and arbitrary) and have concluded that three parameters (the mean line intensity to spacing ratio, the mean line width to spacing ratio, and the effective broadening pressure) are necessary for a complete description of the band absorption. For complete discussions of these models and their limiting forms, one should refer to Tiwari and Batki (1975); Tiwari (1976), Tien (1968); Cess and Tiwari (1972); Edwards (1976); Edwards and Menard (1964); Edwards and Balakrishnon (1972); Edwards et al. (1967), and Felske and Tien (1974). The final form of the total band absorptance relation presented by Edwards and Menard (1964) and Edwards and Balakrishnon (1973) is based on the formulation of the narrow statistical band model. The expression for the transmissivity employed in this model is given by

$$(3.49a) \qquad \tau_j = \prod_{j=1}^{N} \tau_{\omega j} = (1/\Delta\omega) \int_{\omega-(\Delta\omega/2)}^{\omega+(\Delta\omega/2)} [\exp(-\kappa_\omega X)] \, d\omega$$

$$(3.49b) \qquad \tau_j \simeq \exp(-(S_j/d)X/\{1 + [(S_j/d)X/BP_e]\}^{1/2})$$

where B is π times the mean line width to spacing ratio for a dilute mixture at 1 atm pressure, and P_e is an equivalent broadening pressure. Use of Eq. (3.49b) was also made by Felske and Tien (1974) to develop relations for the wide band absorptance from the general statistical band model.

In the present study, four different formulations for the total band absorptance are presented. These are based on various narrow band model relations for absorption. For an exponential wide band, if one assumes that the line intensity is an exponential decaying function of the wave number (Edwards et al., 1967; Edwards and Balakrishnon, 1973), then

$$(3.50) \qquad S_j/d = (S/A_0)(\exp\{[-b_0|\omega - \omega_0|]/A_0\}) = (S/A_0)\zeta$$

where S is the integrated intensity of a wide band, $A_0 = nd$, $b_0 = 2$ for a symmetrical band, and $b_0 = 1$ for bands with upper and lower wave number heads at ω_0. Equation (3.50) is used in the relation for absorption of a narrow band model, and the resulting expression is integrated over the entire band pass to obtain the total absorptance of a wide band as

$$(3.51) \qquad \bar{A}(u, \beta) = A(u, \beta)/A_0 = \int_{\text{wide band}} \bar{A}_N(u, \beta) \, d(\omega - \omega_0)$$

$$= b_0 \int_0^\infty \bar{A}_N(u, \beta, \zeta) \, d\zeta$$

where $u = SX/A_0 = SP\ell/A_0$. The second form of Eq. (3.51) incorporates the relation given by Eq. (3.50).

3.3.2.1. Exponential Wide Band Absorptance from the Elsasser Model. By substituting Eq. (3.50) into Eq. (3.11), the expression for absorptance by the Elsasser band can be written as

$$(3.52) \qquad \bar{A}_N(u, \beta) = 1 - (1/2\pi) \int_{-\pi}^{\pi} \exp(-F_1(u, \beta, z, \zeta)) \, dz$$

where $F_1 = u\zeta \sinh \beta/(\cosh \beta - \cos z)$ and ζ is defined in Eq. (3.50). It should be emphasized here that within a narrow Elsasser band the line intensities are constant, but over a wide spectral interval the line intensity variation is given by the Eq. (3.50). A combination of Eqs. (3.51) and (3.52) results in

$$\bar{A}(u, \beta) = (1/\pi) \int_0^\pi \left[\int_0^1 \zeta^{-1}(F_2(u, \beta, z, \zeta)) \, d\zeta \right] dz \tag{3.53}$$

where $F_2 = 1 - \exp(-F_1)$. The inner integral in Eq. (3.53) can be evaluated in closed form as

$$\int_0^v \{[1 - \exp(-\psi\zeta)]/\zeta\} \, d\zeta = \gamma + \ln(\psi v) + E_1(\psi v) \tag{3.54}$$

where $\psi = F_1/\zeta$, $v = 1$, $\gamma = 0.5772156$ is the Euler's constant, and $E_1(t)$ is the exponential integral of the first order. The final form of the exponential wide band absorptance (based on the narrow Elsasser model) is obtained by combining Eqs. (3.53) and (3.54) as

$$\bar{A}(u, \beta) = \gamma + (1/\pi) \int_0^\pi [\ln \psi + E_1(\psi)] \, dz \tag{3.55}$$

Equation (3.55) can be reduced to several useful limiting forms. One of the important forms is obtained when the weak line approximation for the Elsasser model is valid. Since this approximation is valid for large pressures, the limiting form of the total band absorptance in this case is obtained by letting $\beta \to \infty$ in Eq. (3.55). Alternately, this form can be obtained by combining Eqs. (3.14), (3.50), and (3.51) as

$$\bar{A}(u) = \int_0^1 \zeta^{-1}[1 - \exp(-u\zeta)] \, d\zeta \tag{3.56}$$

$$= \gamma + \ln(u) + E_1(u)$$

Note that in this limit the absorptance is independent of the line structure parameter β. For $u > 10$, $E_1(u) < 10^{-5}$, and Eq. (3.56) reduces to $\bar{A}(u) = \gamma + \ln(u)$, and for $u \leq 0.005$, it is approximately equal to u. Thus, one can write

$$\begin{array}{llll} \bar{A} = u & u \ll 1 & \beta > 1 \\ \bar{A} = \gamma + \ln(u) & u \gg 1 & \beta > 1 \end{array} \tag{3.57}$$

which are the appropriate linear and logarithmic limits of Eqs. (3.55) and (3.56).

When strong line approximation is justified, then the absorption of a narrow Elsasser band is given by the Eq. (3.15). By combining Eqs. (3.15), (3.50), and (3.51), the relation for the strong line approximation for the wide band is obtained as

$$\bar{A}(u, \beta) = \int_0^1 \zeta^{-1}\{\text{erf}[(u\beta\zeta/2)^{1/2}]\}\, d\zeta \tag{3.58}$$

By introducing a new variable $t^2 = u\beta\zeta/2$, Eq. (3.58) can be written as

$$\bar{A}(u, \beta) = 2\int_0^{\sqrt{(u\beta/2)}} t^{-1}\, \text{erf}(t)\, dt \tag{3.59}$$

For all values of t, the series expansion for the error function is given by

$$\text{erf}(t) = (2/\sqrt{\pi}) \sum_{n=0}^{\infty} \{(-1)^n t^{(2n+1)}/[n!(2n+1)]\} \tag{3.60}$$

Upon substituting (3.60) into (3.59), interchanging the summation and integration, and integrating the resulting expression, there is obtained

$$\bar{A}(u, \beta) = (4/\sqrt{\pi}) \sum_{n=0}^{\infty} \{(-1)^n (u\beta/2)^{(2n+1)/2}/[n!(2n+1)^2]\} \tag{3.61}$$

This can be considered a closed form expression for the strong line approximation.

According to the requirements of the square-root limit [see Eq. (3.18)], retaining only the first term in Eq. (3.61) results in

$$\bar{A}(u, \beta) = (4/\sqrt{\pi})(u\beta/2)^{1/2} \tag{3.62}$$

$$= 2(2\beta u/\pi)^{1/2} \qquad \beta \ll 1 \qquad u/\beta \gg 1 \qquad \beta u \ll 1$$

which is the correct square-root limit for the wide band absorption (Cess and Tiwari, 1972; Hsieh and Greif, 1972). Thus, Eq. (3.55) is seen to reduce to correct limiting forms in the linear, square-root, and logarithmic limits.

Since Eq. (3.55) involves double integration, its application in some radiative transfer analyses may require considerably long computational time. This, sometimes, can be avoided by expressing the equation in a series form. In order to do this, Eq. (3.53) is first written as

$$\bar{A}(u, \beta) = \int_0^1 \left(\frac{1}{\pi} \int_0^{\pi} \{1 - \exp[A\zeta/(1 - B\cos z)]\}\, dz\right) \zeta^{-1}\, d\zeta \tag{3.63}$$

where $A = -u \tanh \beta$, and $B = 1/\cosh \beta$. The inner integral in this equation can be evaluated in a series form such that

$$\bar{A}(u, \beta) = \int_0^1 \left\{\sum_{n=1}^{\infty} -(A)^n \zeta^n (\text{SUM}(mn))/[(B+1)^n n!(n-1)!]\right\} \zeta^{-1}\, d\zeta \tag{3.64}$$

where

$$\text{SUM}(mn) = \sum_{m=0}^{\infty} [(n + m - 1)!(2m - 1)! C^m]/[2^m(m!)^2]$$

$$C = 2/(1 + \cosh \beta) = 2B/(B + 1)$$

Integration over ζ in this equation gives a value of $1/n$. Thus, the solution of Eq. (3.53) or Eq. (3.55) in the series representation (which converges rapidly) is given by

(3.65) $$\bar{A}(u, \beta) = \sum_{n=1}^{\infty} \{-(A)^n(\text{SUM}(mn))/[n(B + 1)^n n!(n - 1)!]\}$$

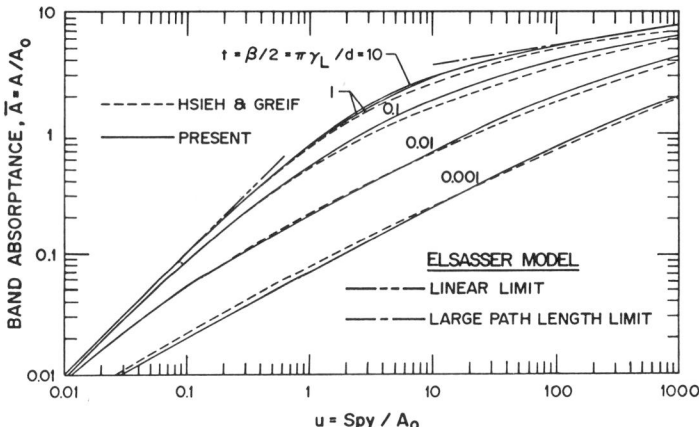

FIG. 19. Wide band absorptance based on narrow Elsasser model.

The solutions of Eqs. (3.55), (3.56), (3.61), and (3.65) are illustrated in Fig. 19 along with the limiting solutions for four different values of the line structure parameter $t = \beta/2 = \pi\gamma/d$. No difference in the results of Eqs. (3.55) and (3.65) were noticed. As such, the solution given by Eq. (3.65) is treated as the "exact" solution. The uppermost curve in the figure represents the solution of Eq. (3.56). For $t \geq 1$, the results of Eqs. (3.55) and (3.56) are identical. For comparison, the theoretical results obtained by Hsieh and Greif (1972) are shown also in Fig. 19. As would be expected, the comparison is not very good at large path lengths.

3.3.2.2. *Exponential Wide Band Absorptance from the Uniform Statistical Model.* Upon combining Eqs. (3.27) and (3.50) an expression for absorption by the uniform statistical model is obtained, which in turn is used in Eq. (4.4) to find the relation for the wide band absorptance as

(3.66) $$\bar{A}(u, \beta) = \int_0^1 [1 - \exp(-\beta L(\eta))]\zeta^{-1} d\zeta$$

where $\eta = u\zeta/\beta$, ζ is defined in Eq. (3.50), and $L(\eta)$ is defined in Eq. (2.28a). Equation (3.66) requires a considerably long time for numerical solution. To save the computational time, the function $L(\eta)$ can be replaced by Goldman's approximation [see Eqs. (2.29e) and (3.28)], and this results in

$$(3.67) \quad \bar{A}(u, \beta) = \int_0^1 (1 - \exp\{-\beta\eta/[1 + (\pi\eta/2)^{5/4}]^{2/5}\})\zeta^{-1}\,d\zeta$$

The numerical results of Eqs. (3.66) and (3.67) are identical and are in good agreement with the theoretical results of Hsieh and Greif (1972) over the entire range of u and β (Tiwari and Batki, 1975).

For sufficiently small η (i.e., for small u/β), the function $L(\eta)$ reduces to η and Eq. (3.66) yields the result for the weak line approximation as

$$\bar{A}(u) = \int_0^1 [1 - \exp(-u\zeta)]\zeta^{-1}\,d\zeta$$

$$= \gamma + \ln(u) + E_1(u)$$

This is exactly the same as Eq. (3.56) and is independent of the line structure parameter β. The linear and logarithmic limits of Eq. (3.66), therefore, are given by the Eq. (3.57).

For large η, $L(\eta) \simeq (2\eta/\pi)^{1/2}$, and Eq. (3.66) reduces to the strong line approximation as

$$(3.68) \quad \bar{A}(u, \beta) = \int_0^1 \{1 - \exp[-(2\beta u\zeta/\pi)^{1/2}]\}\zeta^{-1}\,d\zeta$$

$$= 2\{\gamma + \tfrac{1}{2}\ln(2\beta u/\pi) + E_1[(2\beta u/\pi)^{1/2}]\}$$

For $\beta u \ll 1$, this reduces to the correct square-root limit as given by Eq. (3.62).

It should be emphasized here that the absorptance of a wide band, as obtained from both the narrow Elsasser and uniform statistical models, is the same at sufficiently high pressures. As such, for gases whose spectral behavior could be described either by a narrow Elsasser or a uniform statistical model, the use of Eq. (3.56) should be made in radiative transfer calculations at moderately high pressures ($P \geq 1$ atm).

3.3.2.3. Exponential Wide Band Absorptance from the General Statistical Model. For the sake of completeness and comparison of results, the relation for the wide band absorptance formulated by Felske and Tien (1974) is presented here. This may be obtained by combining Eqs. (2.16), (3.49b), and

(3.50) as

(3.69a) $$\bar{A}(u, \beta) = \int_0^1 \{[1 - \exp(-\rho t)]/[\xi^2 + (2\rho_u)^2]^{1/2}\} \, d\xi$$
$$+ \int_0^1 \{[1 - \exp(-\rho t)]/\xi\} \, d\xi$$

where

(3.69b) $$\rho = [1 - (u\zeta/t)] - 1/[1 + (u\zeta/t)]$$
(3.69c) $$\rho_u = \{(t/u)[1 + (t/u)]\}^{-1/2}$$
(3.69d) $$\xi = \rho/\rho_u \qquad t = \beta/2 = \pi\gamma_L/d$$

In the linear and logarithmic limits, Eqs. (3.69) reduces to the expressions given by Eq. (3.57), and in the square-root limit it reduces to Eq. (3.62). The solutions of Eq. (3.69) are illustrated in Fig. 20 along with the limiting solutions.

3.3.2.4. Exponential Wide Band Absorptance from the Random Elsasser Model. By combining Eqs. (3.42), (3.44), (3.50), and (3.51), the expression for the total band absorptance for this case can be written as

(3.70) $$\bar{A}(u, \beta) = \int_0^1 \left\{1 - \sum_{i=1}^N [1 - (u_i \zeta \sinh \beta_i / G_1(u_i, \beta_i))]\right\} \frac{d\zeta}{\zeta}$$

where

$$G_1(u_i, \beta_i) = [(u_i \zeta \sinh \beta_i + \cosh \beta_i)^2 - 1]$$

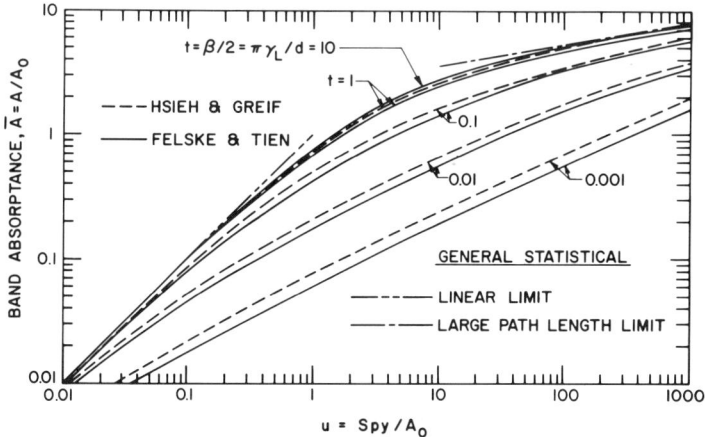

FIG. 20. Wide band absorptance based on the narrow general statistical model.

This is the general result for absorptance derived from the narrow random Elsasser model. The special form of this equation is obtained for $N = 1$ as

$$(3.71) \quad \bar{A}(u, \beta) = \int_0^1 (u \sinh \beta / G_1(u, \beta)) \, d\zeta$$

Note that Eq. (3.71) could have also been obtained by combining Eqs. (3.46), (3.50), and (3.51). The solution of Eq. (3.71) is found to be

$$(3.72) \quad \bar{A}(u, \beta) = \ln[(G_2(u, \beta) + u \sinh \beta + \cosh \beta)/(\sinh \beta + \cosh \beta)]$$

where

$$G(u, \beta) = [(u^2 + 1) \sinh^2 \beta + 2u \sinh \beta \cosh \beta]^{1/2}$$

It can be shown easily that Eq. (3.72) reduces to the correct limiting forms in the linear, square-root, and logarithmic limits.

The relations for the exponential wide band absorptance given by Eqs. (3.55), (3.66), (3.69), and (3.72) are referred to as the exact relations, and the numerical solutions of these equations are termed as the "exact" solutions. These solutions are compared for three different values of the line structure parameter in Fig. 21. It is seen that the absorptance by wide band models follow the same general trend as by narrow band models illustrated in Fig. 18. Once again it should be emphasized that, at larger path lengths, the Elsasser theory predicts higher absorption than the general or uniform

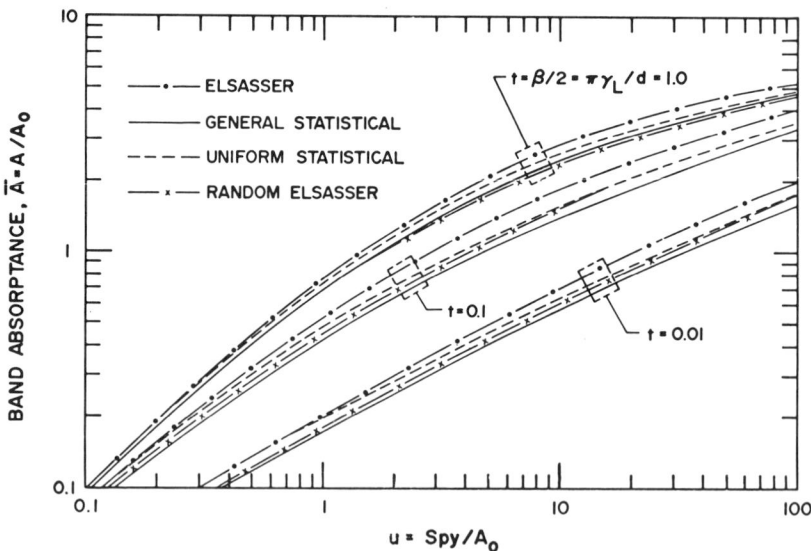

FIG. 21. Comparison of absorptance by wide band models.

statistical model. This fact is clearly evident from the results of Figs. 18 and 21. For $t = \beta/2 = 1$, the absorption by the random Elsasser model in Fig. 21 is seen to be lower than the general statistical model. This does not appear to be physically realistic because the results of the random Elsasser model must fall between the results of the Elsasser and general statistical models.

3.3.3. Axial or Slab Band Absorptance Model. The primary reason for employing the band absorptance models is to represent accurately the absorption-emission characteristics of a vibration-rotation band and consequently eliminate the spectral integration in the radiative flux equations. The angular dependency of radiation usually is not included in the molecular band models. In recent years, however, attempts have been made to incorporate the angular integration of the equation of radiative transfer by introducing the so-called axial or slab band absorptance models (Edwards and Balakrishnon, 1972, 1973; Wassel and Edwards, 1974; Lin and Chan, 1975). By use of these models, in certain cases, the spectral as well as angular integration in the radiative flux equations can be avoided. The expression for the slab band absorptance can, in general, be written as

$$(3.73) \qquad \bar{A}_s(u, \beta) = 2 \int_0^1 \mu \bar{A}(u/\mu) \, d\mu$$

where $\mu = \cos\theta$ and $\bar{A}(u/\mu)$ is the nondimensional total band absorptance. By introducing different relations for the total band absorptance in Eq. (3.73), various expressions for the slab band absorptance can be obtained.

Edwards and Balakrishnan (1972) have proposed that when the pressure is high (i.e., when rotational line structure has smeared out), then the spectral absorption coefficient of many molecular gases can be represented by the exponential-winged band model, and this results in an expression for the slab band absorptance as

$$(3.74) \qquad \bar{A}(u) = \ln(u) + E_1(u) + \gamma + \tfrac{1}{2} - E_3(u)$$

where $E_n(u)$ are the exponential integral functions. As will be shown in the next section, the use of Eq. (3.74) is justified for relatively large path length values.

3.4. Band Absorptance Correlations

The divergence of radiative flux usually involves multiple integrals even for the simple case of energy transfer between a plane-parallel geometry. In order to reduce the mathematical complexities and save the computational time, it often becomes essential to express the integral form of the total band absorptance by fairly accurate continuous correlations. Several continuous relations for total absorptance of a wide band, which are valid over different values of path length and line structure parameter, are available in the literature. A brief description of these correlations is given here in the sequence that they became available in the literature.

The first band absorptance correlation, satisfying the linear, square-root, square-root-logarithmic, and logarithmic limits of a wide band absorptance, was proposed by Edwards and Menard (1964). By comparing the results of correlations in various limits with experimental data over a large range of pressure and temperature, Edwards and co-workers have determined empirically the necessary correlation quantities $S(T)$, $A_0(T)$, and $\beta(T, P_e)$ for the important bands of CO, CO_2, H_2O, and CH_4. These results are summarized by Edwards et al. (1967).

A continuous band absorptance correlation has been proposed by Tien (1968) and Tien and Lowder (1966), and this is of the form

$$(3.75) \qquad \bar{A} = \ln\{uf(t)[(u+2)/(u+2f(t))] + 1\}$$

where

$$f(t) = 2.94[1 - \exp(-2.60t)] \qquad t = \beta/2$$

The choice of Eq. (3.75) was based on the specification of five conditions, and the form of $f(t)$ was chosen so as to give agreement with the correlation of Edwards and Menard. This correlation does not reduce to the correct limiting form in the square-root limit (Cess and Tiwari, 1972). Extensive use of this correlation has been made in various radiative transfer analyses in the past 10 years (see, e.g., Cess and Tiwari, 1972; Tiwari and Cess, 1971; Donovan and Greif, 1975; Cess et al., 1967). The results of this correlation are compared with other correlations in Fig. 22 for $t = \beta/2 = 0.01$ and 1. Comparative results for other t values are available in Tiwari (1976) and Tiwari and Batki (1975). From these results it is concluded that the use of Eq. (3.75) should be restricted to relatively large β values.

Another continuous correlation for band absorptance has been proposed by Goody and Belton (1967), and in terms of the present nomenclature this may be written as

$$(3.76) \qquad \bar{A} = 2 \ln\{1 + u/[4 + (\pi u/4t)]^{1/2}\}$$

Although this correlation satisfies the linear, square-root, and logarithmic limits, its use is restricted to relatively small β values (Cess and Tiwari, 1972).

Tien and Ling (1969) have proposed a simple two-parameter correlation for $\bar{A}(u, \beta)$ applicable under certain thermodynamic conditions, and this is of the form

$$(3.77) \qquad \bar{A}(u) = \sinh^{-1}(u)$$

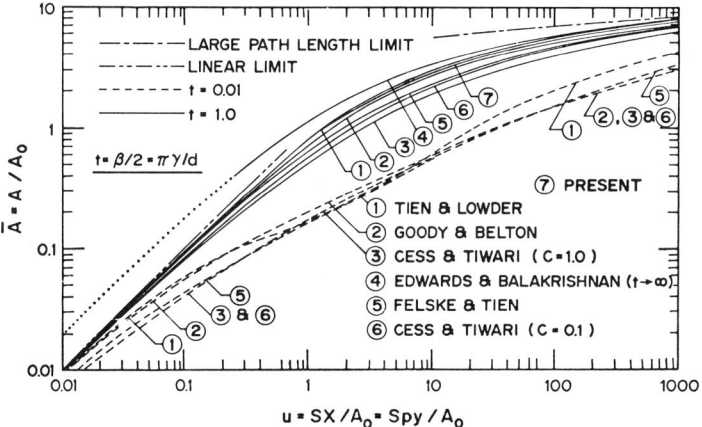

FIG. 22. Comparison of results of band absorptance correlations.

This equation is valid only for the limit of large β. By obtaining the large β limit of Eq. (3.75) as

(3.78) $$\bar{A}(u) = \ln\{2.94u[(u + 2)/(u + 5.88)] + 1\}$$

Tien and Ling have shown that throughout the whole range of u, the maximum deviation of Eq. (3.77) from Eq. (3.78) never exceeds 7%.

A relatively simple continuous correlation for band absorptance, which was introduced first by Cess and Tiwari (1972) and later applied by Cess and Ramanathan (1972), is given by

(3.79) $$\bar{A}(u, \beta) = 2 \ln(1 + u/\{2 + [u(1 + 1/\bar{\beta})]^{1/2}\})$$

where $\bar{\beta} = 4t/\pi = 2\beta/\pi$. It is seen from Fig. 22 (number 3 curves) that this correlation yields lower absorptance than other correlations over the entire range of path length. As such, use of this correlation is justified, at relatively high pressures, for gases whose spectral behavior can be described by the general statistical model.

In the limit of large pressures (i.e., large β limit), the relation for the slab band absorptance proposed by Edwards and Balakrishnon, Eq. (3.74), can be treated as another correlation for the total band absorptance. The results of this correlation are found to be valid only at large path lengths (Tiwari, 1976; Tiwari and Batki, 1975).

Based upon the formulation of the total band absorptance from the general statistical model, as given by Eq. (3.69), Felske and Tien (1974) have proposed a continuous correlation for $\bar{A}(u, \beta)$ as

(3.80) $$\bar{A}(u, \beta) = 2E_1(t\rho_u) + E_1(\rho_u/2) - E_1[(\rho_u/2)(1 + 2t)] + \ln[(t\rho_u)^2/(1 + 2t)] + 2\gamma$$

This correlation is valid for the entire range of the governing parameters. Further discussion of the validity of this correlation is given in the next section.

Another form of the band absorptance correlation is obtained by slightly modifying the original correlation proposed by Cess and Tiwari as

(3.81) $\bar{A}(u, \beta) = 2 \ln(1 + u/\{2 + [u(C + \pi/2\beta)]^{1/2}\})$

where $C \leq 1$. A value of $C = 1$ was suggested in [19]. For $\beta \leq 1$, $C = 0.1$ gives an accurate fit for all path lengths. For $\beta > 1$ and $u \leq 1$, C again is equal to 0.1, but for $\beta > 1$ and $u > 1$ a value of $C = 0.25$ should be used. If it is desired to use only one value of C for all β and path lengths, the value of $C = 0.1$ is recommended. It should be pointed out here that both Eqs. (3.79) and (3.80) reduce to the correct limiting forms suggested in Cess and Tiwari (1972) and Edwards and Menard (1964). The results of this correlation (number 6 curves) are in general agreement with the results of Eq. (3.80) for all u and β values.

The form of the absorptance given by Eq. (3.56) can be treated as another correlation for the total band absorptance. It should be emphasized here that the only restriction in the use of Eq. (3.56) is that the pressure must be sufficiently high. Thus, use of Eq. (3.56) is justified at all path lengths for $t = \beta/2 \geq 1$ (Tiwari, 1976; Tiwari and Batki, 1975).

For gases whose spectral characteristics warrant use of the Elsasser model, the series-form solution given by Eq. (3.65) can be regarded as another correlation for the total band absorptance. As pointed out earlier, the series in Eq. (3.65) converges very rapidly, and the results obtained by this are in excellent agreement with the numerical solution of Eq. (3.55). Thus, use of this correlation in actual radiative transfer problems will provide mathematical flexibilities as well as result in saving of computational time.

3.5. Comparison of Wide Band Absorptance Results

The results of different correlations have been compared with each "exact" solution of the wide band absorptance in this section. In comparing the results of a correlation with an "exact" solution, the limitations of both the particular correlation and the exact solution must be noted. A correlation developed from the general statistical model should not be expected to give good agreement when compared with the exact solution based on the Elsasser model, or *vice versa*.

The wide band absorptance results of various correlations are compared with the wide band "exact" solutions based on the Elsasser and general statistical models in Figs. 23 through 26 for different values of the line structure parameter $t = \beta/2$. Comparative results based on the uniform statistical and random Elsasser models are given in Tiwari (1976) and Tiwari and Batki (1975). As would be expected, maximum errors (in most cases) occur in the intermediate path lengths. This is because most correlations are developed to satisfy at least the linear and logarithmic limits. As discussed earlier, the correlation presented by Edwards and Balakrishnon for large β is seen to be valid only for relatively large u values. In general, the correlation by Tien and Lowder appears to give maximum errors for low β values. The correlation by Goody and Belton gives maximum errors for large u and large β values [because its use is restricted to relatively small values of β (Cess and Tiwari, 1972)].

The maximum error by Felske and Tien's correlation is $+25\%$ (at $t = 0.1$, $u = 0.5$) for the case when it is compared with the exact solution based on the Elsasser model (see Fig. 23). The maximum error by Cess and Tiwari's

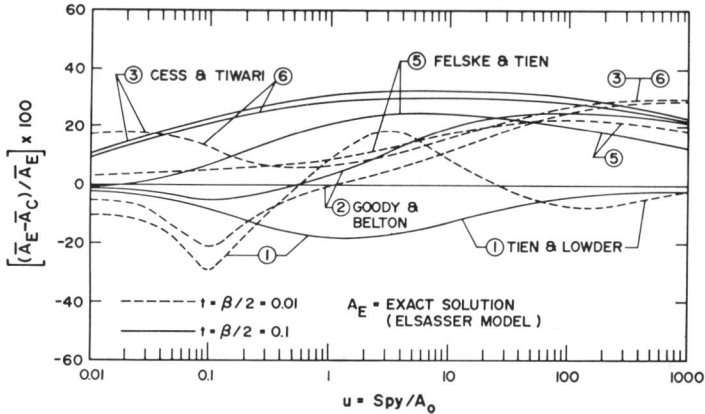

FIG. 23. Errors in the band absorptance correlations when compared with the exact solution (Elsasser model) for $t = 0.01$ and 0.1.

correlation is $+30\%$ (at $t = 0.1$, $u = 10$), and also for the case compared with the exact solution based on the Elsasser model (Fig. 23, curve 6). The advantage in using Eq. (3.81) is that it does not involve any exponential integral and therefore requires significantly less computational time for radiative transfer analyses.

For $t = 1$, the results of correlation given by Eq. (3.56) are within 0.6% of the exact solution based on the Elsasser model and therefore could not be shown in Fig. 24. For $t > 1$, the results of Eqs. (3.55) and (3.56) are identical for all path lengths. When compared with the exact solution based on the uniform statistical model (Tiwari, 1976; Tiwari and Batki, 1975), the results of Eq. (3.56) indicate a maximum error of about 11% for $t = 1$ and 0.7% for

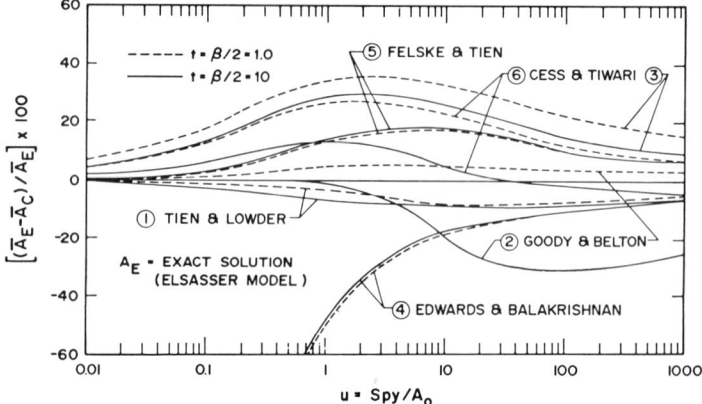

Fig. 24. Errors in the band absorptance correlations when compared with the exact solution (Elsasser model) for $t = 1$ and 10.

$t = 10$. When compared with the exact solution based on the general statistical model (Fig. 26), the results of Eq. (3.56) indicate a maximum error of about 18% for $t = 1$ and 2% for $t = 10$. The use of Eq. (3.56) should, therefore, be made in all cases at sufficiently high pressure ($P = 1$ atm and higher). Equation (3.56) is especially useful for gases whose spectral behavior could be described by the Elsasser and uniform statistical models.

The comparison of results of Felske and Tien's correlation with the numerical solution of Eq. (3.69) indicates excellent agreement (Figs. 25 and 26) for all β and u values. This, however, would be expected because the correlation was obtained from Eq. (3.69) which was derived from the general

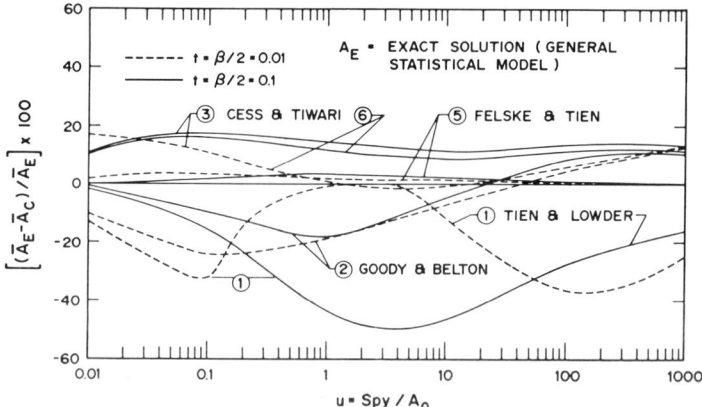

Fig. 25. Errors in the band absorptance correlations when compared with the exact solution (general statistical model) for $t = 0.01$ and 0.1.

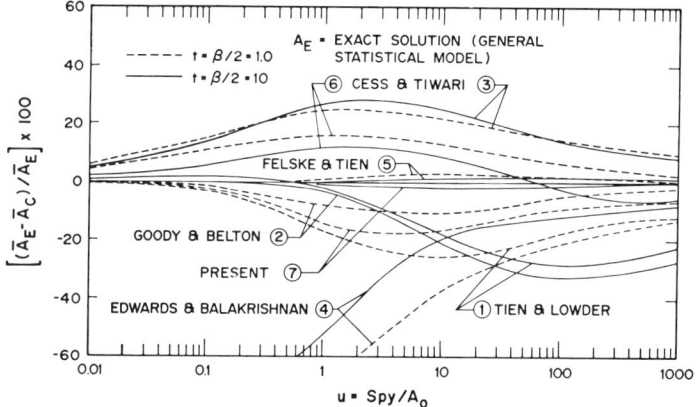

FIG. 26. Errors in the band absorptance correlations when compared with the exact solution (general statistical model) for $t = 1$ and 10.

statistical model. The correlation, therefore, is very useful in radiative transfer analyses involving those molecules whose absorption-emission characteristics can be represented by the general statistical model.

3.6. Band Emissivity (Total Emissivity)

The concept of band emissivity (total emissivity, or, simply, emissivity) has proven to be very important in many radiative transfer analyses. It is particularly useful in calculating atmospheric radiation fluxes and cooling rates. As with the band model correlations, the expressions for emissivity are useful in eliminating the integration over the complicated line structure of the atmospheric spectrum. A large amount of information concerning various relations for emissivity of different gases is available in the literature. Important formulations for emissivity of atmospheric constituents are given in Goody (1964a), Rodgers (1967), Manabe and Wetherald (1967), Taylor and Foster (1974), and Cess (1974). No attempt is made here to summarize all the information available in the literature. Instead, a few fundamental expressions for the emissivity are presented which, along with the band models discussed earlier, can be used in many atmospheric radiative transfer analyses.

For a homogeneous path of absorbing-emitting gases, the spectral emissivity is defined by

(3.82) $$\varepsilon_\omega = A_\omega = 1 - \exp(-\kappa_\omega X)$$

where $A_\omega = \alpha_\omega$ is the spectral absorption defined in Eq. (3.2). The total emissivity (or, simply, emissivity) refers to the emissivity over the entire energy spectrum, and for a homogeneous path of absorber it is defined by

$$\varepsilon(X, p, T) = (\pi/\sigma T^4) \int_0^\infty A_\omega(X, p, T) B_\omega(T)\, d\omega \tag{3.83}$$

where σ is the Stefan–Boltzmann constant.

For a single gas, whose absorption spectrum consists of bands of rotational lines, the column emissivity (Goody, 1964a; Tien, 1968) may be expressed by

$$\varepsilon(X, p, T) = \left[\sum_i B_i(T) A_i(X, p, T)\right] \bigg/ \sum_i B_i(T) \tag{3.84}$$

where $B_i(T)$ is the Planck function evaluated at the band center, $\sum_i B_i(T) = \sigma T^4$, and A_i represents the integrated absorptance of the ith band and is given by Eq. (3.2). For a single band gas, Eq. (3.84) reduces to

$$\varepsilon(X, p, T) = a(T) A_i(X, p, T) \tag{3.85a}$$

where

$$a(T) = \pi B_i(T)/(\sigma T^4) \tag{3.85b}$$

By employing appropriate band absorptance relations for individual bands, theoretical expressions for emissivity of a gas can be obtained from Eq. (3.84). Band absorptance correlations, discussed in Section 3.4, are especially useful for this purpose.

For many atmospheric applications, the process of radiative transfer corresponds to the limit of strong rotational lines (Rodgers, 1967; Cess, 1974), and emissivity can be expressed in terms of a single parameter (see the discussions on the strong line approximation). For example, if one considers the statistical band model consisting of Lorentz lines, then the band absorptance for strong line approximation is given by Eq. (3.32b). A combination of Eq. (3.32b) and Eqs. (3.85) results in

$$\varepsilon(\zeta) = a(T) A_i(\zeta) = a(T)[1 - \exp(-\sqrt{\zeta})] \tag{3.86a}$$

where

$$\zeta = 2\beta^2 x/\pi = 4\gamma_L S_j X/d^2 \tag{3.86b}$$

In the strong nonoverlapping line limit (square-root limit), $A_j(\zeta) = \sqrt{\zeta}$, and Eqs. (3.86) reduces to

$$\varepsilon(\zeta) = a(T)\sqrt{\zeta} \tag{3.87}$$

By making use of Eq. (3.84), Eqs. (3.86) and (3.87) can be extended to the case of multiband gases. The constants appearing in $a(T)$ and ζ can be evaluated for a particular gas (with a single or multiple bands) under varying conditions.

The above relations for emissivity are written for a homogeneous path. In a real atmosphere, the temperature varies along a nonhomogeneous path, and these relations should be appropriately modified. The emissivity for a nonhomogeneous atmosphere (for the path between levels z and z') can be expressed by

$$\varepsilon(z, z') = \int_0^\infty A_\omega(z, z')[B_\omega(z')/B(z')]\, d\omega \tag{3.88a}$$

where

$$B(z') = \int B_\omega(z)\, d\omega = \sigma T^4/\pi \tag{3.88b}$$

Various relations for spectral absorptance can be used in Eqs. (3.88), and the resulting equations can be properly scaled for nonhomogeneous atmospheric applications.

Several relations for emissivity of various gases are available in the literature (Goody, 1964a; Tien, 1968; Rodgers, 1967; Manabe and Wetherald, 1967; Taylor and Foster, 1974; Cess, 1974; Penner, 1973; Hottel, 1954; Hottel and Sarofim, 1967). Early experimental investigations on the emissivity of the gases such as CO, CO_2, H_2O, $CO_2 + H_2O$, SO_2, NH_3, NO_2, and CH_4 are summarized by Hottel (1954) and Hottel and Sarofim (1967). Recent investigations are discussed in Tien (1968), Rodgers (1967), Manabe and Wetherald (1967), Taylor and Foster (1974), Cess (1974), and Penner (1973).

It should be pointed out here that any single relation for the emissivity cannot be expected to apply over a wide range of atmospheric conditions. Furthermore, in radiative flux calculations, not only the expressions for emissivity but also its derivatives are required. Different relations are needed for upward and downward flux calculations. These points are discussed in some detail by Goody (1964a) and Rodgers (1967).

4. Evaluation of Transmittance and Integrated Absorptance of Selected Infrared Bands

In this section, transmittance and integrated absorptance computations are made for several bands of different gases by employing the line-by-line (LBL) and quasi-random band (QRB) model formulations under conditions of pressure and temperature for which experimental measurements are available. The sole motivation for this was to examine the possibility of using the quasi-random band model formulation for transmittance computations as required in surface temperature retrieval and other data reduction procedures (Ludwig et al., 1973; Kunde, 1967; Gupta and Tiwari, 1975, 1976a,b). In some other applications (such as earth radiation budget satellite system studies, etc.), it might be possible to use the wide band correlations for the total band absorptance. For this reason, total absorptances of selected bands were obtained by using the continuous correlations of Tien (1968), Tien and Lowder (1966), Tiwari (1976), Cess and Tiwari (1972), and Felske and Tien (1974). These are compared with the LBL, QRB, and experimental results for a range of physical parameters.

For a homogeneous path of an absorber, the monochromatic transmittance at any wave number location ω is given by

$$\tau(\omega) = \exp[-\kappa(\omega)\bar{u}] \tag{4.1}$$

where $\kappa(\omega)$ is the absorption coefficient at ω per centimeter-atmosphere, and $\bar{u}$ represents the pressure path length of the absorber in centimeter-atmospheres. To calculate the transmittance from Eq. (4.1) it is essential to employ an appropriate spectral model for the absorption coefficient.

4.1. Transmittance Models and Computational Procedures

The line-by-line (direct integration) and quasi-random band models are used in calculating the spectral transmittance. The computational procedures are described in detail in Wyatt et al. (1962), Kunde (1967), Gupta and Tiwari (1975), and Kunde and Maguire (1974).

In the direct integration procedure the entire frequency range of iterest is first divided into a large number of narrow intervals $\Delta\omega$. Each interval is then divided into a variable number of subintervals depending upon the number of lines within the interval. Two very narrow subintervals are created on each side of a line center. The transmittance is computed at four frequency locations in each subinterval and is averaged finally over each interval. The total absorption coefficient at any wave number ω consists of contributions from all the lines in the vicinity and is computed in two parts as

$$\kappa(\omega) = \kappa^D(\omega) + \kappa^w(\omega) \tag{4.2}$$

where $\kappa^D(\omega)$ and $\kappa^w(\omega)$ are called the direct and wing contributions, respectively. Direct contribution originates from lines in very close vicinity (on both sides), and for Lorentz lines this is obtained from

$$\kappa^D(\omega) = \sum_n S_n \gamma_n / \{\pi[(\omega - \omega_n)^2 + \gamma_n^2]\} \tag{4.3}$$

where ω_n refers to the center of the nth contributing line. The wing contribution arises from lines which are farther from ω than the range of direct contribution, and for Lorentz lines this is given by the expression

$$(4.4) \qquad \kappa^w(\omega) = \sum_n S_n \gamma_n / [\Pi(\omega - \omega_n)^2]$$

For complete information on the direct integration procedure, reference should be made to Gupta and Tiwari (1975) and Kunde and Maguire (1974).

In the quasi-random band model, the entire band span Δ is divided into a number of small subintervals of equal spectral width δ. The appropriate number of such intervals is obtained through numerical experimentations. The lines within each subinterval are distributed into five intensity decades, and average intensity for each decade is obtained first. The average transmittance over δ due to a single line of intensity S_n (Gupta and Tiwari, 1975) is given by

$$(4.5) \qquad \bar{\tau}_n(\delta) = \frac{1}{\delta} \int_\delta \exp[-S_n \bar{u} f(\omega, \omega_n)] \, d\omega_n$$

where it should be noted that the variable of integration is the line center location ω_n. If N is the number of lines in an intensity decade, then the average transmittance due to all lines in that decade is given by

$$(4.6) \qquad \bar{\tau}_d(\delta) = \left[\frac{1}{\delta} \int_\delta \exp[-\bar{S}_n \bar{u} f(\omega, \omega_n)] \, d\omega_n \right]^N$$

where $\bar{S}_n$ is the average intensity of all the lines within the decade under consideration. The average transmittance due to all lines in the five intensity decades of the subinterval δ is given by

$$(4.7) \qquad \bar{\tau}_k(\delta) = \prod_{d=1}^{5} \bar{\tau}_d(\delta) = \prod_{d=1}^{5} \left[\frac{1}{\delta_k} \int_{\delta_k} \exp[-\bar{S}_n \bar{u} f(\omega, \omega_n)] \, d\omega_n \right]$$

where subscript k represents the kth spectral subinterval (i.e., δ_k) of the total spectral interval Δ. Equation (4.7) represents the transmittance due to the lines within δ_k. The wings of the lines in the adjacent subintervals also make a significant contribution to the absorption in δ_k. The resultant transmittance over the subinterval δ_k, therefore, is given by

$$(4.8) \qquad \bar{\tau}_k(\delta) = \bar{\tau}_{k-k}(\delta) \prod_{\substack{j=1 \\ j \neq k}}^{K} \bar{\tau}_{k-j}(\delta)$$

where $\bar{\tau}_{k-j}(\delta)$ represents the transmittance in δ_k due to lines in δ_j, and K is the number of adjacent subintervals (on both sides) from which the wing contribution is considered significant. The average transmittance for the entire range Δ is expressed by

$$\bar{\tau}(\Delta) = \frac{1}{K} \sum_{k=1}^{K} \bar{\tau}_k(\delta) \tag{4.9}$$

For other computational details, reference should be made to Kunde (1967) and Gupta and Tiwari (1975).

4.2 Transmittance of Selected Infrared Bands

In this section, transmittances of selected infrared bands are calculated by employing the line-by-line (LBL) and quasi-random band (QRB) models, and the results are compared with the experimental measurements of Burch et al. (1962). The bands selected for comparison are the CO-fundamental (4.6 μm), 4.5 μm N_2O, 4.3 μm CO_2, and 15 μm CO_2 bands. Transmittance results are presented in Figs. 27 through 30 separately for each band. Line-by-line results are shown by the solid lines, QRB results by the histograms, and the experimental results by the broken lines. Integrated absorptance for these bands are compared in the next subsection. The results for the 6.3 μm H_2O band are available in Gupta and Tiwari (1976a).

The values of the computational parameters (such as the number of narrow subintervals for the LBL and QRB models, the ranges of direct and wing contributions, etc.) used for each band and the reasons for using them

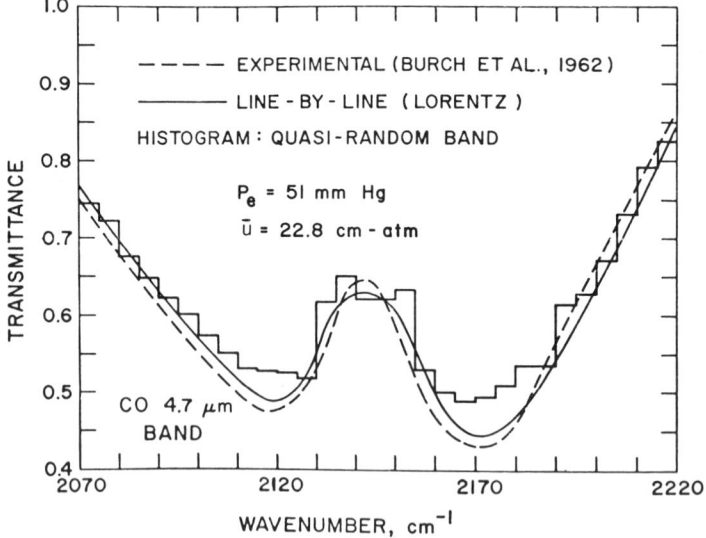

FIG. 27. Comparison of transmittances of the CO-fundamental band.

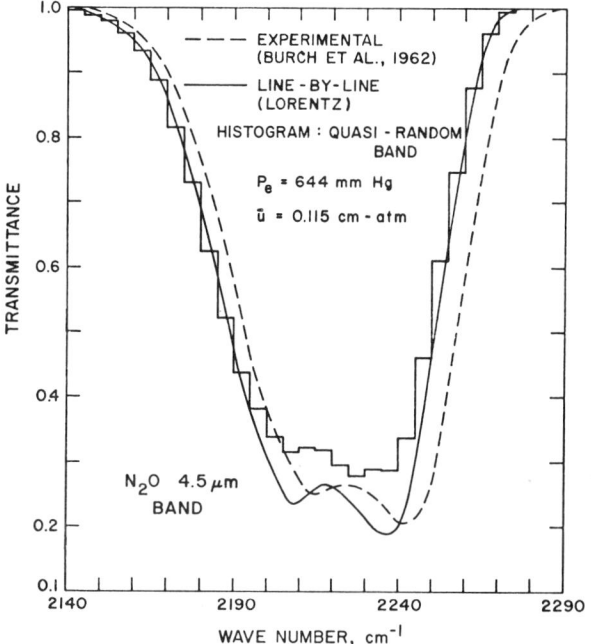

FIG. 28. Comparison of transmittances of the 4.5 μm N_2O band.

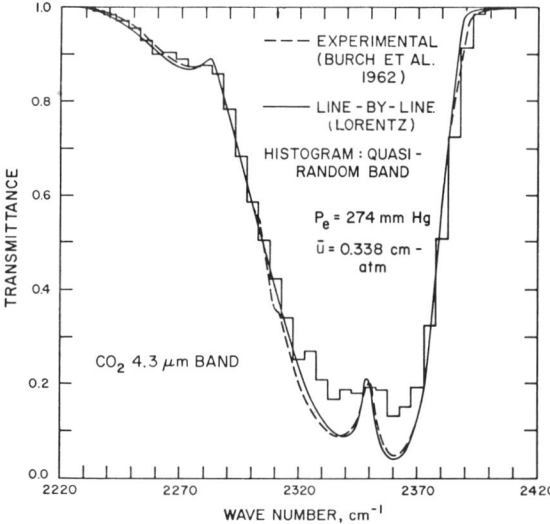

FIG. 29. Comparison of transmittances of the 4.3 μm CO_2 band.

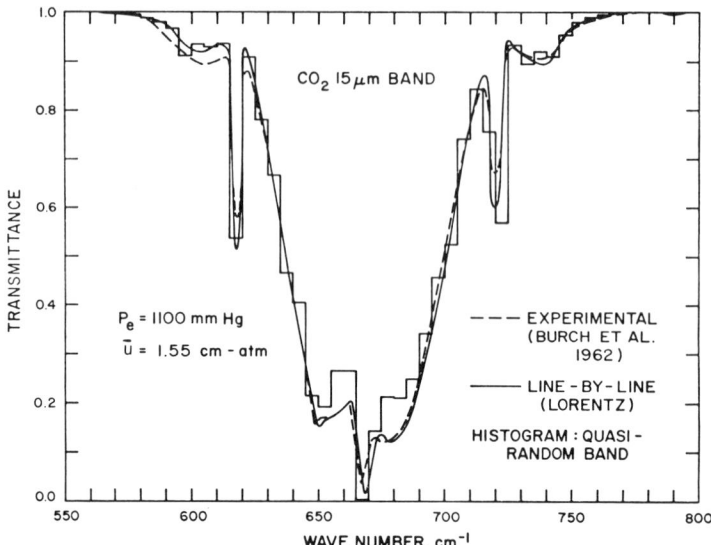

FIG. 30. Comparison of transmittances of the 15 μm CO_2 band.

are discussed in Gupta and Tiwari (1976a). In Burch et al. (1962), the experimental results were obtained by using different effective slit widths for different bands. The theoretical results, therefore, were also degraded with the corresponding slit functions for comparison.

For the CO-fundamental band, transmittances were calculated for the spectral range of $\omega = 2070$ to 2220 cm^{-1}, and these are compared with the experimental results in Fig. 27. The LBL results are seen to be in good agreement with the experimental results, while the QRB results are seen to exhibit appreciable differences (particularly in the P and R branches of the band).

For the 4.5 μm N_2O band, the spectral range selected for transmittance calculations is 2140 to 2290 cm^{-1}, and the results are presented in Fig. 28. With the exception of the fact that the experimental curve is shifted approximately 6 cm^{-1} toward the higher frequency side (with respect to the theoretical curves), the agreement between the LBL and experimental results is very good. The QRB results again show appreciably lower absorption.

For the 4.3 μm CO_2 band, the comparison of theoretical and experimental transmittances is shown in Fig. 29 for the spectral range of 2220 to 2420 cm^{-1}. The agreement between the experimental and the LBL results is seen to be excellent. The QRB results again exhibit a slightly lower absorption. For the 15 μm CO_2 band, the transmittance results are presented in Fig. 30, and the agreement between the three results is seen to be excellent.

From the results presented in this subsection it is concluded that the LBL results are in better agreement with the experimental values than the QRB results. In view of the high accuracy required for the temperature retrieval and other data reduction work (Ludwig et al., 1973; Kunde, 1967; Gupta and Tiwari, 1975, 1976b), it would be desirable to use the line-by-line model. In other atmospheric work, however, use of the quasi-random band model may be justified.

4.3. Integrated Absorptance of Selected Infrared Bands

The total (integrated) absorptances of selected infrared bands were calculated by using the LBL and QRB models and the continuous correlations of Tien and Lowder [Eq. (3.75)], Felske and Tien [Eq. (3.80)], and Cess and Tiwari [Eq. (3.81)]. Results obtained for different pressures, temperatures, and path lengths are compared with available theoretical and experimental investigations in Tiwari and Gupta (1977). Some important results are discussed briefly in this subsection.

Integrated absorptance of the 4.6 μm CO, 4.5 μm N_2O, 4.3 μm CO_2, 15 μm CO_2, and 6.3 μm H_2O bands were calculated by employing the LBL and QRB models for a specific set of temperature, pressure, and path length. For each band, the physical conditions and frequency range were taken to be exactly the same as used in the experimental measurements of Burch et al. (1962). The results are compared in Table I. For the CO-fundamental band, the results show that the LBL absorptance is about 5% lower and the QRB absorptance is about 9% lower than the experimental value. For the 4.5 μm N_2O band, the LBL absorptance is less than 2% lower than the

TABLE I. COMPARISON OF INTEGRATED ABSORPTANCES FOR VARIOUS BANDS[a]

Band identification and frequency range	Experimental absorptance[b]	Line-by-line (LBL) results		Band model (QRB) results	
		Integrated absorptance (cm^{-1})	Percentage difference with measurement	Integrated absorptance (cm^{-1})	Percentage difference with measurement
CO 4.6 μm (1970–2270 cm^{-1})	73.90	70.32	−4.84	67.30	−8.93
N_2O 4.5 μm (2140–2290 cm^{-1})	57.00	56.01	−1.74	51.43	−9.77
CO_2 4.3 μm (2220–2420 cm^{-1})	73.80	74.09	0.39	69.72	−5.53
H_2O 6.3 μm (1200–2100 cm^{-1})	334.0	326.4	−2.28	323.1	−3.26
CO_2 15 μm (550–800 cm^{-1})	66.10	66.57	0.71	65.35	−1.13

[a] $T = 300°K$.
[b] From Burch et al. (1962).

experimental value, while the difference for QRB absorptance is approximately 10%. For the 4.3 μm CO_2 band, the LBL absorptance is within 0.5% of the experimental value, while the QRB absorptance is approximately 5.5% lower. For the 15 μm CO_2 band, the theoretical absorptance results agree with the experimental ones within 1%. These results further indicate that the LBL results are in better agreement with the experimental values than the QRB results.

In order to investigate the validity of various theoretical formulations for an extended range of pressures and path lengths, total absorptance of selected bands were compared with available experimental measurements in Tiwari and Gupta (1977). For brevity, only the results for the 4.6 μm CO and 15 μm CO_2 bands are presented here (see Tables II and III).

For the CO-fundamental band, results obtained at $T = 300°K$ are compared in Table II. The comparison shows that the LBL and QRB results are in good agreement with the experimental results (LBL being slightly better than QRB). Among other theoretical results, the results of Hashemi et al. (1976) provide the best agreement with the experimental and LBL results. Among the results of the correlations, it can be seen that, for the very low pressures, the Tien and Lowder correlation yields higher values of absorptance, and the Cess and Tiwari and Felske and Tien correlations show better agreement. For medium and high pressures, however, Tien and Lowder's correlation yields better agreement while Cess and Tiwari's and Felske and Tien's correlations yield much lower values. One may conclude, therefore, that use of the Tien and Lowder correlation is justified in radiative transfer analyses involving CO at relatively high pressures.

For the 15 μm CO_2-fundamental band, total absorptance results were calculated by employing the LBL and QRB models and the continuous correlations at $T = 300°K$. The QRB results for this band are given also by Kunde (1967) and Young (1964). The results are presented in Table III for six illustrative cases. An inspection of the table shows that the LBL results agree very well with the experimental results. The present QRB results are higher than the experimental values for very low pressures, but they agree well at moderate and high pressures. Young's results are consistently higher, while Kunde's results agree very well. Looking at the results of continuous correlations, it is noted that Tien and Lowder's results are very low at lower pressures but compare better at the higher pressures. The results of Cess and Tiwari's and Felske and Tien's correlation are consistently lower for all cases considered.

Results presented in this section consistently show that the LBL results are in excellent agreement with the experimental values. In most cases QRB results are better than results of any other theoretical formulation. The results of various correlations agree with varying degrees of accuracy

TABLE II. COMPARISON OF ABSORPTANCE RESULTS FOR THE FUNDAMENTAL BAND OF CARBON MONOXIDE[a]

					Absorptance results, A (cm^{-1})					
					Theoretical				Correlation	
Eff. pr. P_e (atm)	Pr. path length $\bar{u} = P_a y$ (atm-cm)	Opt. path $u = Spy/A_0$	Exp. Burch et al. (1962)	Hashemi et al. (1976)	Edwards and Menard (1964); Edwards et al. (1967)	Tiwari and Gupta (1977) LBL[b]	Tiwari and Gupta (1977) QRB[c]	Tien and Lowder (1966)	Cess and Tiwari (1972); Tiwari (1976)	Felske and Tien (1974)
0.0329	0.0110	0.0686	0.84	0.92	1.3	0.737	0.974	1.0132	0.8091	0.8318
0.0345	1.3752	8.5771	14.2	12.6	17.0	12.86	13.00	7.9414	11.9875	11.3233
0.3368	0.0028	0.0175	0.64	0.64	0.68	0.650	0.678	0.6402	0.4906	0.5789
0.3171	0.4366	2.7231	20.5	21.4	22.9	20.55	19.88	22.6158	16.6929	17.1523
0.3118	15.175	94.647	110.0	108.0	119.7	110.24	104.45	113.3707	75.2403	83.3990
3.974	0.0028	0.0175	0.69	0.68	0.68	0.723	0.726	0.6635	0.5925	0.6543
3.908	0.0753	0.4696	16.3	16.5	17.4	16.50	16.11	15.6790	10.2476	12.7278
3.908	1.3752	8.5771	102.6	113.5	97.5	101.95	94.49	100.0165	58.6714	69.7050
3.908	15.175	94.647	183.4	187.6	188.6	185.31	184.43	193.0015	133.2973	155.5740
1.015	47.421	295.765	184.6	183.4	197.7	187.05	186.93	196.6126	138.9117	160.9919

[a] $T = 300°K$. Experimental data of Burch et al. (1962).
[b] Line by line.
[c] Quasi-random band.

TABLE III. COMPARISON OF ABSORPTANCE RESULTS FOR THE 15 μm FUNDAMENTAL BAND OF CARBON DIOXIDE[a]

							Absorptance results, A (cm^{-1})				
						Theoretical				Correlation	
Path length y (cm)	Eff. pr. P_e (atm)	Pr. path length $\bar{u} = P_a y$ (atm-cm)	Opt. path $u = Spy/A_0$	Exp. Burch et al. (1962)	Young (1964) QRB[b]	Kunde (1967) QRB	Tiwari and Gupta (1977)		Tien and Lowder (1966)	Cess and Tiwari (1972); Tiwari (1976)	Felske and Tien (1974)
							LBL[c]	QRB			
400	0.0205	5.73	87.378	34.6	43.3	34.4	35.44	37.49	17.36	16.08	15.72
400	0.0837	5.73	87.378	54.7	67.3	53.6	56.42	57.54	39.11	27.81	28.90
400	0.4000	5.73	87.378	81.9	94.0	79.2	83.23	83.47	69.93	46.25	51.68
400	1.0092	5.73	87.378	95.3	103.9	93.2	86.39	96.91	88.41	58.89	67.85
1,600	0.0050	5.56	84.465	21.7	26.7	20.1	19.88	24.7	5.48	8.57	8.02
1,600	0.0083	9.20	139.804	32.6	39.3	30.4	30.81	34.18	12.58	13.43	12.91

[a] $T = 300°$K. Experimental data of Burch et al. (1962) and Kunde (1967).
[b] Quasi-random band.
[c] Line by line.

5. Upwelling Atmospheric Radiation

It is possible to infer concentrations of various pollutants from an appropriate analysis of upwelling radiation measurements obtained in passive mode experiments (Ludwig *et al.*, 1973; SAI, 1973; Gupta and Tiwari, 1975). Since variables like surface temperature, surface emittance, concentration of water vapor and other more abundant species (e.g., CO_2) affect the upwelling radiance to a greater extent than the less abundant pollutants like CO, a theoretical study of the effects of these variables on the upwelling radiance is essential in order to be able to obtain meaningful information regarding pollutants. In this section, basic equations for calculating the upwelling atmospheric radiance are presented. These account for the various sources of radiation coming out at the top of the atmosphere. The line-by-line and quasi-random band models are used for the evaluation of transmittance and upwelling radiance in the spectral region of the CO-fundamental band (2070 to 2220 cm^{-1}). The theoretical procedure, however, can be extended easily to any other spectral range. Model calculations have been performed to study the effect of different interfering gases, water vapor profiles, surface temperature, and surface emittance on the upwelling radiance and signal change.

As shown in Fig. 31, the radiation emergent from the atmosphere $E(\omega)$

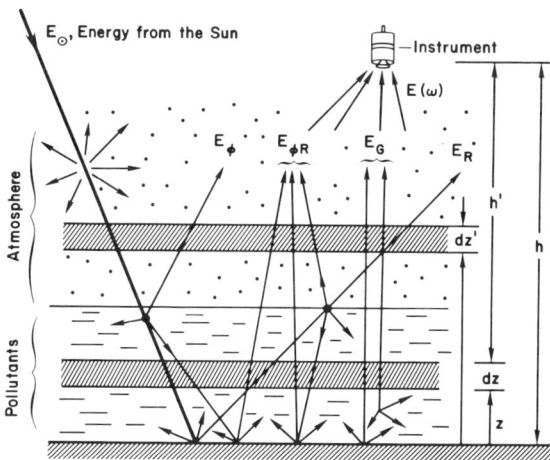

Fig. 31. Radiative energy received by an aircraft- or satellite-mounted instrument.

(Ludwig et al., 1973; Tiwari and Reichle, 1974) may be given by the expression

(5.1) $$E(\omega) = E_G(\omega) + E_R(\omega) + E_\phi(\omega) + E_{R\phi}(\omega)$$

where

$E_G(\omega)$ = Thermal radiation emitted by underlying surface and atmosphere
$E_R(\omega)$ = Incident solar radiation reflected by the surface
$E_\phi(\omega)$ = Radiation scattered by single or multiple scattering processes in the atmosphere without having been reflected from the surface
$E_{R\phi}(\omega)$ = Scattered energy which has undergone a reflection from the surface

In general, these quantities are functions of surface temperature, atmospheric temperature, surface emittance, surface reflectance, sun zenith angle, scattering characteristics of particles, and transmittance of the atmosphere.

In the spectral region of infrared measurements, the effect of scattering is negligible. The incident solar radiation reflected by the surface, however, is important, especially if the surface reflectance is assumed to be high (0.2 and higher).

Upon neglecting the scattering and solar radiation, the expression for thermal radiation emerging from a plane-parallel atmosphere can be written as

(5.2) $$E(\omega) = E_G(\omega) = \varepsilon(\omega)B(\omega, T_s)\tau(\omega, 0) + \int_0^h B(\omega, T(z))\, [d\tau(\omega, z)/dz]\, dz$$

where $\varepsilon(\omega)$ is the surface emittance, $B(\omega, T)$ is the Planck blackbody function, T_s is the surface temperature, $T(z)$ is the temperature at altitude z, and $\tau(\omega, z)$ is the monochromatic transmittance of the atmosphere. The first term on the right-hand side of this equation represents the radiation from the surface, while the second term is the radiation from the atmosphere.

The contribution from sunlight reflected from the surface becomes significant at shorter wavelengths. This contribution is given by the component $E_R(\omega)$ as

(5.3) $$E_R(\omega) = (1/\pi)[1 - \varepsilon(\omega)] \cos \theta H_s(\omega)[\tau(\omega)]^\zeta$$

where θ is the sun zenith angle and $\zeta = 1 + f(\theta)$. The function $f(\theta) = \sec \theta$ for $0 \leq \theta \leq 60°$ and equals $Ch\theta$ for $\theta > 60°$ with $Ch\theta$ denoting the Chapman function. $H_s(\omega)$ is the sun irradiance on top of the atmosphere, and $\tau(\omega)$ is the transmission vertically through the atmosphere.

The total energy emergent from the atmosphere is obtained by integrating either Eq. (5.1) or (5.2) over the specified spectral interval $\Delta\omega$ as

(5.4) $$E_D = E_{\Delta\omega} = \int_{\Delta\omega} E(\omega)\, d\omega$$

The procedure for calculating the upwelling radiance, by employing the line-by-line and quasi-random band models for transmittance, is discussed in detail in Gupta and Tiwari (1975), and computer prrograms are provided. A summary of the procedure is given here.

5.1. Procedure for Calculating the Upwelling Radiance

In radiation modeling for pollution measurement in a nonhomogeneous atmosphere, the upwelling radiation is calculated by dividing the atmosphere into an appropriate number of sublayers. Each sublayer is assumed to be homogeneous in species concentration, temperature, and pressure.

In a specified spectral interval in which a particular pollutant absorbs, the total energy emergent from the atmosphere is obtained from Eq. (5.4). If in this interval, n independent measurements (corresponding to the number of homogeneous layers) could be made to find E_{D1}, E_{D2}, ..., E_{Dn}, then the uniform concentration of the pollutant in each layer (and therefore the concentration profile in the actual atmosphere) could be determined from Eq. (5.4). Because of low concentrations of pollutants in the atmosphere, however, n such measurements are not feasible. Thus, only one independent measurement is usually made, and an average value of the particular pollutant concentration in the atmosphere is obtained. Even if only one value of the pollutant concentration can be obtained from an independent measurement, it is essential to divide the nonhomogeneous atmosphere into several homogeneous layers for the purpose of data reduction. This is because the pressure, temperature, and amount of interfering molecules vary in the atmosphere, and spectroscopic parameters and pressure path lengths are strong functions of these variables.

By employing the Lorentz line-by-line model for atmospheric transmittance, the upwelling radiance at the top of the atmosphere is obtained from Eq. (5.4) for each narrow spectral interval $\Delta\omega$. The exact procedure for doing this is to evaluate the average value of the Planck function for this interval first, then by using the mean value of the transmittance for the interval, evaluate the upwelling radiance at the top of the atmosphere. The total upwelling radiance $(E = \sum E_{\Delta\omega})$ at the top of the atmosphere for the entire spectral range Δ is obtained by summing the radiances of individual intervals.

By employing the quasi-random band model, the total upwelling radiance at the top of the atmosphere can be evaluated from Eq. (5.4). First, the net (integrated) radiance is obtained at the top of the atmosphere for each subinterval by calculating the Planck function and the average transmittance for that subinterval. The total radiance for the entire range Δ is then obtained by summing the integrated radiances of each subinterval.

The signal change $SC = \Delta E$ (in watts per square centimeter per steradian) can be calculated by employing Eq. (5.4) as

$$(5.5) \qquad SC = \Delta E = \int_{\Delta\omega} (E(\omega, \tau_0) - E(\omega, \tau_p))\, d\omega$$

where τ_0 represents the transmittance of a "clean" atmosphere in which the pollutant concentration is zero, and τ_p refers to the transmittance of the atmosphere in the presence of the pollutant. The numerical procedure for evaluating Eq. (5.5) is identical to that described for calculating the upwelling atmospheric radiance.

Science Applications, Incorporated [SAI; NASA contractor, responsible for the development of nondispersive correlation instruments for pollution measurement (Ludwig et al., 1973; SAI, 1973)] has compiled spectral line parameters (position, strength, width, and lower energy level) for lines of the CO-fundamental band and for lines of other molecules which interfere with the CO band. In the calculation of transmittances, these line parameters are directly read from a tape provided by SAI (1973) to NASA-Langley.

Solar irradiances at the top of the atmosphere, at a few selected wave numbers, are also available from the SAI tape (1973). Values obtained from the tape, for the spectral range of present interest, are tabulated in Gupta and Tiwari (1975). In the evaluation of the contribution of the reflected solar radiation to the upwelling radiance, the solar irradiance for each spectral subinterval is obtained from a linear interpolation of the tabulated values (Gupta and Tiwari, 1975).

By employing the Lorentz line-by-line and quasi-random band models for atmospheric transmittance, upwelling radiance and signal change were calculated for several illustrative cases.

5.2. Results of Model Calculations

The results of integrated upwelling radiance at the top of the troposphere (i.e., at 10 km) for different CO concentrations (uniformly distributed throughout the troposphere), in the presence of various interfering molecules, are illustrated in Fig. 32. The solid curves represent the results of the line-by-line model and broken curves the quasi-random band model. As would be expected, the upwelling radiance E decreases with increasing CO concentration and with the inclusion of different interfering molecules. Inclusion of O_3 causes a slight decrease in radiance (not exceeding 0.5 %), and it was difficult to illustrate this decrease in Fig. 32. The agreement between the line-by-line and the quasi-random band model results is seen to be excellent for the case of $CO + H_2O$. The slightly lower radiances for the next two cases is attributed to the overestimation of absorption by the band

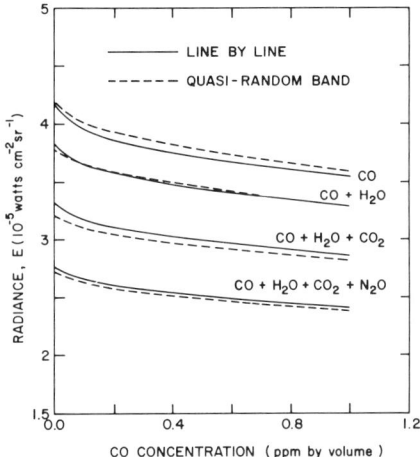

FIG. 32. Upwelling radiance as a function of CO concentration in the presence of interfering molecules; $T_s = 288°K$, $\varepsilon = 0.8$.

model. The reason for this lies in the assumption of random distribution of many lines (in the presence of interfering molecules) in the subintervals of the band model. In the actual spectra, however, the lines are more closely spaced in some regions than in others. The variation in the signal change ΔE with the CO concentration is illustrated in Fig. 33. These results follow the general trend of the results presented in Fig. 32.

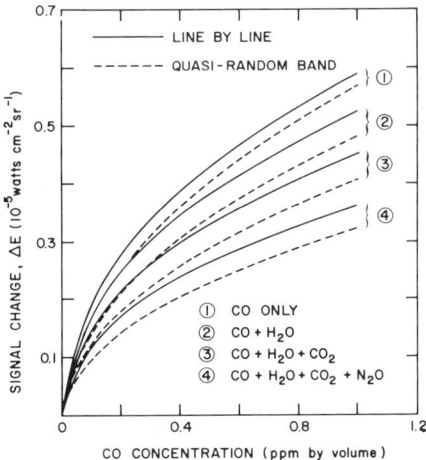

FIG. 33. Signal change as a function of CO concentration in the presence of different interfering molecules; $T_s = 288°K$, $\varepsilon = 0.8$.

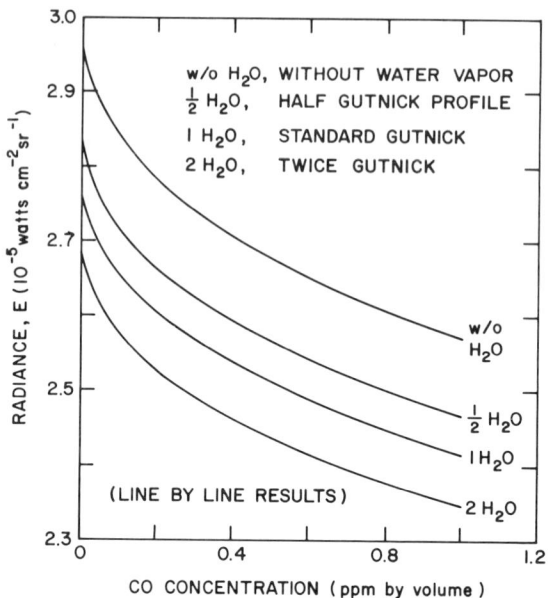

Fig. 34. Upwelling radiance as a function of CO concentration for different water vapor profiles; $T_s = 288°K$, $\varepsilon = 0.8$.

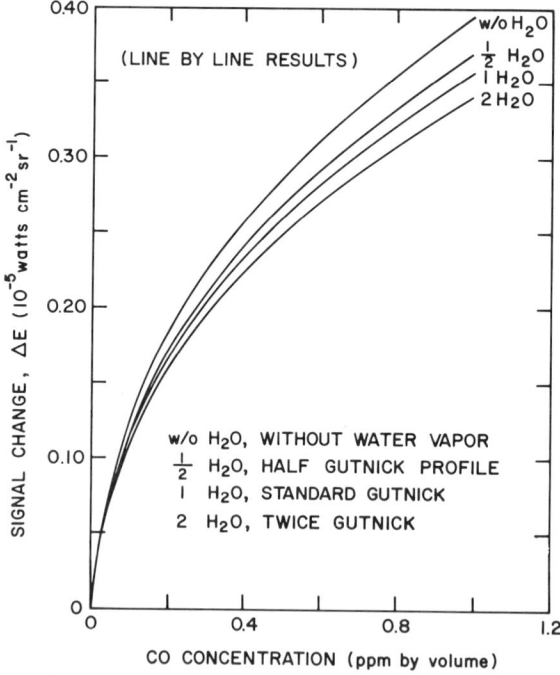

Fig. 35. Effects of water vapor concentration on the signal change; $T_s = 288°K$, $\varepsilon = 0.8$.

The influence of different amounts of water vapor on the upwelling radiance and the signal change is shown in Figs. 34 and 35, respectively. Increased water vapor concentration results in increased absorption in the atmosphere. This, in turn, results in lower values for upwelling radiance and signal change. It should be noted that the effect of CO concentration on the signal change would be relatively small in the presence of a larger quantity of water vapor.

Figure 36 shows the upwelling radiances for surface temperatures of 280, 290, and 300°K and a surface emittance of 0.8. The strong dependence of the upwelling radiance on the surface temperature is obvious from these results. The relatively lower radiance values obtained with the band model are indicative of a slight overestimation of absorption by this model.

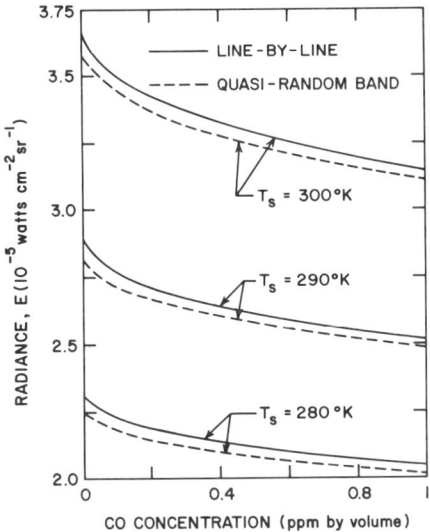

FIG. 36. Upwelling radiance as a function of CO concentration for three different surface temperatures (surface emittance $\varepsilon = 0.8$).

Figure 37 shows the signal change for three different values of surface emittance and for a surface temperature of 288°K. As explained earlier, the radiances obtained from the band model are lower than the line-by-line model because of overestimation of absorption by the band model. The relative increase of the difference for the lower ε values is due to lower total emission from the earth for the small values of surface emittance. In these cases, therefore, the increased absorption by the band model has a greater relative effect on the radiance and signal change.

Figure 38 shows the variation of the upwelling radiance with the surface temperature for a fixed concentration of CO (1 ppm by volume) in the

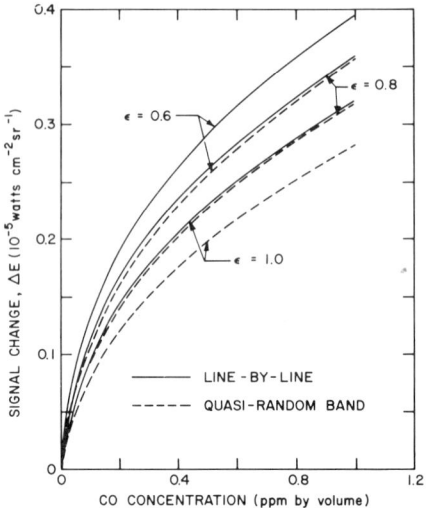

Fig. 37. Signal change as a function of CO concentration for three different surface emittances; $T_s = 288°K$.

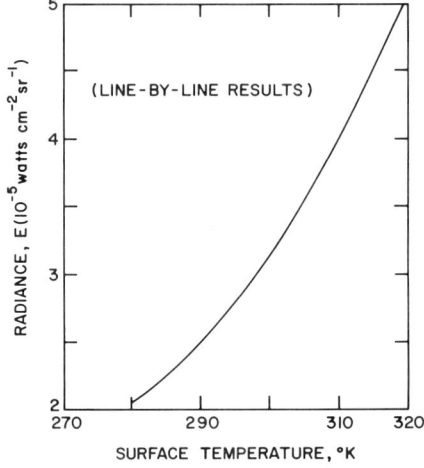

Fig. 38. Upwelling radiance as a function of surface temperature for a fixed CO concentration (1 ppm by volume); $\varepsilon = 0.8$.

atmosphere and for $\varepsilon = 0.8$. The strong dependence of radiance on the surface temperature may be easily explained on the basis of Stefan's law. However, because of the interference from the infrared-active atmospheric molecules, the results obtained here do not exhibit an exact fourth-power relationship.

Figure 39 shows the variation of radiance with the surface emittance for a

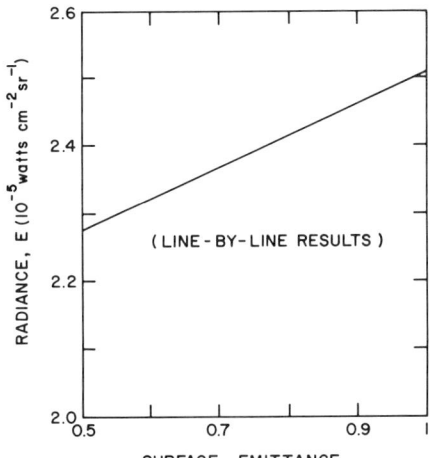

Fig. 39. Upwelling radiance as a function of surface emittance for a fixed CO concentration (1 ppm by volume); $T_s = 288°K$.

fixed CO concentration (1 ppm by volume) and $T_s = 288°K$. As would be expected, the results indicate the linear dependence of radiance on the surface emittance. In general, the ground emittance varies with the wave number. However, for the spectral range of the CO-fundamental band, it was shown in reference [4] that the radiance is not influenced by a significant amount when the wave number-dependent ground emittance is replaced by an averaged value.

Figure 40 shows a comparison of the results obtained from the line-by-line program developed in Gupta and Tiwari (1975) and another line-by-line program (called POLAYER) developed by Science Application Incorporated (1973). In computing the total absorption coefficient at any wave number, the program of Gupta and Tiwari (1975) considers contributions from all the lines up to a fixed wave number location (on both sides) of 45.5 cm^{-1} from the wave number under consideration. This value of 45.5 cm^{-1} for the so-called wing effect was chosen after several numerical experimentations. The POLAYER, on the other hand, considers the effect of a fixed number of 20 lines on each side of the wave number under consideration. This causes the range of wing effect to change depending upon the density of lines in the spectrum. Thus, in some cases, POLAYER will not consider the influence of lines which are only 1 cm^{-1} away from the wave number under consideration. This, of course, will result in underestimation of absorption. This, at least in part, is responsible for the higher integrated radiance (and, therefore, larger signal change) for the POLAYER program.

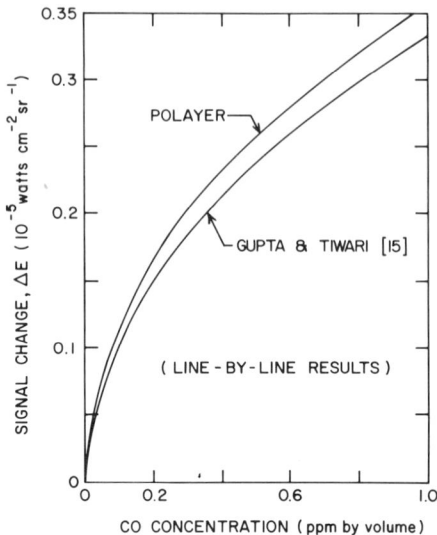

FIG. 40. Comparison of signal change results obtained from the program POLAYER and the present LINBLIN program; $T_s = 288°K$, $\varepsilon = 0.8$. On the average, LINBLIN values are 2 % lower than POLAYER values.

6. Concluding Remarks

The purpose of this study was to review different line and band models for infrared spectral absorption, compare their absorptances and transmittances, indicate their limitations, and establish their usefulness for atmospheric applications.

From the comparison of results of the three line profiles (Lorentz, Doppler, and Voigt), it is concluded that the Voigt line profile should be employed in calculating the transmittance for the middle to upper troposphere and lower stratosphere. The use of the Lorentz line profile is justified for lower tropospheric applications.

From the comparison of wide band absorptance of various band models, it is concluded that the correlation presented by Felske and Tien provides fairly accurate results for all pressures and path lengths. At relatively high pressure, however, the simple correlation given by Tiwari and Batki provides a uniformly better approximation for the total band absorptance.

The homogeneous path transmittances were calculated for the CO-fundamental, 4.5 μm N_2O, 4.3 μm CO_2, and 15 μm CO_2 bands by employing the line-by-line and quasi-random band models. Comparisons of these results with available experimental measurements indicate that in sensitive atmospheric applications (such as surface temperature retrieval and detection of CO concentration), the use of the line-by-line model is desirable.

By employing the line-by-line and quasi-random models, results were obtained to study the effects of different interfering molecules, water vapor profiles, ground temperatures, and ground emittances on the upwelling radiance and signal change in the spectral range of the CO-fundamental band. Physically realistic values of various parameters were used in the model calculations. This information is very useful in interpreting the data obtained from an aircraft- or satellite-mounted instrument to determine the pollutant concentration in the atmosphere.

ACKNOWLEDGMENTS

This work was supported in part by the NASA-Langley Research Center through Grant NSG 1153. Some of the computational work was done by R. R. Batki and S. K. Gupta. I wish to express my appreciation to Dr. H. G. Reichle, Jr. and Dr. S. K. Gupta for helpful discussions and other help in the preparation of this article.

LIST OF SYMBOLS

a	Parameter expressing the ratio of Lorentz to Doppler half-width
A	Band absorptance, cm^{-1}
$A(u, \beta)$	Absorption of a wide band
$\bar{A}$	Nondimensional A
A_j	Absorptance (total absorption) of a single line (jth line)
$\bar{A}_j$	Average value of A_j, dimensionless
$A_{j,\delta}$, $A_{j,D}$, $A_{j,d}$, $A_{j,\infty}$	Absorption of jth line over spectral intervals δ, D, d, and ∞, respectively
A_j^*	Nondimensional A_j, $A_j^* = A_j/A_n$
A_n	Nondimensional constant, $A_n = 2\gamma_D/(\ln 2)^{1/2}$
$A_N(x, \beta)$	Absorption of a narrow band
A_0	Bandwidth parameter, cm^{-1}
A_s	Absorption of a single line in a linear limit, cm^{-1}
$\bar{A}_s(u, \beta)$	Axial or slab band absorptance
$B(\omega, T)$	Planck's function
c	Speed of light
$E(\omega)$	Upwelling radiative energy
E_D	Total upwelling radiative energy in spectral interval $\Delta\omega = D$
$E_G(\omega)$	Thermal radiation of ground and atmosphere
$E_R(\omega)$	Solar radiation reflected from ground
$E_\phi(\omega)$	Solar radiation scattered by the atmosphere without having been reflected by the ground
$E_{R\phi}(\omega)$	Solar radiation reflected from ground outside the field of view and scattered by the atmosphere into the field of view
f_j	Shape factor of jth line
h	Altitude of an aircraft or a satellite
H_s	Sun irradiance at the top of the atmosphere
k	Boltzmann constant
$K(a, v)$	Voigt function
$L(\eta)$	Ladenberg-Reiche function

m — Molecular mass of the absorbing medium
p, P — Partial pressure of the absorbing medium
S — Band intensity or band strength, cm^{-2}
S_j — Line intensity or line strength, cm^{-2}
t — Line structure parameter, $t = \beta/2$
T — Kinetic temperature, °K
T_s — Surface or ground temperature
$T(z)$ — Temperature of atmosphere at altitude z
u — Dimensionless coordinate, $u = Spy/A_0$
$\bar{u}$ — Pressure path length, cm-atm
W_j — Equivalent width of jth spectral line
x — Optical path at the line center
X — Mass of absorbing gas per unit area
β, β_j — Line structure parameter, $\beta = 2t$
γ_j — Line half-width of jth spectral line
ε — Surface emittance
$\varepsilon(X, p, T)$ — Total emissivity
$\varepsilon_j, \varepsilon_1$ — Line overlapping parameter
ε_ω — Spectral emissivity
θ — Sun zenith angle
κ_0 — Absorption coefficient correlation, cm^{-1}
κ_ω — Equilibrium spectral absorption coefficient, cm^{-1}
$\kappa_{\omega j}$ — Absorption coefficient of jth spectral line, cm^{-1}
τ — Radiative transmittance
$\bar{\tau}_{j,D}, \bar{\tau}_{j,\delta}$ — Average transmittance of jth line in the spectral interval D or δ
τ_ω — Spectral transmittance
$\tau_{\omega j}$ — Spectral transmittance of jth line
ω — Wave number, cm^{-1}
ω_j — Wave number the center of jth line
ω_0 — Line center of the main line or band center, cm^{-1}

References

Aida, M. (1975). A statistical method to estimate the vertical transmission through horizontally non-homogeneous media. *J. Quant. Spectrosc. Radiat. Transfer* **15** (6), 503–511.

Allen, L. H. (1963). "Astrophysics—The Atmosphere of the Sun and Stars," 2nd Ed. Ronald Press, New York.

Armstrong, B. H. (1967). Spectrum line profiles: The Voigt function. *J. Quant. Spectrosc. Radiat. Transfer* **7** (1), 61–88.

Armstrong, B. H. (1968). Analysis of Curtis-Godson approximation and radiation transmission through inhomogeneous atmospheres. *J. Atmos. Sci.* **25** (2), 312–322.

Armstrong, B. H., and Nicholls, R. W. (1972). "Emission, Absorption and Transfer of Radiation in Heated Atmosphere." Pergamon, Oxford.

Baranger, M. (1962). Spectral line broadening in plasmas. *In* "Atomic and Molecular Processes" (D. R. Bates, ed.). Academic Press, New York.

Born, M. (1933). "Optik," pp. 482–486. Springer-Verlag, Berlin.

Burch, D. E., Gryvnak, D. A., Singleton, E. B., France, W. L., and Williams, D. (1962). "Infrared Absorption by Carbon Dioxide, Water Vapor, and Minor Atmospheric Constituents," AFCRL-62-698. Air Force Cambridge Res. Lab., Bedford, Massachusetts.

Cess, R. D. (1974). Radiative transfer due to atmospheric water vapor: Global considerations of the Earth's energy balance. *J. Quant. Spectrosc. Radiat. Transfer* **14** (9), 861–871.
Cess, R. D., and Ramanathan, V. (1972). Radiative transfer in the atmosphere of Mars and that of Venus above the cloud deck. *J. Quant. Spectrosc. Radiat. Transfer* **12** (5), 933–945.
Cess, R. D., and Tiwari, S. N. (1972). Infrared radiative energy transfer in gases. *Adv. Heat Transfer* **8**, 229–283.
Cess, R. D., Mighdoll, P., and Tiwari, S. N. (1967). Infrared radiative heat transfer in nongray gases. *Int. J. Heat Mass Transfer* **10** (11), 1521–1532.
Chiarella, C., and Reichel, A. (1968). On the evaluation of integrals related to the error function. *Math. Comput.* **22** (101), 137–143.
Cogley, A. C. (1970). Radiative transport of Lorentz lines in non-isothermal cases. *J. Quant. Spectrosc. Radiat. Transfer* **10** (9), 1065–1075.
Cooper, J. (1966). Plasma spectroscopy. *Rep. Prog. Phys.* **29**, 35–130.
Craig, R. A. (1963). A note on the transmissivity of an Elsasser band. *J. Atmos. Sci.* **20** (1), 66–68.
Curtis, A. R. (1952). Discussion of a statistical model for water vapour absorption. *Q. J. R. Meteorol. Soc.* **78** (338), 638–641.
Donovan, T. E., and Greif, R. (1975). Laminar convection with an absorbing and emitting gas. *Appl. Sci. Res.* **31** (2), 110–122.
Echigo, R., Hasegawa, S., and Miyazaki, Y. (1971). Composite heat transfer with thermal radiation in non-gray medium—Part I. Interaction of radiation with conduction. *Int. J. Heat Mass Transfer* **14** (12), 2001–2015.
Edwards, D. K. (1976). Molecular gas band radiation. *Adv. Heat Transfer* **12**, 116–195.
Edwards, D. K., and Balakrishnan, A. (1972). Slab band absorptance for molecular gas radiation. *J. Quant. Spectrosc. Radiat. Transfer* **12** (10), 1379–1387.
Edwards, D. K., and Balakrishnan, A. (1973). Thermal radiation by combustion gases. *Int. J. Heat Mass Transfer* **16** (1), 25–40.
Edwards, D. K., and Menard, W. A. (1964). Comparison of methods for correlation of total band absorption. *Appl. Opt.* **3** (5), 621–625.
Edwards, D. K., Glassen, L. K., Hauser, W. C., and Tuchscher, J. S. (1967). Radiation heat transfer in nonisothermal non-gray gases. *J. Heat Transfer* **89** (3), 219–229.
Elsasser, W. M. (1942). "Heat Transfer by Infrared Radiation in the Atmosphere," Harvard Meteorological Studies, No. 6. Harvard Univ. Press, Cambridge, Massachusetts.
Ely, R., and McCubbin, T. K., Jr. (1970). The temperature dependence of the self-broadened half-width of the P-20 line in the 001–100 band of CO_2. *Appl. Opt.* **9** (5), 1230–1231.
Felske, J. D., and Tien, C. L. (1974). A theoretical closed form expression for the total band absorptance of infrared-radiating gases. *Int. J. Heat Mass Transfer* **17** (1), 155–158.
Finn, G. D., and Mugglestone, D. (1965). Tables of the line broadening function H (a, v). *Mon. Not. R. Astron. Soc.* **129** (2), 221–236.
Fried, B. D., and Conte, S. D. (1961). "The Plasma Dispersion Function." Academic Press, New York.
Gautschi, W. (1969). Complex error function. *Commun. ACM* **12** (11), 635.
Gautschi, W. (1970). Efficient computation of the complex error function. *SIAM (Soc. Ind. Appl. Math.) J. Numer. Anal.* **7** (1), 187–198.
Gille, J. C., and Ellingson, R. G. (1968). Correction of random exponential band transmission for Doppler effects. *Appl. Opt.* **7** (3), 471–474.
Gilles, S. E., and Vincenti, W. G. (1970). Coupled radiative and vibrational non-equilibrium in a diatomic gas, with application to gas dynamics. *J. Quant. Spectrosc. Radiat. Transfer* **10** (2), 71–97.
Godson, W. L. (1953). The evaluation of infrared radiative fluxes due to atmospheric water vapour. *Q. J. R. Meteorol. Soc.* **79** (341), 367–379.

Golden, S. A. (1967). The Doppler analog of the Elsasser band model. *J. Quant. Spectrosc. Radiat. Transfer* **7** (3), 483–494.

Golden, S. A. (1968). The Doppler analog of the Elsasser band model—II, The integrated emissivity. *J. Quant. Spectrosc. Radiat. Transfer* **8** (3), 877–897.

Golden, S. A. (1969). The Voigt analog of an Elsasser band. *J. Quant. Spectrosc. Radiat. Transfer* **9** (8), 1067–1081.

Goldman, A. (1968). On simple approximations to the equivalent width of a Lorentz line. *J. Quant. Spectrosc. Radiat. Transfer* **8** (2), 829–831.

Goody, R. M. (1964a). "Atmospheric Radiation I: Theoretical Basis." Oxford Univ. Press, London and New York.

Goody, R. M. (1964b). The transmission of radiation through an inhomogeneous atmosphere. *J. Atmos. Sci.* **21** (6), 576–581.

Goody, R. M., and Belton, M. J. S. (1967). Radiative relaxation times for Mars (A discussion of Martian atmospheric dynamics). *Planet. Space Sci.* **15** (2), 247–256.

Griem, H. R. (1964). "Plasma Spectroscopy." McGraw-Hill, New York.

Gupta, S. K., and Tiwari, S. N. (1975). "Evaluation of Upwelling Infrared Radiance from Earth's Atmosphere," TR-75-T14. Sch. Eng., Old Dominion Univ., Norfolk, Virginia.

Gupta, S. K., and Tiwari, S. N. (1976a). "Evaluation of Transmittance of Selected Infrared Bands," TR-76-T7. Sch. Eng., Old Dominion Univ., Norfolk, Virginia.

Gupta, S. K., and Tiwari, S. N. (1976b). "Retrieval of Surface Temperature by Remote Sensing," TR-76-T8. Sch. Eng., Old Dominion Univ., Norfolk, Virginia.

Harris, D. L., III (1948). On the line-absorption coefficient due to Doppler effect and damping. *Astrophys. J.* **108** (1), 112–115.

Hashemi, A., Hsieh, T. C., and Greif, R. (1976). Theoretical determination of band absorptance with specific application to carbon monoxide and nitric oxide. *J. Heat Transfer* **98**, 432–437.

Hottel, H. C. (1954). *In* "Heat Transmission" (W. H. McAdams), Ch. 4. McGraw-Hill, New York.

Hottel, H. C., and Sarofim, A. F. (1967). "Radiative Transfer." McGraw-Hill, New York.

Hsieh, T. C., and Greif, R. (1972). Theoretical determination of the absorption coefficient and the total band absorptance including a specific application to carbon monoxide. *Int. J. Heat Mass Transfer* **15** (8), 1477–1487.

Hummer, D. G. (1964a). "The Voigt Function: An eight-significant-figure table and generating procedure," NBS JILA Rep. No. 24. Univ. of Colorado, Boulder.

Hummer, D. G. (1964b). Expansion of Dawson's function in a series of Chebyshev polynomials. *Math. Comput.* **18** (86), 317–319.

Hummer, D. G. (1965). The Voigt function: An eight significant-figure table and generating procedure. *Mem. R. Astronom. Soc. (G.B.)*, **70**, Part 1, 1–32.

Jansson, P. A., and Korb, C. L. (1968). A table of the equivalent widths of isolated lines with combined Doppler and common broadened profiles. *J. Quant. Spectrosc. Radiat. Transfer* **8** (7), 1399–1409.

Jefferies, J. T. (1968). "Spectral Line Formation." Ginn (Blaisdell), Boston, Massachusetts.

Kaplan, L. D. (1953). Regions of validity of various absorption—coefficient approximations. *J. Meteorol.* **10** (2), 100–104.

Kaplan, L. D. (1954). *Proc. Toronto Meteorol. Conf., R. Meteorol. Soc., 1953* p. 43.

Kielkopf, J. F. (1973). New approximation to the Voigt function with applications to spectral line profile analysis. *J. Opt. Soc. Am.* **63** (8), 987–995.

Kondratyev, K. Y. (1969). "Radiation in Atmosphere." Academic Press, New York.

Kunde, V. G. (1967). Theoretical computations of the outgoing infrared radiance from a planetary atmosphere. *NASA Tech. Memo.* **NASA TN D-4045**.

Kunde, V. G., and Maguire, W. C. (1974). Direct integration transmittance model. *J. Quant. Spectrosc. Radiat. Transfer* **14** (8), 803–817.

Kyle, T. G. (1967). Absorption of radiation by uniformly spaced Doppler lines. *Astrophys. J.* **148** (3), Part I, 845–848.

Kyle, T. G. (1968). Absorption by Doppler-Lorentz atmospheric lines. *J. Quant. Spectrosc. Radiat. Transfer* **8** (8), 1455–1462.

Lin, C. C., and Chan, S. H. (1975). A general slab band absorptance for infrared radiating gases. *J. Heat Transfer* **97** (3), 478–480.

Lindquist, G. H., and Simmons, F. S. (1972). A band model formulation for very non-uniform paths. *J. Quant. Spectrosc. Radiat. Transfer* **12** (5), 807–820.

Ludwig, C. B., Griggs, M., Malkmus, W., and Bartle, E. R. (1973). Air pollution measurements from satellites. *NASA Contract. Rep.* **NASA CR-2324**.

McClatchey, R. A., Fenn, R. W., Selby, J. E. A., Volz, F. E., and Garing, J. S. (1972). "Optical Properties of the Atmosphere," 3rd Ed., AFCRL-72-0497. Air Force Cambridge Res. Lab., Bedford, Massachusetts.

McClatchey, R. A., Benedict, W. S., Clough, S. A., Burch, D. E., Calfee, R. F., Fox, K., Rothman, L. S., and Garing, J. S. (1973). "AFCRL Atmospheric Line Parameters Compilation," AFCRL-TR-73-0096. Air Force Cambridge Res. Lab., Bedford, Massachusetts.

McMillin, L. M., and Fleming, H. E. (1976). Atmospheric transmittance of an absorbing gas: A computationally fast and accurate transmittance model for absorbing gases with constant mixing ratios in inhomogeneous atmospheres. *Appl. Opt.* **15** (2), 358–363.

Malkmus, W. (1967). Random Lorentz band model with exponential-tailed S^{-1} line intensity distribution function. *J. Opt. Soc. Am.* **57** (3), 323–329.

Malkmus, W. (1968). Random band models with lines of pure Doppler shape. *J. Opt. Soc. Am.* **58** (9), 1214–1217.

Manabe, S., and Wetherald, R. T. (1967). Thermal equilibrium of the atmosphere with a given distribution of relative humidity. *J. Atmos. Sci.* **24** (3), 241–259.

Martin, J. K., and Hwang, C. C. (1975). Combined radiant and convective heat transfer to laminar steam flow between gray parallel plates with uniform heat flux. *J. Quant. Spectrosc. Radiat. Transfer* **15** (12), 1071–1081.

Mitchell, A. C. G., and Zemansky, W. M. (1934). "Resonance Radiation and Excited Atoms." Harvard Univ. Press, Cambridge, Massachusetts. (Reprinted, 1961.)

Penner, S. S. (1959). "Quantitative Molecular Spectroscopy and Gas Emissivities." Addison-Wesley, Reading, Massachusetts.

Penner, S. S. (1973). Equilibrium radiation properties of gases. *In* "Handbook of Heat Transfer" (W. M. Rohsenow and J. P. Hartnett, eds.), Sect. 15, Part D, pp. 15–72. McGraw-Hill, New York.

Penner, S. S., and Kavanagh, R. W. (1953). Radiation from isolated spectral lines with combined Doppler and Lorentz broadening. *J. Opt. Soc. Am.* **43** (5), 385–388.

Plass, G. N. (1958). Models for spectral band absorption. *J. Opt. Soc. Am.* **48** (10), 690–703.

Plass, G. N. (1960). Useful representations for measurements of spectral band absorption. *J. Opt. Soc. Am.* **50** (9), 868–875.

Plass, G. N., and Fivel, D. I. (1953). Influence of Doppler effect and damping on line absorption coefficient and atmospheric radiation transfer. *Astrophys. J.* **117** (1), 225–233.

Posener, D. W. (1959). The shape of spectral lines: Table of the Voigt profile. *Aust. J. Phys.* **12** (2), 184–196.

Reiche, F. (1913). *Verh. Dsch. Phys. Ges.* **15**, 3.

Rodgers, C. D. (1967). The use of emissivity in atmospheric radiative calculations. *Q. J. R. Meteorol. Soc.* **93** (395), 45–54.

Rodgers, C. D. (1968). Some extensions and applications of the new random model for molecular band transmission. *Q. J. R. Meteorol. Soc.* **94** (399), 99–102.

Rodgers, C. B., and Williams, A. P. (1974). Integrated absorption of a spectral line with the Voigt profile. *J. Quant. Spectrosc. Radiat. Transfer* **14** (4), 319–323.
Science Application Incorporated (SAI) (1973). "Line-by-Line Program (POLAYER) for Evaluation of Radiance and Signal Change, and Compilation of Spectroscopic Parameters for the Spectral Range of CO Fundamental Band (1973 Version)." (Available at NASA Langley Res. Cent., Hampton, Virginia.)
Seitz, W. S., and Lundholm, D. V. (1964). Elsasser model for band absorption; series representation of a useful integral. *J. Opt. Soc. Am.* **54** (3), 315–318.
Selby, J. E. A., and McClatchey, R. A. (1975). "Atmospheric Transmittance from 0.25 μm: Computer Code LOWTRAN 3," AFCRL-TR-75-0255. Environ. Res., Cambridge Res. Lab., Bedford, Massachusetts.
Simmons, F. S. (1967). Radiance and equivalent widths of Lorentz lines for non-isothermal paths. *J. Quant. Spectrosc. Radiat. Transfer* **7** (1), 111–121.
Taylor, P. B., and Foster, P. J. (1974). The total emissivities of luminous and non-luminous flames. *Int. J. Heat Mass Transfer* **17** (12), 1591–1605.
Tien, C. L. (1966). A simple approximate formula for radiation from a line with resonance contour. *J. Quant. Spectrosc. Radiat. Transfer* **6** (6), 893–894.
Tien, C. L. (1968). Thermal radiation properties of gases. *Adv. Heat Transfer* **5**, 253–324.
Tien, C. L., and Ling, G. R. (1969). On a simple correlation for total band absorptance of radiating gases. *Int. J. Heat Mass Transfer* **12** (9), 1179–1181.
Tien, C. L., and Lowder, J. E. (1966). A correlation for total band absorptance of radiating gases. *Int. J. Heat Mass Transfer* **9** (7), 698–701.
Tiwari, S. N. (1973). "Appropriate Line Profiles for Radiation Modeling in the Detection of Atmospheric Pollutants," TR-73-T3. Sch. Eng., Old Dominion Univ., Norfolk, Virginia.
Tiwari, S. N. (1976). Band models and correlations for infrared radiation. In " Radiative Transfer and Thermal Control" (A. M. Smith, ed.), AIAA Progress in Astronautics and Aeronautics, Vol. 49, pp. 155–182. AIAA, New York.
Tiwari, S. N., and Batki, R. R. (1974). "Infrared Line Models for Atmospheric Radiation," TR-74-T4. Sch. Eng., Old Dominion Univ., Norfolk, Virginia.
Tiwari, S. N., and Batki, R. R. (1975). "Infrared Band Models for Atmospheric Radiation," TR-75-T17. Sch. Eng., Old Dominion Univ., Norfolk, Virginia.
Tiwari, S. N., and Cess, R. D. (1971). Heat transfer to laminar flow of non-gray gases through a circular tube. *Appl. Sci. Res.* **25** (3/4), 155–170.
Tiwari, S. N., and Gupta, S. K. (1977). "Accurate Spectral Modeling for Infrared Radiation," TR-77-NSG-1282. Sch. Eng., Old Dominion Univ., Norfolk, Virginia. (Also ASME Pap. 77-HT-69.)
Tiwari, S. N., and Reichle, H. G. (1974). Application of infrared line models in the detection of atmospheric pollutants. *AIAA/ASME Thermophys. Heat Conf., Boston, Mass.* AIAA Pap. 74-651.
Tubbs, L. D., and Williams, D. (1972). Broadening of infrared absorption lines at reduced temperatures: Carbon dioxide. *J. Opt. Soc. Am.* **62** (2), 284–289.
Van de Hulst, H. C., and Reesinck, J. M. (1947). Line breadths and Voigt profiles. *Astrophys. J.* **106** (1), 121–127.
Walshaw, C. D., and Rodgers, C. D. (1963). The effect of Curtis-Godson approximation on the accuracy of radiative heating-rate calculations. *Q. J. R. Meteorol. Soc.* **89** (379), 122–130.
Wassel, A. T., and Edwards, D. K. (1974). Molecular gas band radiation in cylinders. *J. Heat Transfer* **96** (1), 21–26.
Wassel, A. T., and Edwards, D. K. (1976). Molecular gas radiation in a laminar or turbulent pipe flow. *J. Heat Transfer* **98** (1), 101–107.
Wassel, A. T., Edwards, D. K., and Cotton, I. (1975). Molecular gas radiation and laminar or

turbulent heat diffusion in a cylinder with internal heat generation. *Int. J. Heat Mass Transfer* **18** (11), 1267–1276.

Weinreb, M. P., and Neuendorffer, A. C. (1973). Method to apply homogeneous-path transmittance models to inhomogeneous atmosphere. *J. Atmos. Sci.* **30** (4), 662–666.

Whiting, E. E. (1968). An empirical approximation to the Voigt profile. *J. Quant. Spectrosc. Radiat. Transfer* **8** (6), 1379–1384.

Wyatt, P. J., Stull, V. R., and Plass, G. N. (1962). Quasi-random model of band absorption. *J. Opt. Soc. Am.* **52** (11), 1209–1217.

Yamada, H. Y. (1967). Total radiances and equivalent widths of Doppler lines for nonisothermal paths. *J. Quant. Spectrosc. Radiat. Transfer* **7** (6), 997–1003.

Yamada, H. Y. (1968). Total radiances and equivalent widths of isolated lines with combined Doppler and collision broadened profiles. *J. Quant. Spectrosc. Radiat. Transfer* **8** (8), 1463–1473.

Yamamoto, G., and Aida, M. (1967). Transmission due to overlapping lines. *J. Quant. Spectrosc. Radiat. Transfer* **7** (1), 123–141.

Yamamoto, G., and Aida, M. (1970). Transmission in a non-homogeneous atmosphere with an absorbing gas of constant mixing ratio. *J. Quant. Spectrosc. Radiat. Transfer* **10** (6), 593–608.

Yamamoto, G., Tanaka, M., and Aoki, T. (1969). Estimation of rotational line widths of carbon dioxide bands. *J. Quant. Spectrosc. Radiat. Transfer* **9** (3), 371–382.

Yamamoto, G., Aida, M., and Yamamoto, S. (1972). Improved Curtis-Godson approximation in a non-homogeneous atmosphere. *J. Atmos. Sci.* **29** (6), 1150–1155.

Young, C. (1964). "A Study of the Influence of Carbon Dioxide on Radiative Transfer in the Stratosphere and Mesosphere," Tech. Rep. Dep. Meteorol. Oceanogr., Coll. Eng., Univ. of Michigan, Ann Arbor.

Young, C. (1965a). "Table for Calculating the Voigt Profile," TR-05863-7-T. Coll. Eng., Univ. of Michigan, Ann Arbor.

Young, C. (1965b). Calculation of the absorption coefficient for lines with combined Doppler and Lorentz broadening. *J. Quant. Spectrosc. Radiat. Transfer* **5** (3), 549–552.

Young, S. J. (1975a). Band model formulation for inhomogeneous optical paths. *J. Quant. Spectrosc. Radiat. Transfer* **15** (6), 483–501.

Young, S. J. (1975b). Addendum to: Band model formulation for inhomogeneous optical paths. *J. Quant. Spectrosc. Radiat. Transfer* **15** (12), 1137–1140.

Zachor, A. S. (1967). Absorptance and radiative transfer by a regular band. *J. Quant. Spectrosc. Radiat. Transfer* **7** (6), 857–870.

SCALE ANALYSIS OF LARGE ATMOSPHERIC MOTION SYSTEMS IN ALL LATITUDES

J. Van Mieghem

University of Brussels
Brussels, Belgium

1. Introduction, Generalities, and Statement of the Problem 87
2. Scale Parameters 91
3. Characterization of an Atmospheric Large-Scale Motion System 97
4. Scale Analysis of the Continuity Equation 100
5. Scale Analysis of the Equation of Vertical Motion 102
6. Scale Analysis of the Equations of Horizontal Motion 103
7. Scale Analysis of the Equation of Adiabatic Transformations 106
8. Order of Magnitude of the Divergence Δ of the Horizontal Wind 109
9. Scale Analysis of the Vorticity Equation 113
10. Scale Analysis of the Equation of the Horizontal Wind Divergence 116
11. Discussion of the Basic Scale Equations 117
    11.1 The Middle Latitude Belt 20 to 70° 118
    11.2 The High Latitude Belt 70 to 88° 126
    11.3 The Low Latitude Belt 2 to 20° 126
12. Scale Analysis of the Equation of Balance of Kinetic Energy 128
    References 130

1. Introduction, Generalities, and Statement of the Problem

It is well known that, in certain branches of applied fluid dynamics, special characteristics of the motion under consideration are used to simplify the basic equations so that their integration is facilitated. For example, the fundamental equations of aerodynamics have been considerably simplified by the introduction of the boundary layer approximation and by assuming the incompressibility and homogeneity of the fluid.

Unfortunately, in meteorology one encounters major difficulties. Indeed, atmospheric motions cover a broad spectrum of space and time scales, ranging from the random motions of the air molecules to the planetary zonal vortex involving the atmosphere globally. The classic laws of hydrodynamics and thermodynamics describe and govern all scales of motions except, however, the molecular scale.

It is generally assumed that the short-term evolution of a large-scale atmospheric motion system, such as a weather-producing system, is governed by the hydrodynamic equations for adiabatic, frictionless[1] laminar flow of dry air (ideal gas), that is, by a set of five nonlinear partial differential equations: Euler's equations of motion along the latitude circles, the meridians, and the ascending verticals; and the continuity and adiabatic equations, with five unknown functions: the zonal u, meridional v, and vertical w wind components, the atmospheric pressure p, and air density ρ. The nonlinearity of these basic equations and the existence of a broad spectrum of atmospheric motion systems render their analytical integration impossible, except in very special and simple cases, which are generally without interest from the meteorological point of view. Numerical integration is therefore the only possible escape, and numerical methods have been extensively applied for almost 25 years.

Previously, linearized forms of the fundamental equations of atmospheric dynamics, introduced by V. Bjerknes (1926), were abundantly used to study: (1) the nascent stage of a developing perturbation superimposed on a state of hydrostatic equilibrium in the earth's gravity field or on some simple horizontal and straight shear flow, (2) the stability criterion of the underlying basic state (hydrostatic equilibrium or zonal shear flow) (Bjerknes *et al.*, 1934), and (3) the momentum and energy exchanges between the basic state and a superimposed small perturbation (Van Mieghem, 1973).

In the theory of small perturbations, it is assumed, as a rule, that the general solution, u, v, w, p, ρ of the linearized equations in the infinite half-space $z > 0$ (z is a vertical coordinate) are linear combinations of harmonic wave solutions of the type $A(y, z) \exp m(x - ct)\sqrt{-1}$ for each of the five unknown functions. The function $A(y, z)$ represents the amplitude of the wave superimposed on the basic zonal field $(u_0, 0, 0, p_0, \rho_0)$, m designates the zonal wave number and c the phase velocity of the waves. The coordinates x and y are horizontal: x (the longitudinal coordinate in the direction of the basic zonal flow) is positively oriented to the east and y, positively oriented to the north of the origin of Rossby's β plane (x, y). The known functions u_0, p_0, and ρ_0 are slow-varying with time t and depend mainly upon y and z, the transverse coordinates perpendicular to the basic zonal flow u_0.

In his article entitled "The dynamics of long waves in a baroclinic westerly current," Charney (1947) pointed out that, in the study of atmospheric wave motions, using linearized equations, the analytical integration of these equations is made virtually impossible by the simultaneous existence of a discrete set of wave motions of the above-mentioned type, each of them satisfying the conditions of the problem. The phase speed of the wave will then generally depend not only on the wavelength, but mainly upon

---

[1] It is generally assumed that friction is important only in the surface friction layer.

certain physical parameters characteristic of the problem envisaged. When considering small perturbations of a hydrostatic equilibrium, these parameters are: the earth's gravity g, wave perturbations with a wavelength larger than 100 km, the earth's rotation, and also, although less important, the compressibility of the fluid. When dealing with wave perturbations of a horizontal zonal shear flow on a rotating earth, additional parameters have to be taken into account, namely the horizontal $(\partial u_0/\partial y)$ and vertical $(\partial u_0/\partial z)$ shears of the basic current u_0. The vertical shear $\partial u_0/\partial z$ expresses the baroclinity of the basic flow u_0.

Large-scale atmospheric motions being almost horizontal, the influence of the earth's rotation Ω results mainly from the vertical component Ω_z of Ω and also from the latitudinal variability of this component. Hence, the classic Coriolis parameter ($f = 2\Omega_z = 2\Omega \sin \phi$, $\phi =$ latitude) may be used instead of the earth's rotation $\Omega(0, \Omega_y, \Omega_z)$ and the Rossby parameter ($\beta = (1/r)(df/d\phi) = 2\Omega \cos \phi/r = df/dy$; $f = f_0 + \beta y$, $r =$ distance of an air parcel from the earth's center) must aso be added to the above-mentioned characteristics, at least for very long waves.

Keeping the perturbation equations in their full generality, all theoretically possible wave motions have to be considered at once, such as compressibility waves, short gravity waves with periods much smaller than half a pendulum day ($2\Pi/f$) on which the rotation of the earth has no influence, longer gravity-inertia waves with periods of the order of $2\Pi/f$, cyclone waves with periods larger than $2\Pi/f$, and still longer waves, the Rossby waves, with periods much larger than $2\Pi/f$. The cyclone waves depend mainly on the Coriolis parameter f and the baroclinity $\partial u_0/\partial z$ of the basic zonal flow u_0, and the Rossby waves superimposed on the upper westerlies also depend upon the β parameter. The motions associated with these two last types of waves are quasi-adiabatic, quasi-static, quasi-geostrophic, and gravitationally stable.

For the study of synoptic weather systems in a statically stable[1] atmosphere, only long inertially[2] propagated baroclinic waves are of importance. These waves are the cyclone waves and the Rossby waves. Without simplification of the perturbation equations, one is forced to take into consideration all theoretically possible wave motions, including those of little or no meteorological significance. Moreover, the presence in the linearized

---

[1] We recall that the static stability is defined by $v_s^2 = (g/\theta_0)(\partial \theta_0/\partial z) > 0$, where θ_0 is the potential temperature of the basic flow u_0 and v_s the gyro-frequency of the free vertical gravity oscillations of the air ($v_s \cong 10^{-2}$ s^{-1}, period $2\Pi/v_s \cong 10$ min).

[2] The propagation of baroclinic waves is governed by the absolute vorticity $f - (\partial u_0/\partial y)$ of the zonal shear flow u_0. We recall that the inertial stability is defined by $v_i^2 = f(f - \partial u_0/\partial y) > 0$, where v_i is the gyro-frequency of the free horizontal inertial oscillations of the air ($v_i \cong 10^{-4}$s^{-1}, period $2\Pi/v_i \cong 17h$).

equations of meteorologically unimportant wave motions renders their analytical integration generally impossible. Therefore, this meteorological noise should be filtered out from the perturbation equations. Charney (1947) succeeded in doing so by introducing the geostrophic approximation only in those terms of the perturbation equations in which this can be done rightly. For instance, this cannot be done rightly in the analytical expression of the divergence div $\mathbf{v}_h$ of the horizontal wind velocity $\mathbf{v}_h$, the two terms $\partial u/\partial x$ and $\partial u/\partial y$ both having the same order of magnitude but the sum of these two terms always being of one order of magnitude less. Hence, div $\mathbf{v}_h$ will be strongly affected by observational errors in the horizontal wind components u and v and also by computational errors in the derivatives $\partial u/\partial x$ and $\partial v/\partial y$. In fact, there is equality in order of magnitude of the divergence of the horizontal wind velocity and the error in these two derivatives when they are approximated. In particular, the geostrophic approximation will break down when introduced in div $\mathbf{v}_h$. Thus, div $\mathbf{v}_h$ is an unmeasurable meteorological entity and therefore, as a rule, must be eliminated from the fundamental equations of large-scale atmospheric dynamics (see Section 11.1.1).

This filtering method was extended successfully by Charney (1948) to the more general nonlinear large-scale atmospheric motion systems. Synoptic experience has demonstrated that these motions are slow, quasi-adiabatic (at least when only short-time evolution of weather systems is taken into consideration), quasi-horizontal, quasi-static, and quasi-geostrophic. The three first approximations can be introduced rather easily. As we have shown above, the introduction of the geostrophic approximation, however, offers some serious difficulties, in particular in the equation of continuity. We will see below how this difficulty can be overcome (see Section 11.1.1).

The simplified partial differential equations representing the large-scale nonlinear atmospheric motions must evidently constitute a dynamically consistent system of equations. As a consequence of this requirement, the meteorologically unimportant motions must be filtered out from the fundamental hydrodynamic and thermodynamic equations, on the basis of a set of mathematically consistent principles induced from the experience gained by forecasters.

In this theory of meteorological approximations, Charney has used a kind of scale analysis, similar to that used in the boundary layer theory in aerodynamics, allowing the evaluation of the orders of magnitude of the different terms of the basic equations. All terms of the highest order of magnitude in each of these equations are then kept, only those of at least one order of magnitude less are left out.

Since the meteorologically significant large-scale motions are distinguished from all other types of possible atmospheric motions by large differences in both their three-dimensional space configuration and local lifetime, it is clear that any attempt to evaluate the orders of magnitude and consequently, the relative importance of the different terms in each of the five fundamental equations of atmospheric dynamics and thermodynamics, will be based on the space scales and local lifetimes of the large-scale motions considered.

The scale parameters chosen by Charney (1948) are the horizontal and vertical dimensions S and D of the motion system, the orders of magnitude of the horizontal and vertical wind speeds U and W, and the order of magnitude C of the speed of propagation of the streamline pattern of the horizontal motion. The local lifetime of the motion is then defined by S/C. The introduction of the scale parameter C is justified because the phase speed of the horizontal streamlines of large-scale weather systems does not belong to the interval of phase speeds of the streamline configurations associated with meteorologically unimportant motions. Indeed, the phase speed C is at least one order of magnitude smaller than the corresponding phase speeds associated with motion systems of very little meteorological significance. Among these motion systems we disregard the hypothetical (apparently not yet observed in the atmosphere) slow and long inertial waves (Quinet, 1975). These waves, however, have a vertical scale which is much smaller than the vertical scale of the meteorologically significant motions. Hence, a suitable choice of the vertical scale D will eliminate possible effects of these peculiar inertial waves.

This review paper intends (1) to systematize and synthesize the results of the scaling procedures of Charney (1947, 1948, 1963), Burger (1958), Holton (1969), and Phillips (1963) applied to the equations of continuity (Section 4), vertical and horizontal motion (Sections 5 and 6), adiabatic transformations (Section 7), vorticity (Section 9), and horizontal wind divergence (Section 10) and (2) to extend them to the balance equation of kinetic energy (Section 12).

2. Scale Parameters

To restrict the motion systems under consideration to the large-scale end of the spectrum of existing atmospheric motions, in fact the atmospheric motion systems carrying the bulk of the atmospheric energy, appropriate characteristic dimensional properties of these large-scale systems have been introduced by Charney (1948) into the basic hydrothermodynamic equations in order to separate the atmospheric large-scale motions from all other dynamically possible motions without meteorological significance. These characteristic dimensional properties are listed below.

The spatial dimensions are defined by S, the mean horizontal distance between points at which the velocity components take extreme values, and by D, the corresponding mean vertical distance, so that S and D are the half-wavelengths of the horizontal and vertical flow patterns, respectively.

The velocity field is characterized by U, the average magnitude of the horizontal components u, v of the air velocity $\mathbf{v}$, by W the average magnitude of the vertical component w of $\mathbf{v}$. The mean phase speed of the streamline pattern of the horizontal motion will be represented by C. The separation of the field of motion into a horizontal motion and a vertical motion is justified by the observational fact that the average magnitude of the mean horizontal speed U is at least two orders of magnitude larger than the magnitude of the mean vertical speed W, so that $W \ll U$. Hence the kinetic energy per mass unit $k = \mathbf{v}^2/2$ of the three-dimensional air motion and the kinetic energy per mass unit $\mathbf{v}_h^2/2$ of the horizontal motion of the same unit mass have the same order of magnitude,

(2.1) $$k \cong \frac{u^2 + v^2}{2} \sim U^2$$

where the symbol $\sim$ denotes the equality in orders of magnitude.

We are now in a position to evaluate the orders of magnitude of all entities appearing in the equations of dynamic meteorology, namely the zonal, meridional, and vertical components u, v, and w of the particle velocity $\mathbf{v}$, the pressure p and air density ρ, and the partial space and time derivatives of these quantities. This is done by replacing differentials by finite differences, expressed with the aid of the scale parameters S, D, C, U, and W. Thus, for instance, to evaluate the order of magnitude of $\partial u/\partial s$, u being the zonal velocity component and s a horizontal distance coordinate, we replace $\partial u/\partial s$ by $\Delta u/\Delta s$ and we assume that the flow pattern is such that Δu has the same order of magnitude as u, when $\Delta s = S$, hence $\Delta_S u = U$ so that $(\partial u/\partial s) \sim (U/S)$.

More generally, we have

(2.2) $$\begin{aligned} |\Delta_S u| = |\Delta_S v| &\sim U \\ |\Delta_D u| = |\Delta_D v| &\sim U \\ |\Delta_S w| = |\Delta_D w| &\sim W \end{aligned}$$

hence,

(2.3) $$\left|\frac{\partial u}{\partial x}\right| \sim \left|\frac{\partial u}{\partial y}\right| \sim \left|\frac{\partial v}{\partial x}\right| \sim \left|\frac{\partial v}{\partial y}\right| \sim \frac{U}{S}$$

$$\left|\frac{\partial u}{\partial z}\right| \sim \left|\frac{\partial v}{\partial z}\right| \sim \frac{U}{D}$$

$$\left|\frac{\partial w}{\partial x}\right| \sim \left|\frac{\partial w}{\partial y}\right| \sim \frac{W}{S}$$

$$\left|\frac{\partial w}{\partial z}\right| \sim \frac{W}{D}$$

We recall here that the horizontal streamlines of the westerlies in the free troposphere (roughly between the 800 and 300 mb surfaces) form wave-like patterns along latitude circles, the troughs and ridges of the waves being tilted (generally to the east) with respect to the meridians. The order of magnitude of the distances between any troughline and the ridgelines to the west and to the east defines the horizontal scale length S. It should be stressed that the zonal wind component u tends to be stronger to the east than to the west of the troughlines, while the meridional wind components v tend to be stronger to the west (north wind components, $v < 0$) than to the east (south wind components, $v > 0$) of these lines. The associated vertical motion is generally of the same sign along any given vertical (ascending, $w > 0$) between a troughline and a ridgeline to the east and descending ($w < 0$) between a troughline and the ridgeline to the west), with a maximum intensity at about the midtroposphere decreasing in absolute value above and below and tending to zero both at the earth's surface and the top height of the troposphere (tropopause). Above and below the tropopause level, the vertical velocity w generally has opposite signs. For such a flow pattern the scale conditiotions (2.3) are satisfied.

Equations (2.3) suggest that the flow field, described above, can be identified with the following scale operators

(2.4) $$\left|\frac{\partial \lambda}{\partial x}\right| \sim \left|\frac{\partial \lambda}{\partial y}\right| \sim \frac{\Lambda}{S} \quad \text{and} \quad \left|\frac{\partial \lambda}{\partial z}\right| \sim \frac{\Lambda}{D}$$

where the scale parameter Λ corresponds to the flow parameter λ (such as the components u, v, and w of the wind vector $\mathbf{v}$ and the components of the wind shear tensor $\nabla \mathbf{v}$).

The space derivatives of p and ρ may be evaluated in the same way. In order to simplify the formulation of the theory of meteorological approximations, Charney (1948) has introduced the relative space variations of p and ρ instead of their absolute space variations. In this way the introduction of two characteristic values, one for each of the variables of states p and ρ, can be avoided. To these two variables of states, p and ρ, we must add the absolute air temperature T and the associated potential temperature θ. Pressure p and temperature T are measured variables, while density ρ and potential temperature θ are computed using the ideal gas law $p = R\rho T$ and the classic definition $\theta = T(p_{00}/p)^{R_a/c_{pa}}$ where R is the specific ideal gas constant for moist air, R_a the same gas constant for dry air, c_{pa} the specific heat of dry air at constant pressure, and $p_{00} = 100$ cb. In most cases the approximation $R \cong R_a$ is acceptable.

Let us represent by X any one of the variables of state T, p, ρ, and θ. Taking into account that the pressure and temperature fields have the same horizontal and vertical scales S and D as those of the velocity field in horizontal surfaces (spheres concentric to the earth) and also that the space variations of pressure, density, and temperature are not greater in order of magnitude than their mean values, we may write

(2.5) $$|\Delta_S X| \lesssim \bar{X}_S \qquad |\Delta_D X| \lesssim \bar{X}_D$$

where the symbols $\bar{X}_S$ and $\bar{X}_D$ designate mean values of X over the distances S and D, respectively. Then, by virtue of Eq. (2.5) we have:

(2.6) $$\left|\frac{1}{X}\frac{\partial X}{\partial x}\right| \sim \left|\frac{1}{X}\frac{\partial X}{\partial y}\right| \sim \frac{|\nabla_h X|}{X} \sim \frac{|\Delta_S X|}{\bar{X}_S S} \lesssim \frac{1}{S}$$

$$\left|\frac{1}{X}\frac{\partial X}{\partial z}\right| \sim \frac{|\Delta_D X|}{\bar{X}_D D} \lesssim \frac{1}{D}$$

The symbols $\lesssim$ and $\gtrsim$ denote the inequality in orders of magnitude. For example, $(1/X)(\partial X/\partial s) \lesssim (1/S)$ means simply that in any case the order of magnitude of $(1/X)(\partial X/\partial s)$ is not larger than $1/S$. The symbol ∇_h designates the del operator in a horizontal surface. In Eqs. (2.5) and (2.6), T, p, ρ, and θ may be substituted successively for X.

As usual we designate by H the scale height of the atmosphere,

(2.7) $$H = \frac{R_a T}{g} = \frac{p}{g\rho} \sim 10^4 \quad \text{m}$$

that is, the height of the homogeneous atmosphere of constant density ρ, of surface temperature T, and vertical lapse rate g/R_a. The variability of the scale height in space and time is rather small, so that p and ρ remain nearly proportional along any vertical. In the troposphere, H varies roughly between 6 and 9 km and has the same order of magnitude as the average depth of the troposphere. Large-scale motions in the atmosphere being quasi-hydrostatic, the hydrostatic assumption is valid,

(2.8) $$\frac{1}{\rho}\frac{\partial p}{\partial z} + g = 0 \qquad \text{or} \qquad \frac{1}{p}\frac{\partial p}{\partial z} + \frac{1}{H} = 0$$

pressure p and density ρ along the ascending vertical decrease roughly as the exponential function $e^{-z/H}$. Thus we have

(2.9) $$\left|\frac{1}{p}\frac{\partial p}{\partial z}\right| \sim \left|\frac{1}{\rho}\frac{\partial \rho}{\partial z}\right| \sim \frac{1}{H} \sim 10^{-4} \text{ m}^{-1}$$

and hence,

(2.9') $$\frac{1}{\bar{p}_H}\frac{|\Delta_H p|}{H} \sim \frac{1}{\bar{\rho}_H}\frac{|\Delta_H \rho|}{H} \sim \frac{1}{H}$$

so that

(2.5′) $\qquad |\Delta_H p| \sim \bar{p}_H \qquad |\Delta_H \rho| \sim \bar{\rho}_H$

The vertical increment $|\Delta_H p|$ and $|\Delta_H \rho|$ in p and ρ through the depth H of the troposphere have the same orders of magnitude as those of p and ρ, respectively. Comparing (2.5′) with (2.5) and taking (2.6) into account, we find

(2.10) $\qquad \dfrac{|\Delta_D p|}{\bar{p}_D} \sim \dfrac{|\Delta_D \rho|}{\bar{\rho}_D} \sim \dfrac{D}{H} \qquad \text{with } D \lesssim H$

The order of magnitude of the vertical depth scale D never exceeds the scale height H of the atmosphere.

Synoptic experience has shown that well-developed weather systems extend through the whole depth of the troposphere, so that it is reasonable to assume for large-scale deep motion systems that

(2.10′) $\qquad\qquad D \sim H \qquad \text{with } D \leq H$

Let us now derive with respect to z the logarithmic form of the ideal gas law. Then from (2.9) it follows that

(2.11a) $\qquad \left| \dfrac{1}{T} \dfrac{\partial T}{\partial z} \right| = \left| \dfrac{1}{p} \dfrac{\partial p}{\partial z} - \dfrac{1}{\rho} \dfrac{\partial \rho}{\partial z} \right| \lesssim \dfrac{1}{H}$

the two derivatives on the right-hand side of the equality both being always negative. It is well known that the mean vertical temperature gradient in the troposphere varies roughly between 7 and 8°C km^{-1}, so that effectively

(2.11b) $\qquad \left| \dfrac{1}{T} \dfrac{\partial T}{\partial z} \right| \sim 10^{-5} \text{ m}^{-1} \lesssim \dfrac{1}{H} \sim 10^{-4} \text{ m}^{-1}$

Now using the definition of θ, we find

(2.11c) $\qquad \dfrac{1}{\theta} \dfrac{\partial \theta}{\partial z} = \dfrac{1}{T} \dfrac{\partial T}{\partial z} - \dfrac{R_a}{c_{pa}} \dfrac{1}{p} \dfrac{\partial p}{\partial z} \sim 10^{-5} \text{ m}^{-1} \lesssim \dfrac{1}{H} \sim 10^{-4} \text{ m}^{-1}$

account being taken of the fact that in the troposphere $\partial T/\partial z$ and $\partial p/\partial z$ have, as a rule, the same sign, and of the Eqs. (2.9), (2.11b), and $(R_a/c_{pa}) \sim 0.3$. Large-scale motion systems are stable in the gravity field, so that $(\partial \theta / \partial z) > 0$.

Furthermore, the hydrostatic approximation (Holton, 1972) implies also that

(2.12) $\qquad\qquad \dfrac{\partial}{\partial z}(\Delta_S p) + g(\Delta_S \rho) = 0$

Consequently, we may write $|\Delta_S p|/D \sim g|\Delta_S \rho|$, or finally, by virtue of (2.7), (2.10), and (2.10′),

(2.13) $$\frac{|\Delta_S p|}{\bar{p}_S} \sim \frac{D}{H} \frac{|\Delta_S \rho|}{\bar{\rho}_S} \sim \frac{|\Delta_S \rho|}{\bar{\rho}_S}$$

Hence

(2.13′) $$\frac{|\nabla_h p|}{p} \sim \frac{1}{\bar{p}_S} \frac{|\Delta_S p|}{S} \sim \frac{1}{\bar{\rho}_S} \frac{|\Delta_S \rho|}{S} \sim \frac{|\nabla_h \rho|}{\rho}$$

Replacing $\Delta_S \rho$ in (2.12) by its value deduced from the ideal gas law and recalling the scale relations (2.10′) and the definition (2.7) of the scale height H, we find

(2.14a) $$\frac{|\Delta_S p|}{\bar{p}_S} \sim \frac{D}{H} \frac{|\Delta_S \rho|}{\bar{\rho}_S} \sim \frac{D}{H} \frac{|\Delta_S T|}{\bar{T}_S}$$

Hence, (2.10′),

(2.14b) $$\frac{|\nabla_h p|}{p} \sim \frac{|\nabla_h \rho|}{\rho} \sim \frac{|\nabla_h T|}{T}$$

For the evaluation of the orders of magnitude of the local time derivative $\partial/\partial t$ use may be made of the fact that the streamline pattern moves horizontally together with the associated isobaric and isopycnic patterns at the same phase speed C. The local time variation, however, depends not only on the horizontal translation of the streamline pattern and associated horizontal isobaric and isopycnic configurations, but also on a change in time of the amplitude in these patterns. In this last case the speed C is the resultant of the phase speed of the horizontal streamlines and an additional speed resulting from the change in amplitude of the horizontal streamline configuration. The inclusion in C of this additional speed does not change the order of magnitude of C. With these patterns moving the horizontal distance s in the time s/C, the local change in time is expressed by

(2.15) $$\frac{\partial}{\partial t} \sim C \frac{\partial}{\partial s} \sim C \frac{\Delta_s}{S}$$

As a result of (2.3), (2.6), and (2.15), we have

(2.16) $$\left|\frac{\partial u}{\partial t}\right| \sim \left|\frac{\partial v}{\partial t}\right| \sim C\frac{U}{S} \qquad \left|\frac{\partial w}{\partial t}\right| \sim C\frac{W}{S}$$

and

(2.17) $$\left|\frac{1}{p}\frac{\partial p}{\partial t}\right| \sim \frac{C}{\bar{p}_S}\frac{|\Delta_S p|}{S} \lesssim \frac{C}{S} \qquad \left|\frac{1}{\rho}\frac{\partial \rho}{\partial t}\right| \sim \frac{C}{\bar{\rho}_S}\frac{|\Delta_S \rho|}{S} \lesssim \frac{C}{S}$$

The ratio

(2.18) $$\tau = \frac{S}{C}$$

defines the local lifetime of the motion system envisaged (the half-wave period of the horizontal flow pattern). Denoting by Δ_τ the finite difference over the time interval τ, we have

(2.19) $$\frac{\partial}{\partial t} \sim \frac{\Delta_\tau}{\tau}$$

Hence, by virtue of (2.15),

$$\frac{\partial}{\partial t} \sim C \frac{\Delta_S}{S} \sim \frac{\Delta_\tau}{\tau}$$

and of (2.18),

(2.20) $$\Delta_S \sim \Delta_\tau$$

The space increments of meteorological quantities through the horizontal distance S are equal to their time increments through the period τ.

3. Characterization of an Atmospheric Large-Scale Motion System

Having in mind only large-scale weather-producing motion systems, the scale parameters are chosen to characterize the horizontal and vertical length scales S and D of these systems and the magnitudes of U and W, the average horizontal and vertical wind components. These parameters satisfy the following specifications:

(3.1a)
$$10^6 \text{ m} \leq S \leq 2.10^7 \text{ m}$$
$$D \sim H \quad \text{with } D \leq H \sim 10^4 \text{ m}$$
$$D \ll S \quad W \ll U$$
$$fH \ll U \lesssim 10^2 \text{ m s}^{-1}$$

where f is the Coriolis parameter ($f \gtrsim 10^{-4}$ s^{-1}). More precisely, the inequality $S \gg D$ means that S is at least one order of magnitude larger than D.

The variability ranges we have adopted in (3.1a) for the two basic scale parameters S and U are those proposed by Burger (1958). These ranges are much wider than those introduced by Charney (1948): $S \sim 1000$ km and $U \sim 10$ m s^{-1}. We have shown in the preceding paragraph that the quasi-hydrostatic approximation determines to a certain extent the depth parameter D [see (2.10′)] and we will see that the adiabatic and vorticity equations impose certain limits on the scale parameter W.

The zonal harmonic analysis of the middle latitude westerly flow (Van Mieghem, 1960), has shown that the contributions to the fields of motion and temperature of the harmonic components of zonal wave numbers n larger than 15 are negligibly small. This fact justifies the choice of the lower limit for S: for a latitude circle in middle latitudes having a length of the order of 30,000 km, the shortest possible wavelength is $2S = 30{,}000$ km/15. The length of a great circle represents the longest possible wavelength ($2S = 40{,}000$ km) and thus determines the upper limit of S. The horizontal length scale S and the average value U of the horizontal wind components are much larger than the vertical depth scale D and the average vertical wind speed W, respectively. This results from the fact that for large-scale weather systems, the equation of motion along the vertical reduces to the equation of hydrostatic equilibrium (2.8) with an accuracy of 0.1 %, so that

$$\left(\frac{1}{\rho}\frac{\partial p}{\partial z} + g\right) \sim 10^{-4} g \sim 10^{-3} \text{ m s}^{-2}$$

The great variability of the horizontal scale parameter S suggests the partition of the length interval (3.1a) into two subintervals:

(3.1b)
$$10^6 \text{ m} \leq S < 5 \times 10^6 \text{ m} \quad \text{or} \quad S \sim 10^6 \text{ m}$$
$$5 \times 10^6 \text{ m} \leq S \leq 2 \times 10^7 \text{ m} \quad \text{or} \quad S \sim 10^7 \text{ m} \sim a$$

where a is the mean earth radius. The lengths 10^6 m, 5×10^6 m, and 2×10^7 m are the smallest, the intermediate, and the largest length scales, respectively. The intermediate scale corresponds approximately to the zonal wave with wave number 1 to 3. The first of the two length intervals (3.1b) is characteristic of the weather systems of the synoptic scale, and the second one, of the planetary scale. The analysis of surface and upper air charts has shown that the motion systems of the synoptic scale are most active in the low and middle troposphere and the planetary motion systems in the upper troposphere and low stratosphere of high and middle latitudes.

In addition to (3.1a), we assume that the magnitude of the phase speed C never exceeds the magnitude U of the horizontal wind components, so that

(3.2) $$C \lesssim U \quad \text{with } C \sim 10 \text{ m s}^{-1}$$

The average horizontal wind speed U is an upper limit of the average phase speed C. The propagation of the large-scale motion systems is slower than the air motions. This specification of the magnitude of C precisely distinguishes the meteorologically significant motions [defined by the scale parameters S, D, U, and W, satisfying the scale specifications (3.1a)], from all other theoretically possible atmospheric motions that may possibly exist having the same scale parameters, such as compressibility waves and gravity waves, for example. For these meteorologically unimportant motions, the magnitude of the corresponding phase speed is at least one order of magnitude larger than 10 m s^{-1}. Large-scale motion systems are slow-moving systems in all latitudes.

The characteristic time scale $\tau = S/C$ defining the local lifetime of the motion systems envisaged is equal or larger than the advective time scale S/U. This is a consequence of the characteristic specification (3.2) of the meteorologically important large-scale motion systems. From (3.1a) and (3.2), it follows that the order of magnitude of τ is 10^5 s when $S = 10^6$ m. Thus, the local lifetime for synoptic motion systems is of the order of a few days, and for planetary motion systems of the order of at least 1 week. The time scale τ varies between about 1 day and 3 weeks.

Finally, we wish to draw the attention to the fact that in Charney's theory of meteorological approximations it is implicitly admitted that nonlinear interactions between motions of different scales are negligible. This is a severe limitation.

Table I summarizes the characteristic values of the scale parameters of meteorologically significant weather-producing large-scale motion systems. When instead of one horizontal scale length S, two are chosen: S_x along the parallels and S_y along the meridians, Dickinson (1968) has pointed out that when $(S_y/S_x) \ll 1$, the dynamics of weather systems of this type differ from both Charney's and Burger's scalings.

Finally, we recall the useful practical relations,

$$\frac{c_{pa}}{7} = \frac{c_{va}}{5} = \frac{R_a}{2}$$

among the specific heats and the specific ideal gas constant R_a of dry air, where

$$R_a = 287.05 \text{ J kg}^{-1} \text{ K}^{-1} \quad \text{or} \quad \text{kJ ton}^{-1} \text{ K}^{-1}$$

The largest possible value of the length scale S at a given latitude is defined by $\Pi a \cos \phi$, half the length of the corresponding latitude circle. Table II gives the largest possible values of S as a function of ϕ. From Table II it follows that at low and middle latitudes ($\phi < 70°$), both scale intervals (3.1b) have to be considered. At latitudes higher than 70°, however, the maximum horizontal length scale S is smaller than the length a of the mean earth's radius, so that at these latitudes only the first of the two scale intervals (3.1b) has to be taken into consideration.

TABLE I. Numerical Values of Meteorological Scale Parameters

Mean earth radius	$a = 6{,}370$ km $\sim 5 \times 10^6$ m $\sim 10^7$ m
Synoptic horizontal length scale	10^6 m $\sim S \lesssim a \sim 5 \times 10^6$ m
Planetary horizontal length scale	$S \gtrsim a \sim 5 \times 10^6$ m
Horizontal velocity scale	$fH \ll U \lesssim 10^2$ m s^{-1}
Depth scale	$D \sim H \ll S$ with $D \leq H \sim 10^4$ m
Vertical velocity scale	$W \ll U \lesssim 10^2$ m s^{-1}
Scale of the speed of propagation of the weather systems	10 m s$^{-1} \sim C \lesssim U$
Time scale of synoptic systems	$\tau = S/C \lesssim 5 \times 10^5$ s
Time scale of planetary systems	$\tau = S/C \sim 10^6$ s
Acceleration of gravity	$g \sim 10$ m s^{-2}
Earth's rotation	$\Omega = 0{,}7292 \times 10^{-4}$ s$^{-1} \sim 10^{-4}$ s^{-1}
Shallowness of the atmosphere	$r = a + z \sim a \sim 5 \times 10^6$ m or $z/a \ll 1$
Hydrostatic approximation	$H \ll S$, $W \ll U$
Coriolis parameter $f = 2\Omega \sin \phi$	$\begin{cases} f \sim 10^{-4} \text{ s}^{-1}, \text{ when } \phi > 20° \\ f \sim 10^{-5} \text{ s}^{-1}, \text{ when } 20° \geq \phi > 2° \\ f \lesssim 10^{-5} \text{ s}^{-1}, \text{ when } \phi \leq 2° \end{cases}$
Coriolis parameter $f' = 2\Omega \cos \phi$	$\begin{cases} f' \sim 10^{-4} \text{ s}^{-1}, \text{ when } \phi < 70° \\ f' \sim 10^{-5} \text{ s}^{-1}, \text{ when } 70° \leq \phi < 88° \\ f' \lesssim 10^{-5} \text{ s}^{-1}, \text{ when } \phi \geq 88° \end{cases}$

(From the orders of magnitude of f and f', we deduce the orders of magnitude of the Rossby β parameter, $\beta = f'/r$ and also of $tg\, \phi = f/f'$.)

Static stability v_s^2 in a nonsaturated atmosphere	10^{-4} s^{-2}
Stratification parameters $(-1/\rho)(\partial \rho/\partial z)$ and $(-1/p)(\partial p/\partial z)$	$\sim 10^{-4}$ m^{-1}
Atmospheric pressure	upper limit of $p \sim 10^2$ cb
Air density	upper limit of $\rho \sim 10^{-3}$ to/m^3

TABLE II

ϕ (degree)	0°	5°	10°	20°	30°	40°	50°	60°	70°	75°	80°	85°
$\Pi a \cos \phi$ (km)	20,010	19,935	19,710	18,805	17,331	15,330	12,865	10,005	6,845	5,175	3,475	1,922

4. Scale Analysis of the Continuity Equation

Writing this equation as follows:

$$(4.1\text{a}) \quad \frac{\partial}{\partial t}(\ln \rho) + \left(u \frac{\partial}{\partial x} + v \frac{\partial}{\partial y}\right)(\ln \rho) + \frac{\partial u}{\partial x} + \frac{\partial v}{\partial y} - \frac{v}{r} tg\, \phi$$

$$= -w \frac{\partial}{\partial z}(\ln \rho) - \frac{\partial w}{\partial z} - \frac{2w}{r}$$

where $\partial/\partial x = \partial/r\cos\phi\,\partial\lambda$, $\partial/\partial y = \partial/r\,\partial\phi$, and $\partial/\partial z = \partial/\partial r$, λ and ϕ being the longitude and the latitude, it is clear that:

1. The first three terms on the left-hand side are of the same order of magnitude which is *smaller* than U/S [see Eqs. (2.6), (2.15), and (3.2)].
2. The following two terms of the left-hand side are of the same order of magnitude U/S [see Eqs. (2.3)]. Synoptic experience has shown that these two terms always tend to compensate each other so that the order of magnitude of their sum is smaller than U/S.
3. The last term of the left-hand side never exceeds the value of 10 U/a in the latitude belt 0 to 88°, so that the upper limit of this term in this belt is U/S (see Table I); hence the order of magnitude of the left-hand side never exceeds U/S.
4. The first two terms of the right-hand side are both of the same order of magnitude $(W/D) \sim (W/H)$ [see Eqs. (2.3), (2.9), and (2.10′)] and the third term, of the order of magnitude $(W/a) \ll (W/H)$ (see Table I).

Thus, in the equation of continuity (4.1a), we must keep the terms referred to in (2) and also the first two terms on its right-hand side. Hence the approximate continuity equation assumes the form

$$(4.1b) \qquad \frac{\partial u}{\partial x} + \frac{\partial v}{\partial y} + \frac{1}{\rho}\frac{\partial(\rho w)}{\partial z} = 0$$

Furthermore, we note that

$$(4.2) \qquad \frac{\partial u}{\partial x} \sim \frac{\partial v}{\partial y} \sim \frac{U}{S} \quad \text{and} \quad \left(\frac{\partial u}{\partial x} + \frac{\partial v}{\partial y}\right) \lesssim \frac{U}{S}$$

Consequently, the scale relations

$$(4.3) \qquad \frac{W}{D} \sim \frac{W}{H} \lesssim \frac{U}{S} \lesssim 10^{-4}\,\text{s}^{-1}$$

for deep large-scale weather systems follow from the evaluation of the orders of magnitude of the different terms of the continuity equation (4.1a). When these systems do not extend through the whole depth of the troposphere, we have the scale relations:

$$(4.3a) \qquad \frac{W}{H} \lesssim \frac{W}{D} \lesssim \frac{U}{S} \lesssim 10^{-4}\,\text{s}^{-1} \qquad \text{with } D \lesssim H$$

Furthermore, from (2) and (3), it results immediately that

$$(4.4) \qquad \Delta = \text{div}\,\mathbf{v}_h = \frac{\partial u}{\partial x} + \frac{\partial v}{\partial y} - \frac{v}{r}\,tg\phi \lesssim \frac{U}{S}$$

where Δ designates the divergence of the horizontal velocity field $\mathbf{v}_h$. The divergence of the horizontal wind has an order of magnitude which never exceeds U/S.

5. Scale Analysis of the Equation of Vertical Motion

The equation of motion along the ascending vertical z assumes the form

(5.1) $$\frac{dw}{dt} - \frac{u^2 + v^2}{r} - f'u = -\frac{1}{\rho}\frac{\partial p}{\partial z} - g$$

with $f' = 2\Omega \cos \phi$ and $(\partial p/\partial z) < 0$ always and everywhere in the atmosphere.

The corresponding scale equation defines the order of magnitude of the hydrostatic defect $\chi = (1/\rho)(\partial p/\partial z) + g$. Taking (2.3), (3.2), and (4.3) into account, we have

(5.2a)
$$\chi \sim g - \frac{1}{\bar{\rho}_D}\frac{|\Delta_D p|}{D} \sim \left(\frac{C}{S} + \frac{U}{S} + \frac{W}{D}\right)W + \frac{U^2}{a} + f'U \sim \frac{UW}{S} + \frac{U^2}{a} + f'U$$

where f' represents the Coriolis parameter $2\Omega \cos \phi$. Let us recall here that $f \sim f' \sim 10^{-4}$ s^{-1} in middle latitudes ($20° < \phi < 70°$), $f' \sim 10^{-1} f \sim 10^{-5}$ s^{-1} in high latitudes ($70° \leq \phi < 88°$), and $f' \sim 10 f \sim 10^{-4}$ s^{-1} in low latitudes ($20° \geq \phi > 2°$). Hence, in middle latitudes,

(5.2') $$\frac{1}{g\bar{\rho}_D}\frac{|\Delta_D p|}{D} = \frac{H}{D}\frac{|\Delta_D p|}{p} \sim 1 + \frac{fU}{g} + \frac{U^2}{ga} + \frac{UW}{gS}$$

Now, it is clear that the magnitude of the left-hand side of equation (5.1) is at least one order less than the magnitude of g, when

(5.3a) $$\frac{fU}{g} \lesssim \frac{H}{a} \ll 1 \quad \text{and} \quad \frac{U^2}{gH} \ll 1$$

The table giving the orders of magnitude of the parameters characterizing large-scale motions shows that

$$\frac{fU}{g} < 10^{-3} \quad \frac{H}{a} \sim 10^{-3} \quad \text{and} \quad \frac{U^2}{gH} < 10^{-1}$$

The scale relations (5.3a) are satisfied everywhere in the atmosphere.

Recalling that the symbol $\ll$ indicates that the left-hand side is at most of the order of about one-tenth of the right-hand side, we have, by virtue of (3.1a),

$$\frac{U^2}{ga} \ll \frac{U^2}{gH} \ll 1 \quad \text{and} \quad \frac{UW}{gS} \ll \frac{U^2}{gS} \ll \frac{U^2}{gH} \ll 1$$

The average magnitude W of the vertical wind speed being at least two orders of magnitude smaller than the average magnitude U of the horizontal wind components u and v, and the mean earth radius a being only one order of magnitude larger than the smallest horizontal length scale S, we have

$$(5.3b) \qquad 10^{-4} > \frac{U^2}{ga} > \frac{UW}{gS}$$

The last two terms on the right-hand side of the scale equation (5.2′) may be neglected with respect to $(fU/g) < 10^{-3}$. Finally, assuming with Charney [5] that $U \sim 10$ m s^{-1}, we obtain

$$(5.2b) \qquad \chi \sim fU \text{ m s}^{-2} \sim 10^{-4}g$$

In low latitudes ($\phi \leq 20°$), f must be replaced by f' where this parameter has the same order of magnitude as in the middle latitudes, so that the results obtained above for the middle latitudes may be applied to the low latitudes. In high latitudes ($70° \leq \phi$), f must also be replaced by f', where, however, this parameter is at least one order of magnitude less than f in the middle latitudes so that, in this case, each of the last three terms on the right-hand side of (5.2′) are smaller than 10^{-4}. The hydrostatic defect is negligible in all latitudes. Thus, when the inequalities (5.3a) are satisfied, the scale-equation (5.2a) reduces to (2.10), and the motion envisaged is quasi-hydrostatic.

Let us recall here that the ratio

$$(5.4) \qquad \text{Fr} = \frac{U^2}{gH}$$

is a number, the Froude number of the motion system considered, giving the order of magnitude of the ratio of kinetic energy (2.1) of the unit mass to the potential energy increase of this mass rising through the average depth H of the troposphere. The smallness of the Froude number is a consequence of the gravitational stability of large-scale motion systems. The ratio fU/g is a number representing the order of magnitude of the horizontal Coriolis acceleration fU to the acceleration of gravity g. The quasi-hydrostatic assumption determines the largest value of the vertical dimension of the large-scale motion systems [see Eq. (2.10′)].

6. Scale Analysis of the Equations of Horizontal Motion

The vector equation of motion in horizontal surfaces (spheres concentric to the earth) assumes the form

$$(6.1) \qquad \frac{d\mathbf{v}_h}{dt} + \frac{u\mathbf{k} \times \mathbf{v}_h}{r} tg\phi + \frac{w\mathbf{v}_h}{r} + 2(\Omega \times \mathbf{v})_h = -\frac{\nabla_h p}{\rho}$$

where Ω represents the angular velocity of the earth's rotation and $\mathbf{k}$ the unit vector along the ascending vertical z. The corresponding scale equation defines the order of magnitude of $(-1/\rho)\nabla_h p$. Taking (3.1a), (3.2), and (4.3) into account, we find

$$(6.2) \quad \frac{|\nabla_h p|}{\rho} \sim \left(\frac{C}{S} + \frac{U}{S} + \frac{W}{D}\right)U + \left(\frac{U}{a} + \frac{W}{a}\right)U$$

$$+ fU + f'W \sim \frac{1}{\bar{\rho}_S}\frac{|\Delta_s p|}{S} \sim \left(\frac{U}{S} + f\right)U$$

Indeed, the inequality (4.3) implies that the order of magnitude UW/D of the vertical advection terms of the horizontal accelerations du/dt and dv/dt in the analytical equations of horizontal motion is not larger than the order of magnitude U^2/S of the horizontal advection terms of these accelerations. Moreover, the magnitude of the second term of (6.1) never exceeds $10\ U^2/a$ in the latitude belt 0 to 88°, while in low and middle latitudes ($\phi < 70°$), f' never exceeds $10f$ and U is at least two orders of magnitude larger than W.

Using the definitions (2.7) and (5.4) of the height scale H and the Froude number Fr, the relation (6.2) assumes the form

$$(6.3) \quad \frac{1}{gH}\frac{|\Delta_s p|}{\bar{\rho}_S} \sim \frac{|\Delta_s p|}{p_S} \sim \frac{U^2}{gH}\left(1 + \frac{fS}{U}\right) = \frac{\text{Fr}}{\text{Ro}}(1 + \text{Ro})$$

where

$$(6.4) \quad \text{Ro} = \frac{U}{fS} \lesssim (10^{-4}/f)$$

is the Rossby number of the motion system envisaged. This number represents the ratio of the order of magnitude U^2/S of the terms of the equations of horizontal motion representing the advection of horizontal momentum to the order of magnitude fU of the Coriolis terms.

Assuming that the average horizontal wind speeds never exceed $100\ \text{m s}^{-1}$, we note that

$$\text{Ro} \lesssim 1$$

everywhere in the middle and high latitudes ($f \sim 10^{-4}\ \text{s}^{-1}$), while in tropical latitudes ($f \lesssim 10^{-5}\ \text{s}^{-1}$),

$$\text{Ro} \gtrsim 1$$

Thus, in middle and high latitudes the order of magnitude U^2/S of the horizontal acceleration of the wind never exceeds the order of magnitude fU of the horizontal Coriolis acceleration. The contrary may occur between the tropics.

The smallness of the Rossby number implies the validity of the geostrophic approximation [see Section 8, Eqs. (8.11), (8.12), and (8.13)]. In this case,

$$\frac{U}{S} < f$$

Thus, the deviation from the geostrophic law [see (8.11)] is small when the characteristic frequency U/S of the large-scale midlatitude motion systems is small compared with the characteristic frequency f of a horizontal inertial oscillation.

From (6.3) we deduce the magnitude

(6.5a) $$\frac{|\Delta_S p|}{\bar{\rho}_S} \sim U^2(1 + \text{Ro})\text{Ro}^{-1}$$

of the pressure gradient force. In middle latitudes, we have

(6.5b) $$\frac{|\Delta_S p|}{\bar{\rho}_S} \sim U^2 \text{Ro}^{-1} \quad \text{with Ro} < 1$$

and, in low latitudes,

(6.5c) $$\frac{|\Delta_S p|}{\bar{\rho}_S} \sim U^2 \quad \text{with Ro} > 1$$

Thus, horizontal pressure contrasts $\Delta_S p$ associated with deep large-scale motion systems in the tropics are smaller by a factor Ro (< 1 with $f \sim 10^{-4}$ s^{-1}) than those for midlatitude systems of the same scale. By virtue of (2.20) the same result applies to the time fluctuations $\Delta_S p$.

Finally, from Eqs. (2.14a) and (6.5a), it is easy to deduce the magnitude of thermal contrasts in large-scale motion systems:

(6.6a) $$|\Delta_S T| \sim \frac{1}{R_a} \left| \frac{\Delta_S p}{\bar{\rho}_S} \right| \sim \frac{U^2}{R_a}(1 + \text{Ro})\text{Ro}^{-1}$$

where R_a is the specific ideal gas constant for dry air. In middle latitudes we have

(6.6b) $$|\Delta_S T| \sim \frac{U^2}{R_a} \text{Ro}^{-1} \quad \text{with Ro} < 1$$

and in low latitudes,

(6.6c) $$|\Delta_S T| \sim \frac{U^2}{R_a} \sim 0.3 \text{ K} \quad \text{when } U \sim 10 \text{ m s}^{-1}$$

Thus, horizontal temperature contrasts $\Delta_S T$ associated with deep large-scale motion systems in the tropics are also smaller by a factor Ro (Ro < 1, with $f \sim 10^{-4}$ s^{-1}) than those for midlatitude systems of the same scale. The same result applies to the time fluctuations $\Delta_t T$ [see (2.20)].

This result has an important consequence in atmospheric energetics. The temperature contrasts in large-scale quasi-geostrophic synoptic motion systems (Ro $\sim$ 0.1) of the middle latitudes being one order of magnitude larger than those for large-scale tropical motion systems of the same scale, the available potential energy proportional to the square of these temperature contrasts, is, in the middle latitudes, two orders of magnitude larger than in the tropics. It is indeed well known that the primary energy source for extratropical large-scale motion systems is available potential energy, while near the equator the primary energy source for large-scale motion systems is the release of latent heat in the convective mesoscale cloud systems imbedded in the large-scale motion systems of the tropics. Thus it is likely that there is a strong interaction between the convective mesoscale and large-scale motion systems in the tropics. Unfortunately, an adequate scale analysis of mesoscale motion systems is still lacking.

7. Scale Analysis of the Equation of Adiabatic Transformations

This equation assumes the form

$$(7.1)\quad \frac{c_{va}}{c_{pa}}\left[\frac{\partial}{\partial t}(\ln p) + \mathbf{v}_h \cdot \nabla_h(\ln p)\right] - \left[\frac{\partial}{\partial t}(\ln \rho) + \mathbf{v}_h \cdot \nabla_h(\ln \rho)\right] = -\sigma w$$

where

$$(7.2)\quad \sigma = \frac{c_{va}}{c_{pa}}\frac{\partial(\ln p)}{\partial z} - \frac{\partial(\ln \rho)}{\partial z} = \frac{\partial \ln \theta}{\partial z}$$

defines the static stability ($\sqrt{g\sigma}$ is the gyro-frequency v_S of a free vertical gravity oscillations), and where c_{va} and c_{pa} designate the specific heats of dry air and θ the potential temperature (see Section 2).

Returning to the classic definition of θ, we have the differential relation

$$(7.3)\quad \frac{\nabla\theta}{\theta} = -\frac{\nabla\rho}{\rho} + \frac{c_{va}}{c_{pa}}\frac{\nabla p}{p}$$

between p, ρ, and θ. From this relation and the scale equations (2.13), (2.14a), and (6.3), it follows immediately that

$$(7.4)\quad \frac{|\Delta_S p|}{\bar{p}_S} \sim \frac{|\Delta_S \rho|}{\bar{\rho}_S} \sim \frac{|\Delta_S T|}{T_S} \sim \frac{|\Delta_S \theta|}{\theta_S} \sim \frac{\text{Fr}}{\text{Ro}}(1 + \text{Ro})$$

From the scale relations (7.4) it follows that the horizontal and temporal fluctuations Δ_s and Δ_τ [see (2.20)] are small in all latitudes (Charney, 1963).

Noting that the two terms in square brackets on the left-hand side of (7.1) have the same order of magnitude and are not necessarily of the same sign, and using the scale Eqs. (2.13), (2.15), (3.2), and (6.3), the scale equation

(7.5a) $$\sigma W \sim f\,\text{Fr}(1 + \text{Ro})$$

may be immediately deduced from (7.1). When in large-scale motion systems the static stability decreases, the magnitude W of the vertical velocities increases.

It should be noted that $f(1 + \text{Ro})$ has in middle and high latitudes ($\varphi > 20°$), the same order of magnitude as f and, in low latitudes ($\varphi > 20°$), as U/S, so that the scale equation (7.5a) reduces to

(7.5b) $$\sigma W \sim \frac{fU^2}{gH} = f\,\text{Fr} = \frac{U}{S}\,\text{Fr}\,\text{Ro}^{-1} \quad \text{with Ro} < 1$$

in middle latitudes (Ro $\lesssim$ 1), and to

(7.5c) $$\sigma W \sim \frac{U}{S}\,\text{Fr} \sim 10^{-8}\,\text{s}^{-1} \quad \text{when } U \sim 10\,\text{m s}^{-1}$$

in low latitudes (Ro $\gtrsim$ 1).

Hence, for deep ($D \sim H$) large-scale motion systems of low and middle latitudes having the same scale parameters U, S, and $\sigma \sim 10^{-5}\,\text{m}^{-1}$, the magnitude W of the vertical velocity is larger in extratropical latitudes than in tropical latitudes by a factor Ro^{-1} (with $f \sim 10^{-4}\,\text{s}^{-1}$). Assuming $U \sim 10\,\text{m s}^{-1}$ and $S \sim 10^6\,\text{m}$, we have $W \sim 10^{-2}\,\text{m s}^{-1}$ in middle latitudes and $W \sim 10^{-3}\,\text{m s}^{-1}$ in the tropics. In the tropical belt, this applies only outside the precipitation areas. In these areas, the static stability $g\sigma$ is much smaller than $10^{-4}\,\text{s}^{-2}$. Therefore, the vertical motions are much more important in precipitating tropical motion systems than outside them. In fact they are one order of magnitude larger, and moreover $\partial w/\partial z$ is also larger, especially in the upper troposphere.

In moist air, the static stability is defined by

$$\sigma = \frac{1}{\theta}\frac{\partial \theta}{\partial z} + \frac{L_v}{c_{pa}T}\frac{\partial r_v}{\partial z} \quad \text{with} \quad \frac{\partial r_v}{\partial z} < 0$$

where L_v is the heat of evaporation (released at condensation) and r_v the mixing ratio of water vapor. When the atmosphere is saturated, the saturation mixing ratio must be substituted to r_v. In a cumulus cloud cell, σ is negative in the low troposphere (in the layer 1000 to 600 mb). In a large-scale motion system, however, σ is positive. Therefore, there are in fact two vertical wind fields: (1) the forced vertical motion associated with the large-scale motion systems ($\sigma > 0$) and (2) the convective ascent in the cumulus cloud cells ($\sigma < 0$) and the associated descent between cumulus clouds.

In the tropics the convective cloud cells are deep, their local lifetime is of a few hours ($D \sim H$), and their horizontal dimensions are small ($10^3 < S < 10^4$ m). They are organized in mesoscale lines or rings with horizontal dimensions smaller than 10^5 m ($10^4 < S < 10^5$ m). Mesoscale convective motion systems in turn are grouped in cloud clusters with characteristic horizontal size: $10^5 < S < 10^6$ m. Their lifetime is of a few days. These three convective cloud systems are coupled with the large-scale tropical motion systems. For middle latitude quasi-geostrophic (Ro $\sim 10^{-1}$) motion systems there is one order of magnitude difference between the vertical velocities of middle latitude and low latitude motion systems of the same scale.

Returning to the definition (7.2) of σ and taking into account that both $\partial p/\partial z$ and $\partial \rho/\partial z$ are always and everywhere negative, we find

$$(7.6) \qquad \sigma \sim \frac{c_{pa} - c_{va}}{c_{pa}} \frac{1}{H} \sim \frac{0.3}{H}$$

so that

$$(7.7) \qquad \sigma \lesssim \frac{1}{H}$$

Combining (7.5a), (7.7), and (4.3), we obtain the scale relations

$$(7.8) \qquad f\,\mathrm{Fr}(1 + \mathrm{Ro}) \sim \sigma W \lesssim \frac{W}{H} \lesssim \frac{U}{S} \lesssim 10^{-4}\ \mathrm{s}^{-1}$$

From the second inequality (7.8), it follows that

$$(7.9a) \qquad \frac{fU}{g} \lesssim \frac{H}{S}(1 + \mathrm{Ro})^{-1}$$

This inequality should be compared with the first two inequalities (5.3a). In middle and high latitudes, (7.9a) reduces to

$$(7.9b) \qquad \frac{fU}{g} \lesssim \frac{H}{S} \lesssim 10^{-2} \qquad \text{or} \qquad \mathrm{Ro} \gtrsim \mathrm{Fr}$$

and in low latitudes to the last inequality (5.3a).

The ratio S/H of the horizontal and vertical space dimensions is Prandtl's size ratio of the motion system considered.

Equations (7.4), (2.12), and (2.13) suggest that the fields of the physical state variables of the atmosphere are identified with scale operators as follows:

(7.10)
$$\frac{\Delta_S X}{\bar{X}_S} \sim \frac{\text{Fr}}{\text{Ro}}(1 + \text{Ro})$$

for the fields of the physical variables of state ρ, p, T, and θ;

(7.11)
$$\frac{\Delta_D X}{\bar{X}_D} \sim \frac{D}{H}$$

for the mass field variables ρ and p; and

(7.12)
$$\frac{\Delta_D X}{\bar{X}_D} \lesssim \frac{D}{H}$$

for the temperature field variables T and θ.

8. Order of Magnitude of the Divergence Δ of the Horizontal Wind

We have already shown that, (4.4),

(8.1)
$$\Delta = \frac{\partial u}{\partial x} + \frac{\partial v}{\partial y} - \frac{v}{r} tg\phi \sim \frac{\partial u}{\partial x} + \frac{\partial v}{\partial y} \lesssim \frac{U}{S}.$$

An alternative approximate form of Δ is provided by (4.1b).

Rewriting the continuity equation (4.1a) as follows:

$$\left(\frac{\partial}{\partial t} + \mathbf{v}_h \cdot \nabla_h\right)(\ln \rho) + w\frac{\partial(\ln \rho)}{\partial z} + \Delta + \frac{\partial w}{\partial z} + \frac{2w}{r} = 0$$

and using the scale equations (2.3), (2.9), (2.13), (2.15), (3.1a), (3.2), (6.3), and (7.4), we find the scale equation

(8.2)
$$\Delta \sim f \text{Fr}(1 + \text{Ro}) + \frac{W}{D} \sim \frac{W}{D} \sim \frac{W}{H}$$

by virtue of (2.10′) and the second of the scale relations (7.8) and (2.8).

Introducing the components

(8.3)
$$u_g = \frac{-1}{f\rho}\frac{\partial p}{\partial y} \qquad v_g = \frac{1}{f\rho}\frac{\partial p}{\partial x} \qquad W_g \equiv 0$$

of the geostrophic wind $\mathbf{v}_g$ into the horizontal wind divergence (8.1), we find the analytical expression for the divergence of the geostrophic wind,

(8.4)
$$\Delta_g = \text{div } \mathbf{v}_g = \frac{p}{f\rho}\left[\frac{1}{\rho}\frac{\partial \rho}{\partial x}\frac{\partial p}{\partial y} - \frac{1}{\rho}\frac{\partial \rho}{\partial y}\frac{\partial p}{\partial x} - \left(\frac{\beta}{f} + \frac{tg\phi}{r}\right)\frac{1}{p}\frac{\partial p}{\partial x}\right]$$

where

(8.5) $$\beta = \frac{df}{dy} = \frac{1}{r}\frac{\partial f}{\partial \phi} = \frac{2\Omega \cos \phi}{r} = \frac{f'}{r}$$

represents the Rossby parameter.

Using the scale equations (7.4), we obtain after some transformations the order of magnitude of Δ_g

(8.6a) $$\Delta_g \sim (1 + \text{Ro})\left[f\,\text{Fr}(1 + \text{Ro}) + \left(\frac{f'}{f} + \frac{f}{f'}\right)\frac{U}{a}\right]$$

The largest value of the ratios f/f' and f'/f is 10 in the latitude belts $70° \leq \phi < 88°$ and $2° < \phi \leq 20°$, respectively, so that the sum $(f'/f) + (f/f')$ never exceeds 10 in the latitude belt 2 to 88°. Hence the scale formula (8.6a) reduces to

(8.6b) $$\Delta_g \sim \left(\frac{fU}{gH} + \frac{1}{a}\right)U \sim \frac{U}{a}$$

in the middle latitudes ($20° < \phi < 70°$), to

(8.6c) $$\Delta_g \sim \left(\frac{fU}{gH} + \frac{10}{a}\right)U \sim \frac{10U}{a}$$

in the high latitudes ($70° \leq \phi < 88°$), and to

(8.6d) $$\Delta_g \sim (1 + \text{Ro})\left[\frac{U^2}{gH} + 10\frac{S}{a}\right]\frac{U}{S} \sim 10(1 + \text{Ro})\frac{U}{a}$$

in the low latitudes ($20° \gtrless \phi > 2°$). In the scale equations (8.6b) and (8.6c) account has been taken of the first inequality (5.3a). The largest value of the first term in the brackets of formula (8.6d) is 10^{-1}, and of the second term is 10. In the low latitude belt the maximum value of Ro is 10.

We may now relate the Richardson number, proportional to the static stability $g\sigma$ and inversely proportional to the square of the vertical windshear:[1]

(8.7) $$\text{Ri} = \frac{\dfrac{g}{\theta}\dfrac{\partial \theta}{\partial z}}{\left|\dfrac{\partial \mathbf{v}_h}{\partial z}\right|^2} = \frac{g\sigma}{\left|\dfrac{\partial \mathbf{v}_h}{\partial z}\right|^2}$$

---

[1] We recall that the Richardson number is proportional to the ratio of the rate $\rho K_H(1/\theta) \times (\partial \theta/\partial z)$ at which work has been done by the small-scale eddies against gravity to the rate of supply $\rho K_M(\partial \mathbf{v}_h/\partial z)^2$ of energy by the Reynolds' shearing stresses, where K_H and K_M are the coefficients of eddy conductivity and eddy viscosity.

of the motion system considered, to its Rossby number defined in (6.4). Taking into account the definition (7.2) of σ, and the scale formula (7.8), we find the order of magnitude of Ri:

(8.8) $\quad \text{Ri} \sim g\sigma \dfrac{D^2}{U^2} \sim \dfrac{fD}{W}(1+\text{Ro})\dfrac{D}{H} \sim \dfrac{fH}{W}(1+\text{Ro}) \gtrsim \dfrac{fS}{U}(1+\text{Ro})$

by virtue of (2.10′) and (4.2); hence, (6.4),

(8.9) $\quad\quad\quad\quad\quad\quad\quad\quad \text{RiRo} \gtrsim 1 + \text{Ro}$

Another useful expression for Ri is given by

(8.7′) $\quad\quad\quad\quad\quad\quad\quad\quad \text{Ri} = \dfrac{\dfrac{H}{\theta}\dfrac{\partial \theta}{\partial z}}{\text{Fr}}$

The Richardson number Ri will now be used in order to relate the scale parameter W/H to U/S in a more precise way than does the scale formula (4.3). Combining the definition (8.7) and (6.4) and the scale equations (7.5a) and (8.8), we obtain, after some elementary transformation,

(8.10a) $\quad\quad\quad\quad\quad \dfrac{W}{D} \sim \dfrac{W}{H} \sim \dfrac{U}{S}(1+\text{Ro})(\text{RoRi})^{-1}$

This relation between the ratios U/S and W/D of the velocity scale parameters U and W and the corresponding scale parameters of the space dimensions simplifies to

(8.10b) $\quad\quad\quad\quad\quad\quad \dfrac{W}{D} \sim \dfrac{W}{H} \sim \dfrac{U}{S}(\text{RiRo})^{-1}$

in middle and high latitudes, and to

(8.10c) $\quad\quad\quad\quad\quad\quad \dfrac{W}{D} \sim \dfrac{W}{H} \sim \dfrac{U}{S}\text{Ri}^{-1}$

in low latitudes (Charney, 1963).

The geostrophic assumption implies, by definition, that

(8.11) $\quad\quad\quad\quad\quad\quad\quad \dfrac{1}{\bar{\rho}_s}\dfrac{|\Delta_s p|}{S} \sim fU$

Daily synoptic weather analysis has shown that this approximation is justified in middle and high latitudes above the friction layer. Substituting (6.3) into (8.11) we obtain

(8.12) $\quad\quad\quad\quad\quad\quad\quad\quad \text{Ro} \ll 1$

This is the Rossby criterion for quasi-geostrophic motion. When this criterion is verified, we have

$$\frac{U^2}{S} \ll fU \tag{8.13}$$

so that, in this case, the horizontal acceleration of the wind may be neglected with respect to the horizontal Coriolis acceleration. Moreover, when the condition (8.12) is satisfied (Charney, 1963), we have

$$\frac{|\Delta_s p|}{\bar{p}_s} \sim \frac{|\Delta_s \rho|}{\bar{\rho}_s} \sim \frac{|\Delta_s \theta|}{\bar{\theta}_s} \sim \frac{\text{Fr}}{\text{Ro}} \tag{8.14}$$

account being taken of (7.4).

By virtue of (8.9), the Rossby criterion (8.12) is equivalent to the Eady criterion (1949),

$$\text{Ri} \gg 1 \tag{8.15}$$

for quasi-geostrophic motion in middle and high latitudes.

In low latitudes, however, the scale relation (7.4) (Charney, 1963) reduces to

$$\frac{|\Delta_s p|}{\bar{p}_s} \sim \frac{|\Delta_s \rho|}{\bar{\rho}_s} \sim \frac{|\Delta_s \theta|}{\bar{\theta}_s} \sim \text{Fr} \tag{8.16}$$

Considering the magnitudes assumed by Ro as a function of latitude ϕ and the horizontal length and velocity scales S and U, it is easy to formulate the following propositions.

1. In middle and high latitudes ($\phi > 20°$), where $f \sim 10^{-4}$ s^{-1}, the motion systems of the synoptic ($S \sim 10^6$ m) and planetary ($S \sim 10^7$ m) length scales are quasi-geostrophic when $U \sim 10$ m s^{-1} [Charney's (1948) case]; when, however, the horizontal velocity scale belongs to the velocity interval ($50 \leq U < 100$ m s^{-1}), then the motion systems of the planetary scale only are quasi-geostrophic.
2. In the low latitude belt ($20° \leq \phi < 2°$), where $f \sim 10^{-5}$ s^{-1}, motion systems are quasi-geostrophic when and only when $U \sim 10$ m s^{-1} and $S \sim 10^7$ m.
3. In any case, in the equatorial belt (2°S to 2°N), the geostrophic assumption is unacceptable.
4. Finally, comparing the Rossby number (6.4) and the Froude number (5.4), it is immediately seen from (3.1a) that, for large-scale motion systems,

$$\text{Ro} \gg \text{Fr} \quad \text{or} \quad \frac{H}{S} \gg \frac{fU}{g} \tag{8.17}$$

except for motion systems of the planetary scale ($S \sim a \sim 10^7$ m), when the scale parameter U approaches its upper limit ($U = 100$ m s^{-1}). In this last case

(8.18) $\qquad \text{Ro} \sim \text{Fr} \sim 10^{-1} \quad \text{or} \quad \dfrac{H}{S} \sim \dfrac{fU}{g} \sim 10^{-3}$

These results are consistent with (7.9b).

9. Scale Analysis of the Vorticity Equation

This equation assumes the form

$$(9.1) \quad \left(\frac{\partial}{\partial t} + \mathbf{v}_h \cdot \nabla_h + w\frac{\partial}{\partial z}\right)(f+\xi) + (f+\xi)\left(\Delta + \frac{2w}{r}\right) + \frac{\partial w}{\partial x}\left(\frac{\partial v}{\partial z} + \frac{v}{r}\right)$$
$$- \frac{\partial w}{\partial y}\left(\frac{\partial u}{\partial z} + \frac{u}{r} + f'\right) = \frac{p}{\rho}\left(\frac{1}{p}\frac{\partial p}{\partial y}\cdot\frac{1}{\rho}\frac{\partial \rho}{\partial x} - \frac{1}{p}\frac{\partial p'}{\partial x}\cdot\frac{1}{\rho}\frac{\partial \rho}{\partial y}\right)$$

where β has been defined in (8.5) and where the relative vorticity ξ is expressed by

$$(9.2) \quad \xi = \frac{\partial v}{\partial x} - \frac{\partial u}{\partial y} + \frac{u}{r}\,\mathrm{tg}\phi$$

Noting that

$$(9.3) \quad \beta = \frac{f'}{a} \quad \text{and} \quad \xi \sim \frac{U}{S}$$

it is easy to find, by virtue of (3.1a), the scale equation corresponding to (9.1):

$$(9.4\mathrm{a}) \quad \frac{f'}{a}U + \left(\frac{C}{S} + \frac{U}{S} + \frac{W}{D}\right)\frac{U}{S} + \left(f + \frac{U}{S}\right)\left(\frac{W}{D} + \frac{W}{a}\right) + \frac{W}{S}\left(\frac{U}{D} + \frac{U}{a}\right)$$
$$+ \frac{W}{S}\left(\frac{U}{D} + \frac{U}{a} + f'\right) \sim \frac{gH}{S^2}\frac{\mathrm{Fr}^2}{\mathrm{Ro}^2}(1+\mathrm{Ro})^2$$

account being taken of (2.13), (6.3), and (8.2).

It is immediately seen from (2.7), (2.10), (2.10'), (3.1a), (3.1b), (4.3), and Table I (see Section 3), that

$$(9.5) \quad \frac{C}{S} \lesssim \frac{U}{S} \qquad \frac{W}{a} \ll \frac{W}{D} \sim \frac{W}{H} \qquad \frac{U}{a} \ll \frac{U}{D} \qquad f' \ll \frac{U}{D}$$

and

$$(9.6) \quad \left| w\frac{\partial \xi}{\partial z} + \xi\Delta + \left(\nabla_h w \times \frac{\partial \mathbf{v}_h}{\partial z}\right)\mathbf{k} \right| \sim \frac{U}{S}\frac{W}{D}$$

so that the scale equation (9.4a) reduces to

(9.4b) $$\frac{U^2}{S^2} + \frac{f'U}{a} + f\frac{W}{D}(1 + \text{Ro}) \sim f^2\text{Fr}(1 + \text{Ro})^2$$

The first term on the left-hand side of (9.4b) represents the horizontal advection of vorticity, the second term represents the β term, the term fW/D represents the combined effect of the earth's rotation and the divergence Δ of the horizontal wind on the change in vorticity, and the term $fW\text{Ro}/D = WU/DS$ represents the terms of the vorticity equation on the left-hand side of (9.6), that is, the vertical advection of vorticity, the vorticity change due to the divergence of the horizontal wind, and the twisting terms of the vorticity equation. The solenoidal terms of this equation are represented on the right-hand side of (9.4b).

The second scale relation (7.8) and the first scale relation (4.2) show that the solenoidal term on the right-hand side of (9.4b) is smaller than the third term on the left-hand side, so that neglecting the solenoidal term and rearranging the terms of the scale equation (9.4b), we find

(9.7a) $$f\frac{W}{D}(1 + \text{Ro}) \sim \frac{U^2}{S^2} + \frac{f'U}{a} \sim \frac{U^2}{S^2}\left(1 + \frac{S}{a\text{R}'\text{o}}\right) = \frac{f'U}{aS^2}\left(\frac{aU}{f'} + S^2\right)$$

where $\text{R}'\text{o}$ is the Rossby parameter associated with the Coriolis parameter $f' = 2\Omega \cos \phi$,

(9.8a) $$\text{R}'\text{o} = \frac{U}{f'S}$$

with

(9.8b) $$\text{R}'\text{o} \lesssim 1$$

in low and middle latitudes ($\phi < 70°, f' \sim 10^{-4} \text{ s}^{-1}$), and

(9.8c) $$\text{R}'\text{o} \gtrsim 1$$

in higher latitudes ($\phi > 70°, f' \sim 10^{-5} \text{ s}^{-1}$).

From this scale equation, it results that there exists a critical intermediate horizontal length S_* (Burger, 1958),

(9.9) $$S_* = a\text{R}'\text{o} \quad \text{or} \quad S_* = \left(\frac{aU}{f'}\right)^{1/2}$$

in the range of horizontal length scales of large atmospheric disturbances. This critical scale length is smaller than a in low and middle latitudes

($\phi < 70°$), and larger than a in higher latitudes ($\phi > 70°$). At these latitudes, the order of magnitude of the largest possible horizontal length scale does not greatly exceed 10^6 m. In low and middle latitudes ($70° > \phi$) we have

(9.10a) $$10^6 \text{ m} \leq S_* \leq 3.10^6 \text{ m}$$

but in higher latitudes,

(9.10b) $$3.10^6 \text{ m} \leq S_* \leq 10^7 \text{ m}$$

At these latitudes motion systems of the planetary scale do not exist.

In the latitude belt 20 to 70°, both Coriolis parameters f and f' have the same order of magnitude (10^{-4} s^{-1}); however, f' in higher latitudes and f in lower latitudes are at least one order of magnitude less, so that the scale equation (9.7a) assumes the following simplified forms:

(9.7b) $$\frac{W}{D} \sim \frac{U}{S}\left(\text{Ro} + \frac{S}{10a}\right) \quad \text{with Ro} \lesssim 1$$

when $88° > \phi \geq 70°$, where $f \sim 10f' \sim 10^{-4}$ s^{-1},

(9.7c) $$\frac{W}{D} \sim \frac{U}{S}\left(\text{Ro} + \frac{S}{a}\right) \quad \text{with Ro} \lesssim 1$$

when $70° > \phi > 20°$, where $f \sim f' \sim 10^{-4}$ s^{-1}, and

(9.7d) $$\frac{W}{D} \sim \frac{U}{S}\left(1 + \frac{10S}{a\text{Ro}}\right) \quad \text{with Ro} \gtrsim 1$$

when $20° \geq \phi > 2°$, where $10f \sim f' \sim 10^{-4}$ s^{-1}. Equation (9.7c) is Burger's (1958) key scale relation.

Let us now take again the general form (9.4b) of the scale equation corresponding to the vorticity equation and substitute in both sides of (9.4b) $(U/S)\text{Ro}^{-1}$ for f; we find immediately the form

(9.10) $$\frac{U^2}{S^2} + \frac{f'U}{a} + (1 + \text{Ro}^{-1})\frac{UW}{SD} \sim \frac{U^2}{S^2}\text{Fr}(1 + \text{Ro})^2\text{Ro}^{-2}$$

or, by virtue of (8.10a) and (9.9),

(9.11) $$1 + (1 + \text{Ro})^2\text{Ro}^{-2}(\text{Ri}^{-1} + \text{Fr}) \sim \frac{S}{a\text{R'o}} \sim \frac{S^2}{S_*^2}$$

In Charney's (1948) case, we have: $\text{Fr} \sim 10^{-3}$, $\text{Ri}^{-1} \sim 10^{-2}$, and $S_* \sim 10^6$ m.

10. Scale Analysis of the Equation of the Horizontal Wind Divergence

This equation assumes the form

$$(10.1) \quad \left(\frac{\partial}{\partial t} + \mathbf{v}_h \cdot \nabla_h + w\frac{\partial}{\partial z}\right)\Delta + \Delta\left(\Delta + \frac{2w}{r}\right) + 2\left(\frac{\partial v}{\partial x}\frac{\partial u}{\partial y} - \frac{\partial u}{\partial x}\frac{\partial v}{\partial y}\right)$$

$$+ \frac{\partial w}{\partial x}\left(\frac{\partial u}{\partial z} + \frac{u}{r} + f'\right) + \frac{\partial w}{\partial y}\left(\frac{\partial v}{\partial z} + \frac{v}{r}\right) + 2\frac{tg\phi}{r}\left(u\frac{\partial u}{\partial y} + v\frac{\partial v}{\partial y}\right)$$

$$+ \frac{u^2 + v^2}{r} - f\xi + \beta u$$

$$= \frac{p}{\rho}\left(\frac{1}{\rho}\frac{\partial \rho}{\partial x} \cdot \frac{1}{p}\frac{\partial p}{\partial x} + \frac{1}{\rho}\frac{\partial \rho}{\partial y} \cdot \frac{1}{p}\frac{\partial p}{\partial y}\right) - \frac{1}{\rho}\left(\frac{\partial^2 p}{\partial x^2} + \frac{\partial^2 p}{\partial y^2} - \frac{tg\phi}{r}\frac{\partial p}{\partial y}\right)$$

where the divergence $\Delta \simeq \text{div } \mathbf{v}_h$ of the horizontal wind $\mathbf{v}_h$ has been defined in (8.1), and the vorticity ξ and the Rossby β parameter in (9.2).

Using the scale relations (8.2), (9.3) and (2.7), (2.13), (2.15), (3.1a), (3.1b), (3.2), (4.3), and (7.4), it is easy to deduce a scale equation from (10.1). The order of magnitude of the left-hand side of (10.1) is represented by

$$(10.2') \quad f(1 + \text{Ro})\frac{U}{S}$$

It should be noted here that the term $f'U/a$ has been deleted because its order of magnitude never exceeds fU/S in the latitude belt (2 to 88°). The order of magnitude of the right-hand side of (10.1) is given by

$$(10.2'') \quad gH\frac{|\nabla_h^2 p|}{p} + \frac{gH}{S^2}\left(\frac{\text{Fr}}{\text{Ro}}\right)^2 (1 + \text{Ro})^2$$

By virtue of (7.8), the second term of (10.2") never exceeds (10.2') in order of magnitude, so that the scale equation corresponding to the equation of the divergence of the horizontal wind assumes the simple form

$$(10.3) \quad gH\frac{|\nabla_h^2 p|}{p} \sim f(1 + \text{Ro})\frac{U}{S}$$

where $\nabla_h^2 p$ represents the Laplacian of pressure p in horizontal surfaces (spheres concentric to the earth).

In middle and high latitudes, the right-hand side of (10.3) reduces to fU/S and the equation of the horizontal wind divergence to its geostrophic approximation (Burger, 1958),

$$(10.4) \quad -\frac{1}{\rho}\nabla_h^2 p + f\xi \cong 0$$

In low latitudes, however, the right-hand side of (10.3) simplifies to U^2/S^2 and the equation of the horizontal wind divergence to

(10.5) $\qquad \rho(\nabla_h u \times \nabla_h v)\mathbf{k} = \tfrac{1}{2}\nabla_h^2 p$

11. Discussion of the Basic Scale Equations

For convenience, we will rewrite the equations that have been established in Sections 7, 8, 9, and 10 as follows:

(7.4) → (11.1a) $\qquad \dfrac{|\Delta_S p|}{\bar{p}_S} = \dfrac{|\Delta_S \rho|}{\bar{\rho}_S} = \dfrac{|\Delta_S \theta|}{\bar{\theta}_S} \sim \dfrac{\text{Fr}}{\text{Ro}}(1 + \text{Ro})$

(7.11) → (11.1′) $\qquad \dfrac{|\Delta_D p|}{\bar{p}_D} = \dfrac{|\Delta_D \rho|}{\bar{\rho}_D} \sim \dfrac{D}{H} \sim 1 \qquad \text{with } D \leq H$

(7.12) → (11.1″) $\qquad \dfrac{|\Delta_D \theta|}{\bar{\theta}_D} \lesssim \dfrac{D}{H} \sim 1 \qquad \text{or} \qquad \sigma \lesssim \dfrac{1}{H} \qquad \text{with } D \leq H$

(7.5a), (8.9), and (8.10a) → (11.2a) $\quad \sigma W \sim f\,\text{Fr}(1 + \text{Ro}) \qquad$ or

$\qquad\qquad \dfrac{W}{D} \sim \dfrac{W}{H} \sim \dfrac{U}{S}(1 + \text{Ro})(\text{RoRi})^{-1} \qquad \text{with RoRi} \gtrsim 1 \dotplus \text{Ro} \gtrless 1$

(8.2) → (11.3) $\qquad \Delta \sim \dfrac{W}{D} \sim \dfrac{W}{H} \qquad \text{with } D \sim H \text{ and } D \leq H$

(9.7a) → (11.4a) $\qquad \dfrac{W}{D} \sim \dfrac{W}{H} \sim \dfrac{U}{S}\left(\text{Ro} + \dfrac{f'\,S'}{f\,a}\right)(1 + \text{Ro})^{-1} \qquad$ or

$\qquad\qquad\qquad (1 + \text{Ro})^2(\text{RoRi})^{-1} \sim \left(\text{Ro} + \dfrac{f'\,S}{f\,a}\right)$

(10.3) → (11.5a) $\qquad \dfrac{|\nabla_h^2 p|}{p} \sim \dfrac{fU}{g}\dfrac{1 + \text{Ro}}{HS}$

Combining (7.5a) and (9.5), we obtain Prandtl's size ratio S/D of the horizontal length scale to the depth scale of the motion system,

(11.6) $\qquad \dfrac{S}{D} \sim \dfrac{S}{H} \sim \left(\dfrac{v_s}{f(1+\text{Ro})}\right)\left(1 + \dfrac{S}{a\text{R}'\text{o}}\right)^{1/2}$

where $v_s^2 \equiv g\sigma$ is the static stability, v_s being the gyro-frequency of free vertical oscillations in the gravity field.

Furthermore, we have

(11.7')
$$f \sim 10^{-5}\, s^{-1} \qquad f' \sim 10^{-4}\, s^{-1} \qquad \text{Ro} \gtrsim 1 \qquad \text{R'o} \lesssim 1 \qquad f(1+\text{Ro}) \sim \frac{U}{S}$$

in the low latitude belt 2 to 20°,

(11.7'') $\quad f \sim f' \sim 10^{-4}\, s^{-1} \qquad \text{Ro} \sim \text{R'o} \lesssim 1 \qquad f(1+\text{Ro}) \sim f$

in the middle latitude belt 20 to 70°, and

(11.7''')
$$f \sim 10^{-4}\, s^{-1} \qquad f' \sim 10^{-5}\, s^{-1} \qquad \text{Ro} \lesssim 1 \qquad \text{R'o} \gtrsim 1 \qquad f(1+\text{Ro}) \sim f$$

in the high latitude belt 70 to 88°.

11.1. *The Middle Latitude Belt 20 to 70°*

In this belt, scale equations (11.1a) and (11.2a) assume the following simplified forms [(8.14) and (7.5b)]:

(11.1b) $\qquad \dfrac{1}{\bar{p}_s}\dfrac{|\Delta_s p|}{S} \sim \dfrac{1}{\bar{\rho}_s}\dfrac{|\Delta_s \rho|}{S} \sim \dfrac{1}{\bar{\theta}_s}\dfrac{|\Delta_s \theta|}{S} \sim \dfrac{fU}{gH} \lesssim 10^{-7}$

and

(11.2b)
$$\sigma \sim f\,\text{Fr} \sim \frac{fU}{gH}\frac{U}{W} \quad \text{or} \quad \frac{W}{D} \sim \frac{W}{H} \sim \frac{U}{S}(\text{RoRi})^{-1} \quad \text{with RoRi} \gtrsim 1$$

and Burger's key relation (9.7c) or (11.4a), the form

(11.4b) $\qquad \dfrac{W}{D} \sim \dfrac{W}{H} \sim \dfrac{U}{S}\left(\text{Ro} + \dfrac{S}{a}\right) \qquad \text{or} \qquad (\text{RoRi})^{-1} \sim \text{Ro} + \dfrac{S}{a}$

with Ro $\lesssim$ 1. Burger's key relation (11.4b) demonstrates that the magnitude of W depends uniquely on the magnitude of S and U, so that, by virtue of (11.2b), the magnitude of σ is also uniquely determined by the magnitude of the same scale parameters S and U.

Assuming that $S \lesssim a$, we have

(11.5b) $\qquad \dfrac{W}{D} \sim \dfrac{W}{H} \lesssim \dfrac{U}{S} \qquad \text{or} \qquad \dfrac{S}{D} \sim \dfrac{S}{H} \lesssim \dfrac{U}{W}$

so that, in this case, the vertical advection term $w(\partial v_h/\partial z)$ is not more important than the horizontal advection term $v_h \cdot \nabla_h v_h$ in Eq. (6.1) of horizontal motion, and consequently, the smallness of the Rossby parameter (Ro $\ll$ 1) is sufficient to validate the geostrophic approximation (8.7).

Burger was the first to point out that two possibilities may occur: the horizontal length scale S may be smaller ($S \lesssim a$Ro) or larger ($S \gtrsim a$Ro) than aRo, where Ro < 1, or, what amounts to the same thing,

$$S \lesssim S_* \quad \text{or} \quad S \gtrsim S_* \quad \text{with } S_* = (aU/f)^{1/2} \lesssim 3.10^6 \text{ m}$$

We will in turn consider the two cases Burger (1958): (1) The moderately large-scale motions [$S < S_*$ or $1 > $ Ro $> (S/a)$] belonging to the synoptic scale range, and (2) the very large-scale motions [$S > S_*$ or $1 > (S/a) > $ Ro] of the planetary scale and the large-scale end of the synoptic scale range.

11.1.1. Moderately Large-Scale Motion Systems. In this case, we have

(11.6) $\quad S \lesssim a\text{Ro} = a\dfrac{U}{fS} \quad$ or $\quad S \lesssim S_* \lesssim 3.10^6$ m $\quad$ with $\quad S_* = (aU/f)^{1/2}$

and Burger's key relation (11.4a) reduces to

(11.7) $\quad f\dfrac{W}{D} \sim \dfrac{U^2}{S^2}\left(1 + \dfrac{S}{a\text{Ro}}\right) \quad$ or $\quad \dfrac{W}{H} \sim \dfrac{U\text{Ro}}{S} = \dfrac{1}{f}\dfrac{U^2}{S^2} \lesssim 10^{-4}$ s^{-1}

account being taken of (3.1); hence,

(11.7') $\quad\quad\quad\quad\quad\quad W \lesssim 1$ m s^{-1}

Now, eliminating W between (11.2b) and (11.7), we find the Prandtl's size ratio associated with moderately large-scale motions,

(11.8) $\quad\quad\quad\quad\quad\quad \dfrac{S}{H} = \dfrac{\sqrt{g\sigma}}{f} = \dfrac{v_s}{f}$

expressing in this case the ratio of the horizontal to the vertical scale lengths S and H. From (11.7) it is seen that W increases with increasing magnitude of the horizontal velocity components and decreases with the horizontal scale length. Rewriting (11.8) and taking (11.6) into account, we find

(11.9) $\quad\quad\quad\quad\quad\quad \sigma H \sim \dfrac{F^2 S^2}{gH} \lesssim \dfrac{a}{H}\dfrac{fU}{g} \lesssim 1$

Since $U \lesssim 10^2$ m s^{-1}, the last inequality is consistent with (7.7). The first scale relation (11.9) shows a rapid increase of the static stability $g\sigma$ with increasing horizontal dimension S of the motion systems.

In Charney's case (1948),

(11.10) $\quad S \sim 10^6$ m $\sim S_*\quad (S/a) \ll 1\quad$ and $\quad \sigma H \sim 10^{-1}$

Thus, in this case the inequality (11.9) reduces to

$$\text{(11.11)} \qquad \frac{f^2 S^2}{gH} \ll 1$$

or multiplying both members of the inequality (11.11) by U^2,

$$\text{(11.12)} \qquad \frac{f^2 U^2}{gH} \ll \frac{U^2}{S^2}$$

The left-hand side of (11.12) represents, in the case envisaged here, the magnitude of the right-hand side of equation (9.1), in other words the magnitude of the solenoidal terms of the vorticity equation. Thus, the inequality (11.12) shows that the solenoidal terms are negligibly small with respect to the advection U^2/S^2 of vorticity U/S.

Returning to (11.2b) and (11.7) and combining these scale relations with (11.12), we may write (Burger, 1958)

$$\text{(11.13)} \qquad \frac{fU^2}{gH} \ll \frac{U^2}{fS^2} \sim \frac{W}{H}$$

These last scale relations demonstrate that in the continuity equation (4.1a) the individual time derivative $[(\partial/\partial t) + \mathbf{v}_h \cdot \nabla_h]$ of $\ln \rho$ along the horizontal motion of the air parcels (ρ = air density) is negligibly small with respect to the magnitude W/H of the term $\rho\, \partial w/\partial z$. Indeed, we have

$$\text{(11.14)} \qquad \left| \frac{1}{\rho} \frac{\partial \rho}{\partial t} + \frac{\mathbf{v}_h}{\rho} \cdot \nabla_h \rho \right| \sim \frac{fU^2}{gH} \ll \frac{W}{H}$$

account being taken of (2.3), (2.10), (2.10′), (2.15), (3.2), and (11.1b). This conclusion is consistent with the result of Section 4.

The choice of orders of magnitude of the scale parameters U and S,

(11.15) $\quad\quad S \sim 10^6$ m $\quad$ and $\quad U \sim 10$ m s^{-1}

made by Charney (1948) implies that Ro = $(U/Sf) \ll 1$, so that the criterion of the validity of the geostrophic approximation is satisfied. Hence (11.7),

$$\text{(11.16)} \qquad \frac{W}{H} \ll \frac{U}{S} \quad \text{or} \quad \frac{S}{H} \ll \frac{U}{W}$$

Moreover, the last scale relation (11.13) implies that, in Charney's case (11.15),

(11.13′) $$W \sim 10^{-2} \text{ m s}^{-1}$$

Finally, returning to the definition (8.4) of the Richardson number Ri, we find, by virtue of (11.7), that

$$\text{RiRo}^2 \sim 1$$

in agreement with (8.5). Therefore, the criterion of the validity of the geostrophic approximation Ro ≪ 1 is equivalent to the criterion Ri ≫ 1, the second form of the criterion being less restrictive than the first because the Rossby number occurs squared in $\text{RiRo}^2 \sim 1$.

The scale relations (11.10), (11.12), (11.13), and (11.15) allow us to omit certain terms from the basic equations of dynamic meteorology. The second scale relation (11.10) shows that metrical terms in $1/r$ may be neglected: (1) in the expressions of the horizontal wind divergence Δ and the vorticity ξ,

$$\Delta \cong \frac{\partial u}{\partial x} + \frac{\partial v}{\partial y} \quad \text{and} \quad \xi \cong \frac{\partial v}{\partial x} - \frac{\partial u}{\partial y} \quad \text{with } \Delta \ll \xi$$

by virtue of (8.2) and (11.16) (see also Section 4), and (2) in the equations of horizontal motion and in the continuity and vorticity equations. Moreover, the deletion of terms containing w in this last equation is justified by the scale relation (11.16).

Concluding, we may state that under Charney's assumptions, the equation of vertical motion (5.1) reduces to the hydrostatic equation (2.8), the equations of horizontal motion (6.1) to the equation of geostrophic motion $\rho f(\mathbf{k} \times \mathbf{v}_g) + \nabla_h p = 0$, the continuity equation (4.1a) to the simplified equation (4.1b), while the adiabatic equation (7.1) remains unchanged (Burger, 1958).

The disappearance of the local time derivative in the equations of motion and continuity shows the quasi-stationary character of the air flow, which is confirmed by synoptic experience. This important fact facilitates the use of the primitive equations for diagnostic purposes but, unfortunately, prohibits their application for prognostic purposes. However, as a consequence of the assumption $C \sim U$ [see (3.2)], the local time derivative does not disappear in the adiabatic equation nor in the vorticity equation which, by virtue of (11.2a) and (11.16), reduces to the simplified equation

(11.17) $$\left(\frac{\partial}{\partial t} + \mathbf{v}_h \cdot \nabla_h\right)\xi + \beta v + f\Delta = 0$$

account being taken of (11.7).

Eliminating the unmeasurable quantity Δ between (11.17) and (4.1b), we find for the vorticity equation the simplified form

$$(11.17') \quad \left(\frac{\partial}{\partial t} + \mathbf{v}_h \cdot \nabla_h\right)\xi + \beta v = \frac{f}{\rho}\frac{\partial(\rho w)}{\partial z}$$

where, on the left-hand side we may now introduce the geostrophic approximation $\xi \cong \xi_g \cong \nabla_h^2 p/f\rho$. The introduction of the geostrophic approximation is not allowed in (11.17) but is allowed in (11.17') [see Section 1, and Section 4, Item (2)].

In the middle latitude belt (20 to 70°), the geostrophic horizontal wind divergence Δ_g satisfies the scale relation (8.6b) so that (8.2),

$$(11.18) \quad \Delta \sim \frac{W}{H} \gtrsim \frac{U}{a} \sim \Delta_g$$

by virtue of (5.3a) and (11.4b). Indeed, in the case of moderately large-scale motion systems $(S/a) <$ Ro, so that $W/H > U/a$. Even if the geostrophic approximation were applicable to Δ, the geostrophic value Δ_g is an unacceptable approximation of Δ, except in Charney's case (11.4a), in which case $\Delta \sim \Delta_g$. Only for large values of the Rossby number (U/fS), not exceeding S/a, however [that is, for small values of S ($=10^6$ m) and also for large values of U (> 50 m s^{-1}) for which the geostrophic approximation becomes questionable] does Δ_g appear small compared to Δ (Burger, 1958).

11.1.2. Very Large-Scale Motion Systems. In this case, we have

$$(11.19) \quad S \gtrsim a\text{Ro} = \frac{aU}{fS} \quad \text{or} \quad S \gtrsim S_*$$

$$\text{with } S_* = \left(\frac{aU}{f}\right)^{1/2} \quad \text{and} \quad 10^6 \leq S_* < 3 \times 10^6 \text{ m}$$

and Burger's key relation (11.4a) reduces to

$$(11.20) \quad \frac{W}{D} \sim \frac{W}{H} \sim \frac{U}{a} \lesssim 10^{-5} \text{ s}^{-1}$$

account being taken of (3.1a); hence,

$$(11.20') \quad W \lesssim 10^{-1} \text{ m s}^{-1}$$

The scale relations (11.20) replace the scale relations (11.7) of the preceding case.

Combining the scale relations (11.2b) and (11.20), we find:

$$(11.21) \quad \sigma H \sim \frac{fUa}{gH} \lesssim 1$$

consistent with (7.7). The scale relation (11.21) replaces the scale relation (11.9).

The scale relations (11.20) and (11.21) show that the scale parameter W of the vertical velocity and the static stability $g\sigma$ are both independent of S and increase linearly with U, so that in this case there is no relation to replace Prandtl's relation (11.8) of the preceding case. The moderately large and very large motion systems behave quite differently. This difference in behavior becomes more and more evident when motion systems of the planetary scale are considered. At this scale,

(11.22) $$S \sim a \sim 10^7 \text{ m}$$

The Rossby number of motion systems of the planetary scale is very small,

(11.23) $$\text{Ro} \sim \frac{U}{Sf} \ll 1$$

account being taken of (3.1a) and (11.22), so that here the geostrophic approximation may be used.

Combining (11.23) and (11.22), we find

(11.24) $$\frac{U}{S} \ll \frac{fS}{a} \quad \text{or} \quad \frac{U^2}{S^2} \ll \frac{fU}{a}$$

This inequality implies that in the vorticity equation, the terms

$$(\partial/\partial t + \mathbf{v}_h \cdot \nabla_h)\xi$$

are negligibly small against the β term. Hence at the planetary scale, the vorticity equation can no longer be used for prognostic purposes. With decreasing smallness of Ro, that is to say with increasing values of U, the less quasi-stationary is the air flow.

Rewriting (11.20), we have, in contrast with (11.15) in Charney's case,

(11.25) $$\frac{W}{H} \sim \frac{U}{S} \lesssim 10^{-5} \text{ s}^{-1} \quad \text{with } S \sim a$$

This scale equation implies that horizontal (U^2/S^2) and vertical (UW/SH) advection terms of vorticity equal the magnitude of the twisting terms $[(\partial w/\partial x)(\partial v/\partial z) - (\partial w/\partial y)(\partial u/\partial z)]$.

By virtue of (8.2), we find that

(11.26) $$f\Delta \sim \frac{fU}{a} \sim \beta U$$

Thus, on the left-hand side of the vorticity equation (9.1), the term $f\Delta$ is of the same order of magnitude as the β term (fU/a) and must be retained. All other terms on this left-hand side may be deleted against βU and $f\Delta$, account being taken of (5.3a), (11.22), (11.24), and (11.25). The order of magnitude of the solenoidal terms on the right-hand side of the vorticity equation is by virtue of (11.1b), of the order of magnitude

$$\text{(11.27)} \qquad \text{solenoidal terms} \sim \frac{f^2 U^2}{gH} \gg \frac{U^2}{a^2} \sim \frac{U^2}{S^2} \qquad \text{with } S \sim a$$

because the coefficient $f^2/gH(10^{-13}\ \text{m}^{-2})$ is one order of magnitude larger than $1/a^2(10^{-14}\ \text{m}^{-2})$ (see Table I in Section 3). Hence the scale relation (11.27) is independent of the magnitude of U. Returning to (7.9b), we find

$$\text{(11.28)} \qquad \frac{f^2 U^2}{gH} \lesssim \frac{fU}{S} \qquad \text{with } S \sim a$$

account being taken of the first inequality (5.3a). From (11.27) and (11.28), it is seen that the solenoidal term $f^2 U^2/gH$ never exceeds the β term fU/a but always predominates over the horizontal advection U^2/S^2 of vorticity U/S.

In Charney's case, when $U \sim 10\ \text{m s}^{-1}$, the symbol $\lesssim$ in (11.28) must be replaced by $\ll$ so that, for the planetary scale ($S \sim a$) with $U \sim 10\ \text{m s}^{-1}$, the vorticity equation reduces to the very simple equation.

$$\text{(11.29)} \qquad \beta v + f\Delta = 0$$

This equation is evidently equivalent to the geostrophic approximation (8.6b). Thus, convergence of the horizontal wind implies a northward displacement of the air parcels, and divergence to the contrary implies a southward displacement.

According to the theory of Rossby waves, very long waves have a very fast retrogression to the west. From (11.29), it is seen that this is not the case when $U \sim 10\ \text{m s}^{-1}$ and $S \sim a \sim 10^7\ \text{m}$ (Burger, 1958).

For larger values of U ($10\ \text{m s}^{-1} < U \lesssim 100\ \text{m s}^{-1}$) however, the solenoidal terms of the vorticity equation become increasingly important and should be retained on the right-hand side of the vorticity equation when U is approaching its upper limit. Therefore, these terms should be inserted on the right-hand side of (11.29) (Burger, 1958). Moreover, for increasing values of U, the horizontal and vertical advection of vorticity (U/S with $S \sim a$) and the twisting terms become more and more important and should be kept in the vorticity equation together with the $f\Delta$ and βU terms. For large values of U, the approximate vorticity equation (Burger, 1958) assumes the form

$$\text{(11.30)} \qquad \left(\frac{\partial}{\partial t} + \mathbf{v}_h \cdot \nabla_h + w\frac{\partial}{\partial z}\right)\xi + \beta v + f\Delta + \left(\nabla_h w \times \frac{\partial \mathbf{v}_h}{\partial z}\right)\mathbf{k} = \left(\nabla_h p \times \nabla_h \frac{1}{\rho}\right)\mathbf{k}$$

Although the local time derivative of ξ has a magnitude smaller than the horizontal advection of ξ, this time derivative is retained in the vorticity equation in order to preserve its use for prognostic purposes. When the magnitudes of the horizontal wind components approach their upper limit ($U = 100$ m s^{-1}), the approximate form (11.30) of the vorticity equation is not very much simpler than its full form. Moreover, the metrical terms in the expressions (4.4) and (9.2) of Δ and ξ may no longer be deleted, although the distance r from the earth's center may be replaced by a, the mean radius of the earth.

Furthermore, at the planetary scale (11.22), as a result of (11.23) and (5.3a), the equations of horizontal motion may be replaced by their geostrophic approximations and the equation of vertical motion by its hydrostatic approximation. The adiabatic equation remains unchanged as in the preceding subsection. Things are a bit more complicated regarding the continuity equation: the simplified form (4.1b),

$$\Delta + \frac{1}{\rho}\frac{\partial(\rho w)}{\partial z} = 0$$

is acceptable provided $U \sim 10$ m s^{-1}. If U increases so that (8.13) ceases to hold, namely when at the planetary scale, U approaches its upper limit, then (8.14),

$$\frac{fU}{g} \sim \frac{H}{S} \quad \text{or} \quad \frac{fU^2}{gH} \sim \frac{U}{S} \sim \frac{W}{H} \sim \Delta$$

account being taken of (11.25) and (8.2). In this case the individual horizontal time derivative of the air density ceases to be negligible in the continuity equation for large values of S ($S \sim a \sim 10^7$ m). Indeed, in this case, we have

$$(11.31) \qquad \left|\frac{1}{\rho}\frac{\partial \rho}{\partial t} + \frac{\mathbf{v}_h}{\rho}\cdot \nabla_h \rho\right| \sim \frac{fU^2}{gH} \sim \frac{U}{S} \sim \frac{W}{H} \sim \Delta$$

so that we have to retain here the full form of Eq. (4.1a).

In Eq. (10.1) of the divergence of the horizontal wind we must now evidently include the terms βu and $f\xi$ [see Eq. (11.26)]. On the right-hand side of Eq. (10.1), the first two terms may no longer be omitted, except in the case $U \sim 10$ m s^{-1}, for the same reason we retained the solenoidal terms in the vorticity equation. Not only the terms containing density derivatives but also most of the terms containing w should be kept on the left-hand side. Only the following terms:

$$\frac{2\Delta w}{r} \qquad \frac{\partial w}{\partial x}\left(\frac{u}{r}+f'\right) \qquad \frac{\partial w}{\partial y}\frac{v}{r}$$

may be neglected. All the terms of the right-hand side must be retained (Burger, 1958).

11.2. The High Latitude Belt 70 to 88°

In this belt Burger's key relation (11.4a) assumes the form

$$(11.4c) \qquad \frac{W}{D} \sim \frac{W}{H} \sim \frac{U}{S}\left(\text{Ro} + \frac{S}{10a}\right)$$

that is, the form (11.4b) that may be used in the middle latitude belt, in which, however, a must be replaced by $10a$. Hence Burger's threshold value $S_* = (10aU/f)^{1/2}$ of S has here a higher upper limit ($S_* \lesssim 10^7$ m). All the other scale relations used in the middle latitude belt may be applied in the high latitude belt so that the moderately large motion systems predominate in this belt (see, in Section 3, Table II).

11.3. The Low Latitude Belt 2 to 20°

In this belt, scale equations (11.1a) and (11.2a) assume the following simplified forms [(8.16), (8.10c), and (7.5c)]:

$$(11.1c) \qquad \frac{|\Delta_s p|}{\bar{p}_s} \sim \frac{|\Delta_s \rho|}{\bar{\rho}_s} \sim \frac{|\Delta_s \theta|}{\bar{\theta}_s} \sim \text{Fr} = \frac{U^2}{gH} \lesssim 10^{-1}$$

and

$$(11.2c) \qquad \sigma W \sim \frac{U}{S}\text{Fr} \quad \text{or} \quad \frac{W}{D} \sim \frac{W}{H} \sim \frac{U}{S}(\text{Ri})^{-1} \quad \text{with Ri} \gtrsim 1$$

and Burger's key relation (9.7a) or (11.4a), the form

$$(11.4d) \qquad \frac{W}{D} \sim \frac{W}{H} \sim \frac{U}{S}\left(1 + \frac{S}{a\text{Ro}}\right) \sim \frac{U}{S}\left(1 + \frac{S^2}{S_*^2}\right) \quad \text{or} \quad \text{Ri}^{-1} \sim 1 + \frac{S^2}{S_*^2}$$

The threshold value S_* of S is given by

$$S_* = \left(\frac{aU}{f'}\right)^{1/2} \lesssim 3 \times 10^6 \text{ m}$$

When $U \sim 10$ m s^{-1}, then $S_* \sim 10^6$ m. When $U \sim 10$ m s^{-1} and $S \sim a \sim 10^7$ m, then Ro $= U/fS \ll 1$, and the planetary motion is geostrophic. In all other cases, the geostrophic approximation may not be used, and also, in any case, the geostrophic approximation may not be used in the equatorial belt (2° S to 2° N).

In Sections 6 and 7, we have shown that in the tropics, outside the precipitation areas, the vertical component of the air velocity, the horizontal pressure and temperature contrast, and the corresponding fluctuations in time have magnitudes much smaller than those in extratropical large motions of the same scale. Thus, in the absence of condensation, motions in the tropics are quasi-horizontal, and the horizontal windfield is very nearly nondivergent ($\Delta \sim 0$) with a very weak coupling in the vertical. In Charney's case: $U \sim 10$ m s^{-1} and $S \sim 10^6$ m, $W \sim 10^{-3}$ m s^{-1}, and $\Delta \sim 10^{-7}$ s^{-1} in deep tropical synoptic systems.

In this case, the scale equation (9.4a) outside the precipitation areas reduces to (11.4d),

(11.32) $$\frac{U^2}{S^2} + \frac{f'U}{a} \sim 0 \quad \text{or} \quad S \sim S_* = \left(\frac{aU}{f'}\right)^{1/2}$$

and the vorticity equation (9.1) to the barotropic vorticity equation

(11.33) $$\left(\frac{\partial}{\partial t} + \mathbf{v}_h \cdot \nabla_h\right)(\xi + f) = 0$$

representing the free Rossby waves.

Thus, deep large-scale motion systems outside the precipitation areas in the tropics are barotropic. Therefore, no conversion between available potential energy and kinetic energy occurs in the tropics outside the precipitation areas, so that the air motion must be driven by lateral coupling with extratropical and precipitating tropical motion systems (Charney, 1963; Holton, 1972). The quasi-horizontal nondivergent flow, free of energy sources outside the precipitation areas, transfers laterally the influence from the extratropical large-scale motions to the intertropical convergence zone, a narrow zone of strong cumulus convection. Hence, the position of this zone seems to be determined by the extratropical circulation in both hemispheres (Charney, 1963). The zone itself may be considered as an internal boundary layer between the quasi-horizontal nondivergent motions on both sides of the zone. Tropical and extratropical large-scale motion systems behave in quite different ways.

In tropical latitudes, the ratio of the horizontal Coriolis acceleration fU to the acceleration of gravity g is at least one order of magnitude smaller than in extratropical latitudes, where the large-scale motion systems are quasi-geostrophic. This ratio decreases with decreasing latitude and is zero at the equator. Thus, we may expect in the equatorial latitudes the existence of large-scale motion systems with a much stronger gravitational character. Indeed, much of the energy in the tropical atmosphere is contained in planetary waves characterized by a strong vertical coupling in the gravity field, namely: (1) internal gravity waves of the Kelvin type, propagating eastward ($C \sim 30$ m s^{-1}) and (2) mixed Rossby gravity waves propagating westward. Both planetary wave motions propagate downward, thus carrying energy upward and their vertical dimension is only a few kilometers ($D \ll H$) (Holton, 1969, 1972).

In shallow ($D \ll H$) planetary motion systems, the divergence $\Delta \sim W/D$ of the horizontal wind may be large, even when the vertical velocity remains small. The full primitive equation must be used to represent adequately the entire spectrum of large-scale motion systems in the tropics (Holton, 1969).

In precipating tropical systems, the vertical velocity w (>0) and its vertical gradient $\partial w/\partial z$ in the intertropical convergence zone are much larger, especially in the upper half of the troposphere. The air motion in these systems cannot be described by the barotropic vorticity equation. Most of the terms of the vorticity equation (9.1) must be retained here. Moreover, the interaction between mesoscale convective cells and large-scale motion systems must be taken into account. Indeed, horizontal momentum U and thus vorticity U/S are transported vertically in these cells.

The rising motions are large and concentrated in the intertropical convergence zone meandering along the equator.

The release of latent heat in this narrow zone is the primary source for the maintenance of the tropical large-scale motion systems outside the zone where the static stability $(g/\theta)(\partial \theta/\partial z)$ is positive and the subsiding motions small but spread over much larger areas. A cumulus cloud system is statically unstable:

$$\frac{1}{\theta}\frac{\partial \theta}{\partial z} + (L_v/c_{pa}T)\frac{\partial r_v}{\partial z} < 0 \quad \text{with} \quad \frac{\partial r_v}{\partial z} < 0$$

Such a system results from the release of this instability in a saturated tropical atmosphere with a lapse rate larger than the moist adiabatic lapse rate. The mass convergence in the frictional boundary layer of the trade wind zones supplies not only the water vapor to the intertropical convergence zone but also initiates in this zone the lifting of the saturated equatorial air masses.

In order to understand the dynamics of the tropical atmosphere, we must improve our knowledge of the different scales of convective motion systems, the interactions between these motion scales ($10^3 < S < 10^4$ m, $10^4 < S < 10^5$ m, 10^5 m $< S < 10^6$ m), and between these scales and the larger scale motion systems.

12. Scale Analysis of the Equation of Balance of Kinetic Energy

Multiplying the vector equation (6.1) of horizontal motion scalarly with $\mathbf{v}_h$ and the equation (5.1) of vertical motion with w, adding them, and taking frictional forces into account, we obtain (Van Mieghem, 1973) by virtue of the continuity equation (4.1a),

(12.1) $$\frac{\partial \rho k}{\partial t} + \text{div}(\rho k \mathbf{v} + p\mathbf{v} + \mathbf{A}) = p \, \text{div} \, \mathbf{v} - g\rho w - \rho \Delta$$

where $k = \mathbf{v}^2/2$ is the kinetic energy of the mass unit, Δ the rate of the frictional heating of the same unit of mass, and $\mathbf{A}$ the flux of kinetic energy, through the boundary of the volume unit, due to friction. As a rule, $\mathbf{A}$ is negligible against $p\mathbf{v}$.

The right-hand side of the equation of balance represents, per unit volume, the rate of production of kinetic energy. This production results from the conversion into kinetic energy of gravitational potential energy at the rate $-g\rho w$ and the conversion of internal energy at the rate $p \, \text{div} \, \mathbf{v} - \rho \Delta$, with $\Delta > 0$.

The flux of kinetic energy $\mathbf{F}(k)$ through the boundary surface of the unit volume is given by the terms in parenthesis on the left-hand side of (12.1),

(12.2a) $$\mathbf{F}(k) = \rho k \mathbf{v} + p\mathbf{v} + \mathbf{A} \cong \rho k \mathbf{v} + p\mathbf{v}$$

The corresponding scale relation assumes the form

(12.2b) $$|\mathbf{F}(k)| \sim p(\text{Fr} + 1)|\mathbf{v}| \sim |p\mathbf{v}|$$

or

(12.2c) $$|\mathbf{F}_h(k)| \sim pU \quad \text{and} \quad |\mathbf{F}_z(k)| \sim pW$$

by virtue of (5.3a) and (5.4). Thus, for large-scale motion systems, the flux $\mathbf{F}(k)$ of kinetic energy k is well represented by $p\mathbf{v}$.

The propagation velocity of kinetic energy is defined by

(12.3a) $$\frac{\mathbf{F}(k)}{\rho k} = \left(1 + \frac{p}{\rho k}\right)\mathbf{v} + \frac{\mathbf{A}}{\rho k} \cong \left(1 + \frac{p}{\rho k}\right)\mathbf{v}$$

The corresponding scale equation may be written as follows

(12.3b) $$\frac{|\mathbf{F}(k)|}{\rho k} \sim (1 + \text{Fr}^{-1})|\mathbf{v}| \gg |\mathbf{v}|$$

or

(12.3c) $$(|\mathbf{F}_h(k)|/\rho k) \sim \text{Fr}^{-1} U \gg U \quad \text{and} \quad \left(\frac{|\mathbf{F}_z(k)|}{\rho k}\right) \sim \text{Fr}^{-1} W \gg W$$

again by virtue of (5.3a) and (5.4). Thus, for large-scale motion systems, the velocity of kinetic energy propagation is at least one order of magnitude larger than the air velocity.

The kinetic energy produced in the atmosphere through conversion of total potential energy into kinetic energy is rapidly dispersed into the whole atmosphere, entailing rather slow large-scale air motions ($U \lesssim 10^2$ m s^{-1}).

Acknowledgment

I wish to thank Dr. A. Quinet for his interest in the subject and his valuable remarks and suggestions.

References

Bjerknes, V. (1926). Die atmosphärischen Störungsgleichungen. *Beitr. Phys. Atmos.* **13**, 1–14.

Bjerknes, V., Bjerknes, J., Solberg, H., and Bergeron, T. (1934). " Hydrodynamique Physique avec Applications à la Météorologie Dynamique," Vol. 23. Presses Universitaires de France, Paris.

Burger, A. P. (1958). Scale consideration of planetary motions of the atmosphere. *Tellus* **10** (2), 195–205.

Charney, J. G. (1947). The dynamics of long waves in a baroclinic westerly current. *J. Meteorol.* **4**, 135–162.

Charney, J. G. (1948). On the scale of atmospheric motions. *Geofys. Publ.* **17** (2), 1–17.

Charney, J. G. (1963). A note on large-scale motions in the tropics. *J. Atmos. Sci.* **20** (6), 607–609.

Dickinson, R. E. (1968). A note on geostrophic scale analysis of planetary waves. *Tellus* **20**, 548–550.

Eady, E. T. (1949). Long waves and cyclone waves. *Tellus* **1**, 33–52.

Holton, J. R. (1969). A note on scale analysis of tropical motions. *J. Atmos. Sci.* **26** (4), 770–771.

Holton, J. R. (1972). "An Introduction to Dynamic Meteorology," International Geophysics Series, Vol. 16. Academic Press, New York.

Phillips, N. A. (1963). Geostrophic motion. *Rev. Geophys.* **1** (2), 123–176.

Quinet, A. (1975). Sur la simulation en laboratoire des mouvements troposphériques de grande échelle. *Publ. Inst. R. Meteorol. Belg.*, Ser. *B* No. 78, 1–28.

Van Mieghem, J. (1960). Zonal harmonic analysis of the northern hemisphere geostrophic windfield. *Monogr. Int. Union Geod. Geophys.* No. 8, 1–57.

Van Mieghem, J. (1973). "Atmospheric Energetics," Oxford Monographs on Meteorology. Oxford Univ. Press (Clarendon), London and New York.

SYMMETRIC CIRCULATIONS OF PLANETARY ATMOSPHERES

E. L. Koschmieder

Atmospheric Science Group
University of Texas, Austin, Texas

1. Introduction 131
2. Convection on a Nonuniformly Heated Sphere 134
   2.1 The Academic Case 137
   2.2 The Case of Nonuniform Heating 143
   2.3 The Case of Uniform Heating with Rotation 149
   2.4 Nonuniform Heating with Rotation 151
   2.5 Symmetric Baroclinic Instability 156
   2.6 The Circulation of Jupiter's Atmosphere 157
   2.7 Convection in the Atmosphere on Earth 161
3. Azimuthal Disturbances 163
   3.1 The Rotating Annulus Experiment 164
   3.2 Finite Azimuthal Disturbances 170
4. Nonlinearity 170
   4.1 Numerical Studies 172
5. Summary and Conclusions 175
   List of Symbols 176
   References 177
   Note Added in Proof 181

1. Introduction

Over the last 20 years an extraordinary effort has been made to understand through numerical analysis the general circulation of the atmosphere of the Earth. As is well known, numerical models of the general circulation have been developed which reproduce satisfactorily the major characteristics of the irregular, time-dependent flow of the atmosphere. Comparatively little attention has been paid during these years to the problem of the circulation of the atmosphere over an idealized Earth with a homogeneous surface, where one might expect to find an axisymmetric circulation. The disinterest in this problem is, in part, due to the excitement over the possibility of solving numerically the real problem of the complex circulation on the real earth, and in part due to the belief that an axisymmetric circulation would not occur even on an Earth with a perfect surface. The classic studies of Hadley (1735) and Ferrel (1859) had assumed the existence or at least the possibility of a symmetric circulation. This tradition was broken in the 1930s, and the notion was introduced that the circulation of the atmosphere has necessarily to be asymmetric. In particular, Jeffrey's (1926) study is in this context often referred to as the turning point. In modern terms one would say that baroclinic instability makes it unlikely for a symmetric circulation to occur.

Nevertheless, it is important to make sure about this point and to gather more information about symmetric circulations. In particular, we will not be able to state that the general circulation is necessarily asymmetric, if we do not know the solution of the symmetric case and study its stability against azimuthal disturbances. If the ultimate goal of the exploration of the general circulation is its analytical description, there is actually no choice but to begin with the symmetric case. In a review of the nature of the global circulation of the atmosphere Lorenz (1969) outlined many essential points regarding symmetric circulations. We quote Lorenz (1969, p. 4): "For many purposes the real atmosphere may be approximated by an ideal atmosphere, where the incoming solar energy varies only with latitude, and the underlying surface of the earth is uniform in elevation and composition. There then exists at least one particularly simple atmospheric circulation pattern which is compatible with the heat sources and the surface geography; this circulation is completely symmetric with respect to the earth's axis and does not fluctuate with time." Lorenz refers to this circulation as the ideal (Hadley) circulation. He continues: (Lorenz, 1969, p. 5) "There is no simple argument eliminating the possibility of a single direct cell, with or without the earth's rotation." And (Lorenz, 1969, p. 8): "In working with an idealized atmosphere it would appear reasonable to choose a rather simple heat source. However, the resulting ideal Hadley circulation must depend critically upon this choice, conceivably one choice might lead to a circulation like Hadley's, while another might lead to one like Thomson's and Ferrel's." And finally (Lorenz, 1969, p. 9): "Perhaps one can show that the ideal Hadley circulation satisfies conditions for instability. To the best of our knowledge this has not been done."

In spite of the absence of a systematic investigation of symmetric circulations, a consensus seems to have been reached as to what kind of symmetric circulation should occur on Earth. Several textbooks maintain that the atmosphere would circulate in a single direct cell if the Earth did not rotate, had a warm equator and a cold pole, and a surface temperature independent of longitude. This circulation is referred to as the Hadley circulation without rotation. A schematic picture of this circulation is shown in Fig. 1a. The term direct cell means that air rises over the warmest area of the bottom (the tropics) and sinks over the coldest area, the arctics. The existence of this circulation is deduced from experimental evidence, according to which a fluid layer over a plane plate with a horizontal temperature gradient circulates in one direct cell. The second statement made in textbooks concerns the circulation on the rotating Earth. It is maintained that *with* rotation, a warm

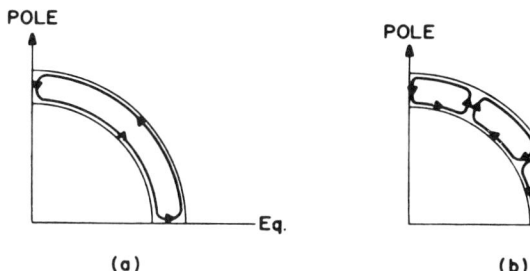

FIG. 1 (a). Hadley circulation (schematic); (b) three-cellular, meridional circulation (schematic).

equator, a cold pole, and axisymmetric temperatures the circulation of the atmosphere would be three-cellular (Fig. 1b). It is often said that conservation of angular momentum makes it impossible for the one-cellular or Hadley circulation to occur on the rotating Earth. There is no theoretical proof that this is really correct for a viscous, heat-conducting fluid layer. There is also laboratory evidence which supports the view that a Hadley circulation on a rotating sphere is possible.

Actually, the three-cellular picture of the idealized circulation on the rotating Earth has been deduced from atmospheric data, after they have been averaged over a long time. One cannot, of course, expect to find a steady symmetric circulation on the surface of the Earth which has such gross irregularities as the continents and the oceans, the mountain ranges, the temperature distortions caused by the ocean currents, and the daily and seasonal temperature changes. When the real atmospheric motions are averaged there is either just one direct Hadley circulation or the three-cellular circulation. The data do not seem to be really conclusive. However, the trade winds and the westerlies in midlatitudes support the three-cellular model strongly. All relevant facts concerning these aspects of the general circulation and previous theoretical studies of the general circulation have been discussed in detail in Lorenz's (1967) review. Streamlines of the mean meridional circulation for four 3-month periods can be found in Newell et al. (1969), and monthly mean streamlines are given by Oort and Rasmusson (1971).

The notion of a three-cellular circulation of the atmosphere on Earth is supported by unambiguous evidence that similar circulations occur on Jupiter and Saturn, although that is rarely ever mentioned. As a matter of fact, Fig. 2 is clear proof that axisymmetric atmospheric circulations do indeed exist. This leads to a final point regarding symmetric circulations. We should study not only and specifically the general circulation of the Earth, but rather the circulation on any nonuniformly heated planet. A solution of the

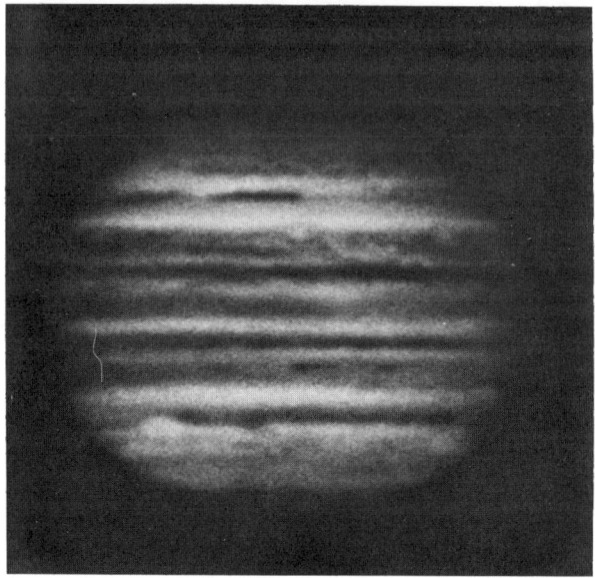

Fig. 2. Circulation on Jupiter (Jan. 25, 1968), ultraviolet photograph. (After Fountain and Larson, 1973.)

more general problem would, of course, include a solution of the general circulation on Earth. By specifying particular constants in the general solution, such as radius, gravity, angular velocity, solar constant, etc., for one planet, one should obtain the type of circulation characteristic for the planet. Just being able to do so would be the most rigorous verification of the validity of the solution of the general circulation problem.

In this chapter we will focus the discussion on information concerning axisymmetric circulations that has been obtained from fluid dynamics investigations. We shall see, in particular, that motions in the form of axisymmetric rings or cells occur quite naturally if heating from below is involved in the problem. In Section 2.6 we will apply the results of these investigations to the circulation of Jupiter's atmosphere.

2. Convection on a Nonuniformly Heated Sphere

The circulation on a rotating, nonuniformly heated planet is, in technical terms, a problem of convection on a sphere. The term convection refers to fluid motions caused by temperature differences or unequal heating. However, the meaning of the term convection is not unique, since convection can be due to *horizontal* temperature differences, to which the fluid responds in every case; or to *vertical* temperature differences, to which the fluid might not respond at all if the stratification is stable, or might respond if a critical negative vertical gradient is exceeded. Convection due to horizontal temperature differences which are constant and independent of the fluid motion will be referred to as externally "forced convection." This notation varies from the term "free convection" or "natural convection" used particularly in the engineering literature to describe convection caused by either horizontal or vertical temperature differences. Convection caused by heating from below or cooling from above is traditionally referred to as Bénard convection. Forced convection is characterized by a circulation making one big loop with fluid rising over the warm area and sinking over the cooler area (a direct cell). Bénard convection is characterized by a roll or cellular type of flow, with rolls of opposite circulation adjacent to each other. For Bénard convection to occur it is *necessary* that the fluid has an unstable vertical temperature distribution before onset of convection. That means in the case of the atmosphere that the potential temperature has to decrease with height. However, the average stratification in the atmosphere of the Earth is usually close to neutral. Hence, Bénard convection would seem to be of no significance for the general circulation. But, this observation is misleading. We measure the temperature in the atmosphere *after* the atmospheric motions have come into full swing. The initial state of the atmosphere, namely the state in which the air would be at rest, is characterized by radiative equilibrium. And the vertical temperature distribution in radiative equilibrium can be highly unstable, as is shown in Fig. 3, taken from Goody's (1964) book. The degree of vertical instability will usually be a function of latitude. In general, conditions are such that forced convection as well as Bénard convection should occur simultaneously on the planets.

Convection on an axisymmetrically heated, rotating sphere is relevant to the circulations on Earth, Jupiter, and Saturn, and possibly on Mars. Axisymmetric convection on a nonrotating sphere may be relevant to the circulation on Venus also. We note that axisymmetric heating on Earth means neglecting or time-averaging the daily heating cycle, as well as introduction of a homogeneous surface. Details of the atmospheres of the other planets cannot be discussed here; for a review of this topic the reader is referred to Goody (1969). We mention here only that Jupiter differs from Earth insofar as Jupiter (as well as Saturn) releases a substantial amount of internal heat (about two-thirds of the energy emitted from Jupiter is of internal origin). Jupiter's atmosphere is therefore essentially heated from below, so that in this case Bénard convection should, by all means, be important for the explanation of the circulation. For much data about the circulation on Jupiter see Peek's (1958) book, and for Saturn see Alexander's (1962) book. The Earth's atmosphere is also essentially heated from below due to the transparency of the atmosphere to solar radiation.

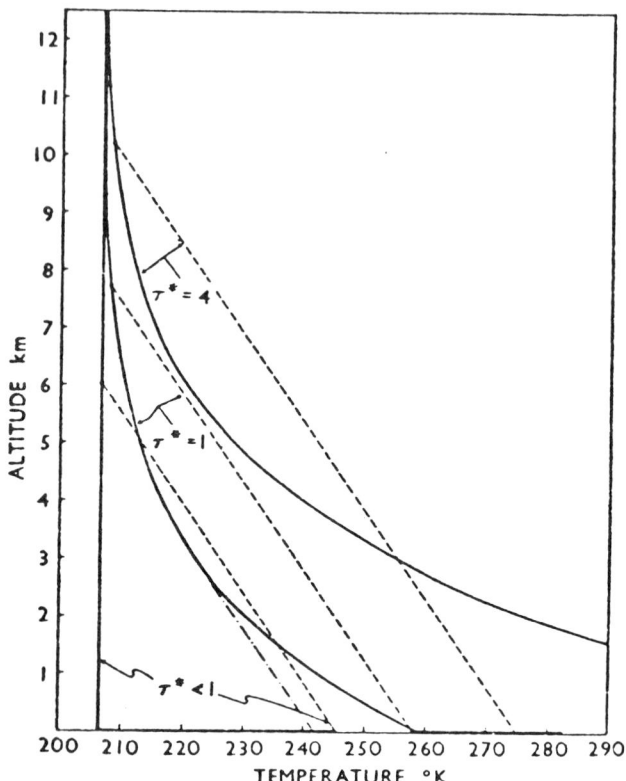

Fig. 3. Temperature profiles under radiative equilibrium (full lines) and as modified by tropospheric convection (broken lines). (After Goody, 1964.) (Copyright by Oxford University Press. Used with permission.)

Convection on a rotating, axisymmetrically heated sphere is such a complex problem that its solution can only be the end of a line of reasoning. We will therefore strip the problem successively of its difficulties in order to develop a solution from the solutions of the more simple subproblems. We can omit the horizontally nonuniform heating first and look at convection on a uniformly heated rotating sphere. We can then do away with the mathematically most cumbersome aspect, namely rotation. The problem left is convection on a horizontally nonuniformly heated resting sphere. This is what one would expect to take place on the surface of Venus, provided the 234-day rotation period of Venus is negligible. Finally, if we eliminate the nonuniform heating in the resting case, we arrive at convection on a resting, uniformly heated sphere. Although this is clearly an academic problem, it is the foundation on which we have to build. We shall now proceed with the simplest case first, and build up to a discussion of the full problem of nonuniform heating with rotation.

2.1. The Academic Case

Following earlier work of Wasiutinsky (1946) and Jeffreys and Bland (1951, 1952), the convective motions in a sphere heated from *within* have been studied in a series of articles by Chandrasekhar (1952, 1953a, 1957). The results of these investigations are summarized in Chandrasekhar's (1961) book. The motivation for these studies was astrophysical problems. Although obviously applicable to atmospheric problems, these papers do not seem to have created much meteorological interest. However, it was realized by Koschmieder (1959) that Chandrasekhar's theory can be modified to apply to the problem of convection in a spherical shell heated uniformly from *below*. The motions of a viscous, heat-conducting, incompressible fluid following the linearized Navier–Stokes equations and the linearized equation of thermal conduction are then determined by the equation

$$\Delta[r^3\Delta^2(rv_r)] = \frac{C}{r^3}L(rv_r) \tag{2.1}$$

where v_r is the radial speed, Δ is the Laplace operator in spherical polar coordinates, C is a constant representing an eigenvalue of the equation, and L is an operator representing the angular parts of the Laplace operator. The approximations made in this study correspond exactly to the approximations made in the linear theory of Bénard convection on a plane plate. Special solutions of Eq. (2.1) are obtained with the product solution

$$v_r = F(r)P_n^m(\theta)e^{\pm im\phi} \tag{2.2a}$$

where $P_n^m(\theta)$ is an associated Legendre polynomial. It was found by Koschmieder that the flow field which corresponds to any of the special solutions (2.2a) can be determined qualitatively from the maxima of v_r. Figure 4 shows schematically some possible axisymmetric flows (m = 0), which occur in the form of *axisymmetric rings*. To each solution belongs a specific eigenvalue C_n, or in practical terms a specific temperature difference across the fluid layer. There is, as a matter of fact, an infinite number of such special solutions of the type (2.2a), or modes as they are called. Each axisymmetric mode of degree n is accompanied by $2n + 1$ modes which are dependent on the azimuth. However these solutions are unrealistic, as will be explained later. For the case of a shallow layer the eigenvalues C_n of the first 10 modes have been determined by Koschmieder (1965).

The validity of the obtained solutions in the form of axisymmetric rings has to be verified somehow. They actually contradicted the belief held at that

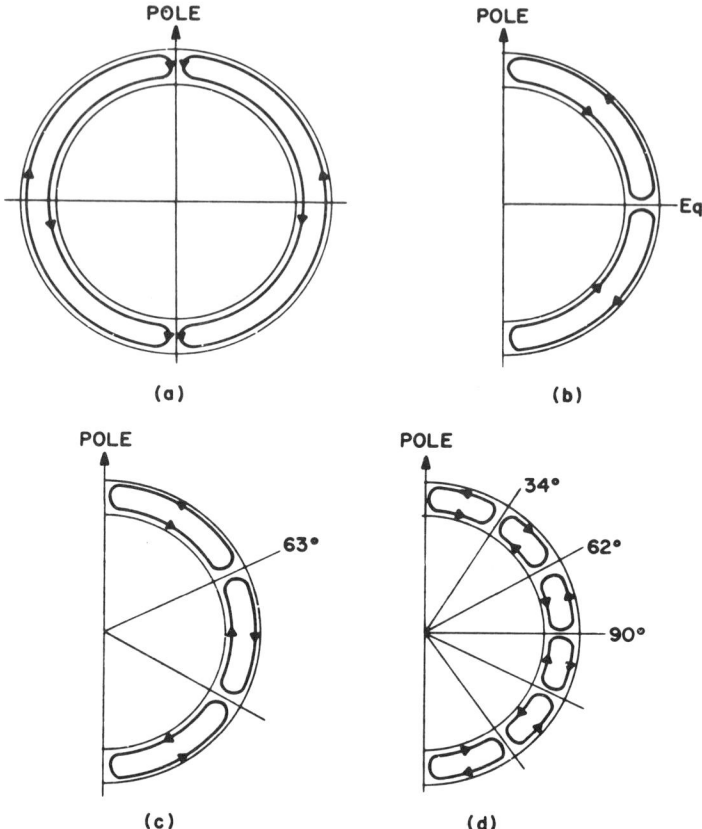

Fig. 4. Several possible convective motions on a uniformly heated nonrotating sphere: (a) state P_1, (b) state P_2, (c) state P_3, (d) state P_6. (After Koschmieder, 1959.)

time that hexagonal Bénard cells would be the proper solution of the convection problem. Since it is not possible to simulate in a laboratory a gravitational field on a sphere which is constant and points perpendicular to the surface of the sphere, we will have to rely on models where convection takes place on a plane plate, where g is constant and perpendicular to the surface. It has been shown by Zierep (1958) that the linear equations describing convection in a viscous, heat-conducting, incompressible fluid on a plane heated uniformly from below do not have only solutions in the form of straight parallel rolls or hexagonal cells, etc. (as discussed in detail in Chandrasekhar, 1961), but also an axisymmetric solution. This is the ring cell, given by the equation

$$w(x, y, z) = AI_0\left(\frac{a}{h}r\right)F(z) \qquad (2.3\text{a})$$

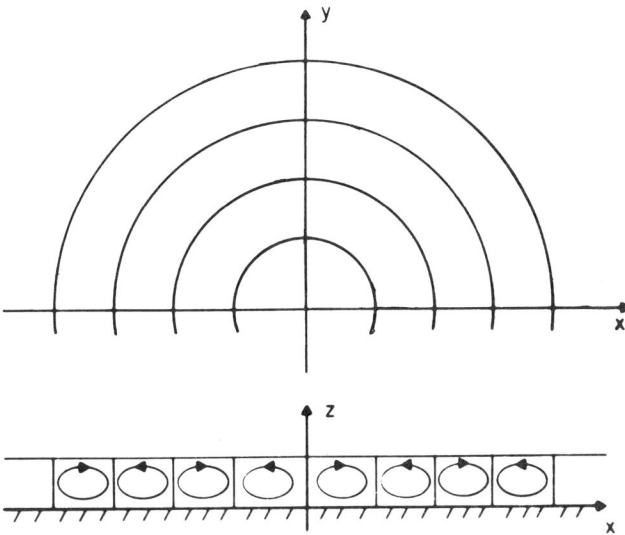

FIG. 5. Schematic view of the ring cell. (After Zierep, 1958.)

where w is the vertical velocity component, A is a constant, I_0 is the Bessel function of order zero, and a the so-called wave number of the flow. A schematic presentation of the ring cell is shown in Fig. 5. The ring cell should form on an infinite plane around a single disturbance if the critical Rayleigh number is reached. The Rayleigh number is a nondimensional measure of the applied vertical temperature difference and is given by the formula

(2.4a) $$R = \frac{\alpha g \Delta T}{\nu \kappa} d^3$$

where α is the volume expansion coefficient of the fluid, g the gravitational acceleration, ΔT the adverse temperature difference across the fluid, d the fluid depth, ν the kinematic viscosity, and κ the thermal diffusivity. The existence of the ring cell has been verified experimentally by von Tippelskirch (1959).

Using the fact that the nodal surfaces of the Legendre polynomials of infinite order coincide with the nodal surfaces of the Bessel function of zeroth order, Koschmieder (1959) has shown that the ring cell is the asymptotic solution of the convective motions on a sphere in the limit of a very shallow layer. Koschmieder (1965) has shown, furthermore, that in a shallow layer, if the width of the cells is the same as the depth of the fluid (as is approximately the case for the ring cells), the eigenvalue C_n for onset of convection on a sphere approaches the critical Rayleigh number for onset of convection on a plane plate of infinite extent. Thus, the ring cell indeed

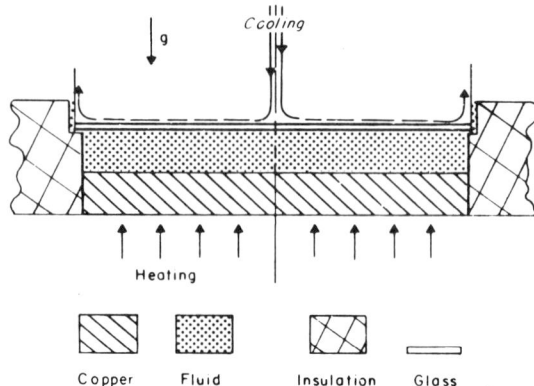

Fig. 6. Schematic view of a convection apparatus with uniform heating from below.

properly models the convective motions on a sphere. Instead of using strictly the ring cell, we model the convective motions on a sphere by the convective motions in a shallow fluid layer on a plane *circular* plate heated uniformly from below and cooled uniformly from above. A schematic section through an apparatus with which such an experiment is made is shown in Fig. 6. If the fluid is in touch with a lid on top of the fluid, then a pattern of circular concentric rings forms as was shown by Koschmieder (1966a), see Fig. 7. Between 1 and 13 perfect concentric rings have been produced in this way. The critical temperature difference for onset of convection agrees within the error of measurement with the value which follows from the critical Rayleigh number. Likewise, the so-called wavelength λ of the flow (defined as the ratio of the width of two cells divided by the fluid depth) agrees, within a small experimental uncertainty, with the theoretical prediction. The theoretical value is $\lambda_c = 2.0$, which means that the cells are as wide as they are deep. For a thorough theoretical investigation of the convective motions in circular containers see Charlson and Sani (1970) and Jones et al. (1976). We note that axisymmetric rings in circular fluid layers have been maintained at supercritical Rayleigh numbers up to about $5R_c$. That means that these rings exist not only under conditions which are described by linear theory, but that they exist also in the nonlinear range.

As stated above, the solutions for convection on a sphere which depend on azimuth seem to be unrealistic. These solutions are of the form

(2.2b) $$v_r = F(r)P_n^m(\theta) \cos m\phi \qquad m \neq 0$$

In the plane case the corresponding flows are of the form

(2.3b) $$w = DJ_m\left(\frac{a}{h}r\right) \cos m\phi \qquad m \neq 0$$

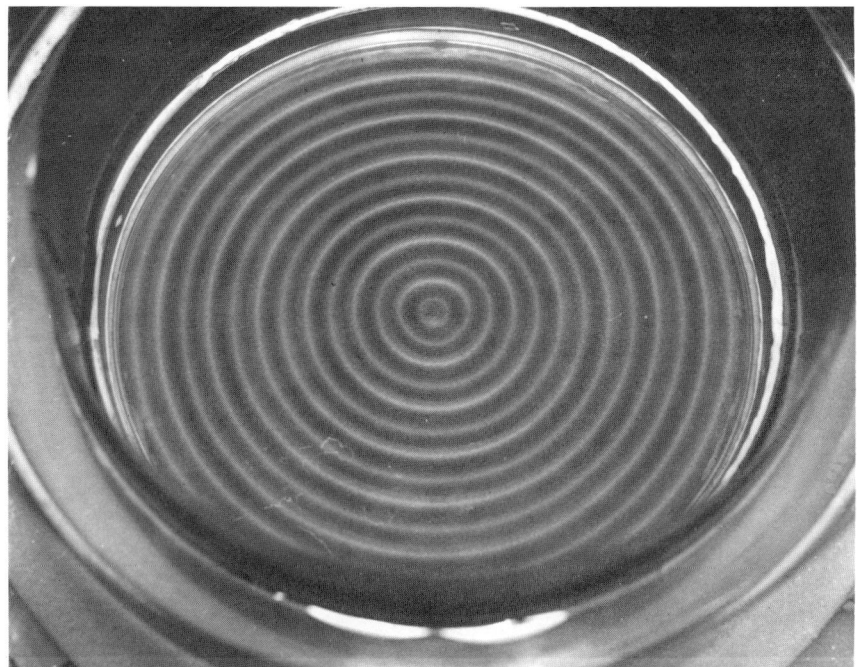

Fig. 7. Circular concentric convection rolls on a uniformly heated nonrotating plane and under a uniformly cooled glass plate. Fluid silicone oil. Visualization with aluminum powder. Dark lines indicate locations of vertical fluid motion, bright rings indicate location of predominantly horizontal flow. (After Koschmieder, 1966a.)

according to Zierep (1959). As seen from above the flow should then look like Fig. 8. The straight, diagonal lines are the nodal planes of $\cos m\phi$, delineating the azimuthal cell boundaries. However, the azimuthal wavelength would not be constant according to Fig. 8, but rather a function of the distance from the center and would soon approach very large values as compared with the radial wavelength. All evidence obtained with flow in circular containers in which azimuthal disturbances are present indicates that, if an azimuthal wavelength develops, the azimuthal wavelength is of the same order as the radial wavelength and independent of r. A good example of such a flow can be seen in the second ring on Fig. 14.

The question will be asked: for what reason does Fig. 7 not show the familiar hexagonal Bénard convection cells, an example of which is reproduced in Fig. 9. The absence of the hexagons in Fig. 7 is solely due to the glass lid which is in touch with the top of the fluid layer. The fluid then corresponds exactly to the assumptions made in theory, namely uniform

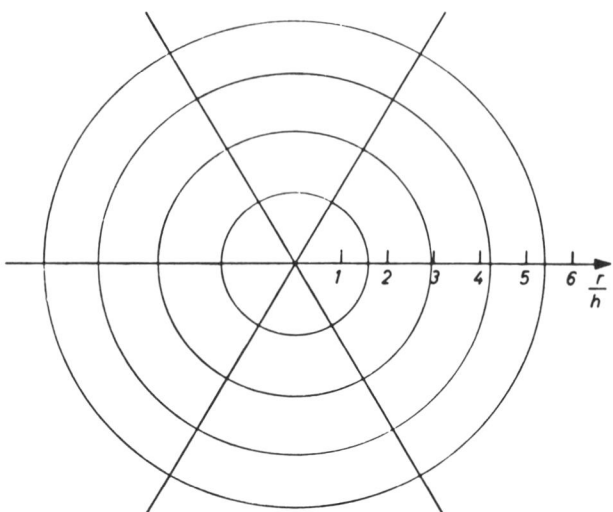

Fig. 8. Ring cells with an azimuthal dependence. (After Zierep, 1959.)

Fig. 9. Hexagonal convection cells on a uniformly heated nonrotating plane and under an air surface. Fluid silicone oil. Visualization with aluminum powder. (After Koschmieder, 1974.) (Copyright by John Wiley and Sons. Used with permission.)

temperature at the bottom, and a different, lower, but likewise uniform temperature on top. The circular wall makes the problem axisymmetric and, as long as linear conditions prevail, one would expect an axisymmetric solution. Indeed, circular rolls develop. However, if the upper surface is free (air), surface tension, or more precisely the variation of surface tension with temperature, is involved. It has been established *unambiguously* that this is the cause for the hexagonal cells. Surface tension driven hexagons can also form if the temperature stratification is stable (*i.e.*, when warm fluid is on top of cold fluid). The necessary condition for the formation of hexagonal cells is that the critical value of the Marangoni number is exceeded. The Marangoni number is defined by the equation

$$(2.5) \qquad M = \frac{dS}{dT} \frac{\Delta T d}{\rho \nu \kappa}$$

where S is the surface tension coefficient of the fluid. Details cannot be discussed here; for a review of the present knowledge of the Bénard convection problem in fluids the reader is referred to Koschmieder (1974). Since surface tension effects are not involved in atmospheric motions, one wonders about the cause for the well-known hexagonal or more precisely polygonal cloud patterns. We would like to suggest that the polygonal cells in the atmosphere are created by random, finite nonuniformities of the bottom temperature, in contrast to the assumed perfect uniform bottom temperatures in theory and proper model experiments. Such random nonuniformities do undoubtedly exist over land, but also over sea due to convective circulation in the thermocline. It is very easy to produce a polygonal convection pattern in the laboratory, if only the bottom plate is heated irregularly. As a matter of fact it has been shown by Koschmieder (1975) that almost any arbitrary pattern can be produced by disturbances at the bottom. For a possible explanation of the puzzling "open" and "closed" atmospheric convection cells see Krishnamurti (1975).

Let us summarize the results concerning the academic case. It is an entirely unambiguous result of linear theory that convective motions in a shallow fluid layer on a uniformly heated resting sphere can form a system of axisymmetric rings. This result is fully supported by experimental evidence. There is but one restriction to this result, namely the assumption of the validity of linear theory. The consequences following from this simplification will be discussed later.

2.2. The Case of Nonuniform Heating

The next logical step is the exploration of the fluid motions caused by the *simultaneous* presence of horizontal and vertical temperature differences. Very little theoretical work has been done to analyze this topic. Laboratory experiments seem to provide an easier approach to the problem. Imagine a plane circular plate heated axisymmetrically, so that, for example, the rim is warm and the center cool, with the temperature decreasing steadily from the rim to the center. This plate is covered with a fluid layer of uniform depth. The fluid layer is cooled uniformly on top through a glass lid. As the temperature at the rim is gradually increased warm fluid rises at the rim, moves inward along the lid, sinks at the cold center, and returns along the bottom. This circulation is due to the horizontal temperature difference in the bottom plate and will be referred to as the density circulation. Whenever the vertical temperature distribution is *stable*, that means whenever the temperature on top of the fluid is everywhere warmer than at the bottom, the density circulation is one big loop, extending from the rim to the center. This is to be expected from all previous investigations of steady flow due to horizontal temperature differences. If, however, the temperature over part of the bottom plate is raised sufficiently to exceed the temperature on top of the fluid, then we deal with the combination of a horizontal temperature difference with an *unstable* vertical temperature distribution. In other words, we deal with a combination of forced convection with Bénard convection. In shallow fluid layers the flow is then no longer just one big loop, but consists of a series of direct and "indirect" cells, as was shown experimentally by Koschmieder (1966b). An example of such a flow is shown in Fig. 10. The streamlines of the flow in Fig. 10 are indicated underneath the photograph. Fluid rises at the wall over the warmest part of the bottom plate and forms a rim roll if the vertical temperature difference there is unstable. The unstable vertical temperature field in Fig. 10 extends from the rim almost to the center of the plate. Adjacent to the rim roll is a roll with a circulation opposite to the anticlockwise circulation of the overall density circulation. Hence, the circulation in this roll is weak and the roll is narrow as compared to the rim roll in which the circulation of the roll is of the same sense as the overall density circulation. The appearance of rolls of alternately larger and smaller sections is characteristic for flows caused by a combination of a horizontal temperature gradient with an unstable vertical temperature gradient. The narrow rolls are here referred to as "counter rolls"; they correspond to what is often called an indirect cell. Inside from the counter roll in Fig. 10 is another wide roll turning anticlockwise, then comes another counter roll, and so on. Near the center, where the vertical temperature gradient is stable, follows finally a regular density circulation. Photographs of similar patterns with rolls and counter rolls can be found in Koschmieder's (1966b) paper. What we observe in these experiments is a clearcut *superposition* of the rolls caused by the vertical instability upon the overall density circulation caused by the horizontal temperature difference.

This result, if it is to be true, must be independent of the direction of the horizontal temperature gradient. And indeed, the motions of a fluid on a

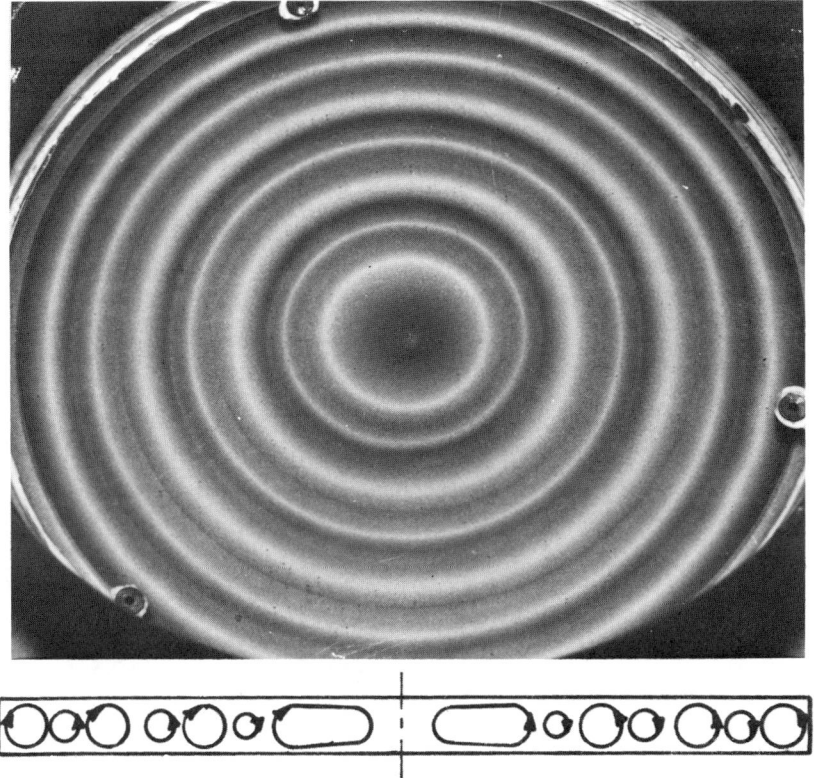

Fig. 10. Convective motions on a nonuniformly heated nonrotating plane, with a positive radial temperature gradient in the bottom. Top plate of glass, cooled uniformly. (After Koschmieder, 1966b.)

circular plate with a warm center and a cold rim combined with an unstable vertical temperature gradient are strictly analogous to those in Fig. 10. An example of the results of such an experiment is shown in Fig. 11. The overall density circulation is now upward at the center and downward at the rim. Superposed on this flow are three rings caused by vertical instability. The second ring from the center is another example of a narrow counter roll. Since the temperature field is axisymmetric, the flow is axisymmetric too and forms circular concentric rolls. There is no reason to doubt that axisymmetric rolls with alternately larger and smaller sections could also occur on an axisymmetrically heated resting sphere with gravity, as long as conditions prevail which correspond to those studied by linear theory. We note that the experimental result does actually also hold for moderately supercritical conditions, where the linear description of convection is, strictly speaking, no longer valid. There seems to be no other experimental investigation of Bénard convection on a plate heated nonuniformly from below.

146 E. L. KOSCHMIEDER

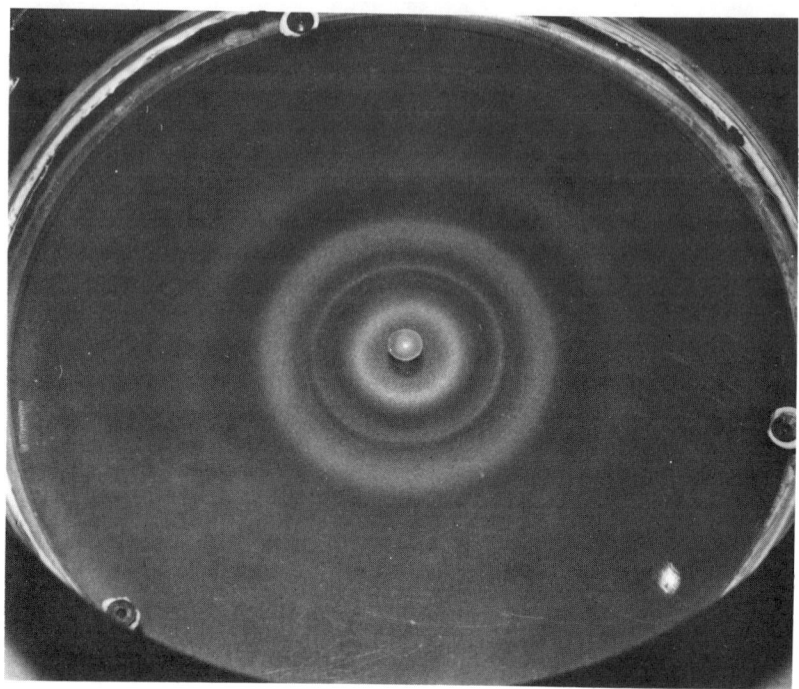

Fig. 11. Convective motions on a nonuniformly heated nonrotating plane, with a negative radial temperature gradient in the bottom. Top plate of glass, cooled uniformly. (After Koschmieder, 1966b.)

Some aspects of the results of these experiments have been confirmed theoretically by Müller (1966). The only other mathematical analysis of Bénard convection on a plane heated nonuniformly from below is Weber's (1973) paper. Two-dimensional atmospheric motions caused by nonuniform heating in a nonrotating system have been studied by Dickinson (1969).

Applying the knowledge obtained from the experiments discussed above to the problem of convection on a sphere, we realize one important point. Namely we see that the commonly held opinion, that the atmospheric circulation on a nonrotating Earth with a warm equator and a cold pole would occur in the form of just one direct cell (Fig. 1a), is not necessarily correct. If there is vertical instability over parts of, or over the entire sphere, cells *must* form and, assuming axisymmetric conditions, rings will be observed. Another essential thing to be learned is how the so-called "indirect" cells come about. Indirect cells are those in which the circulation is such that the air (or fluid) is sinking over the warm part of the bottom and rising over the cold part. The classic example of an indirect cell is the Ferrel cell over the midlatitudes of the Earth, see Fig. 1b. In the meteorological literature the Ferrel cell is explained as a second-order consequence of baroclinic wave motions in a geostrophic zonal flow; see, for example, Phillips (1954), Kuo (1956), and Saltzman and Tang (1972). There is, to my knowledge, no known example in fluid dynamics of a single indirect cell caused by a horizontal temperature difference in a shallow fluid layer. Such a circulation, if it is just a reversed direct cell, seems to contradict the basic thermodynamic principle that heat flows from warm to cold. However, in a vertically unstable layer, neighboring convection cells (Bénard cells) have necessarily opposite circulations and are hence, with regard to an imposed horizontal temperature gradient, either direct or indirect cells.

On a global scale nonuniform axisymmetric heating is relevant, at least in a first approximation, to the circulation of the atmosphere of Venus, provided that the very slow rotation of Venus (234 days per revolution) can be neglected. The axis of symmetry on Venus is the line connecting the subsolar point, as the warmest point, with the antisolar point, the coldest point on the planet's night side. One expects, of course, rising motion at the subsolar point. With a homogeneous surface the air would spread uniformly into all directions from the subsolar point, making one big loop to the antisolar point, provided the stratification is stable everywhere. At the antisolar point the air would sink and then return along the surface to the subsolar point. An example of a similar symmetric circulation is shown in Fig. 12, taken from Goody and Robinson (1966). In this particular study the Venusian atmosphere is considered to be opaque. The motion of the atmosphere is then driven by the nonuniform temperature *on top*. The observation of the Venera 9 and Venera 10 space probes (Keldysh, 1977) that there is sufficient direct light at the surface of Venus to cast shadows, creates doubt as to the validity of the assumption of an opaque Venusian atmosphere. Numerical studies of the circulation on Venus have been made by Mintz (1961), Sasamori (1971), and in particular by de Rivas (1973, 1975). A figure in de Rivas (1973) shows the streamlines of a symmetric circulation, which is quite similar to the result of Goody and Robinson. Stone (1968) has made an analysis of Hadley circulations on rotating and nonrotating planets, which should be relevant to the circulation on Venus. However, the assumptions made in Stone's study are very restrictive, the Rayleigh number being either $< 10^3$ or $\gg 10^3$.

We must also consider the possibility that the vertical temperature distribution on Venus might be locally unstable. Such an unstable temperature distribution is not necessarily due to heating from below but can be caused by internal heating as well, namely by the absorption of radiation within the

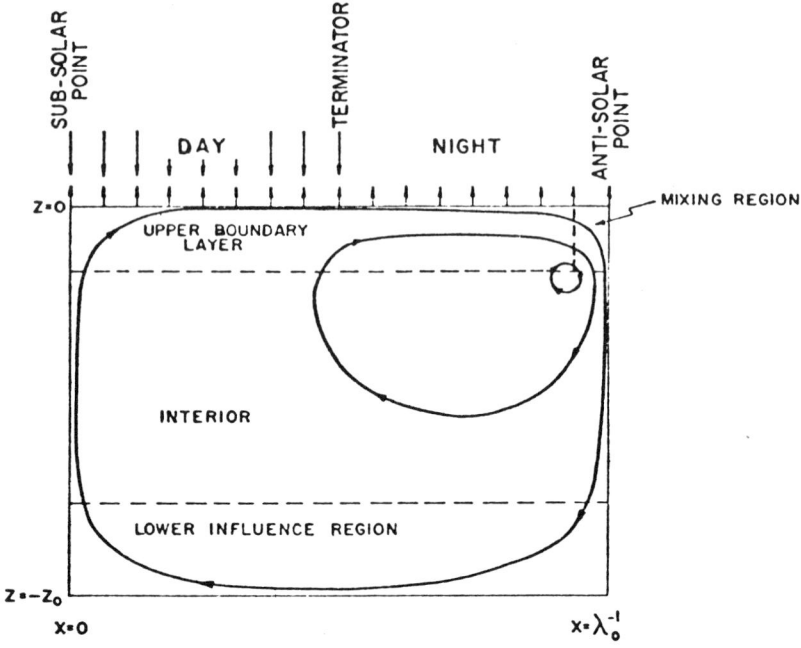

FIG. 12. Axisymmetric theoretical circulation on Venus. (After Goody and Robinson, 1966.)

atmosphere or by the release of latent heat. Convective motions in a spherical shell caused by heating from *within* occur, according to Chandrasekhar's theory, in the form of the same ring patterns as the motions caused by heating from below (Fig. 4), although the eigenvalues differ. In the case of a fluid on a plane plate with a uniform distribution of internal heat sources the usual Rayleigh number is replaced by a corresponding Rayleigh number

(4b) $$\mathrm{Ra} = \frac{g\alpha}{\nu\kappa} \frac{d^3}{32} \left(\frac{Sd^2}{2\kappa} \right)$$

where $S = H/\rho C$ is the volumetric rate of energy production. Linear theoretical investigations of the stability problem for internal heating of plane fluid layers have been made by Roberts (1967) and Thirlby (1970). Experimental investigations have been performed by Tritton and Zarraga (1967), Schwiderski and Schwab (1971), and Kulacki and Goldstein (1972). Although discrepancies between theory and experiment remain, there is no doubt that vertical instability can be caused by internal heating. If vertical instability occurs on Venus, either from heating from below or by internal heating, it would be necessarily nonuniform but in a first approximation axisymmetric. One would therefore expect a circulation in circular concentric rings around the subsolar point, quite similar to the flow shown in Fig. 11.

The thinking about the circulation on Venus has been changed drastically by the spectacular ultraviolet pictures obtained from the flyby of Mariner 10 (Murray *et al.*, 1974). These pictures of cloud fields in the troposphere/lower stratosphere do "display highly symmetrical motions relative to the *rotational axis*" (emphasis added). No "evidence of structures similar to large-scale cyclonic eddies" has been observed. Polygonal cells which appear around the subsolar point seem to make it quite likely that the vertical temperature field is there indeed unstable. The entire cloud pattern rotates retrograde with a puzzling 4-day period. Information of the overall circulation of the deep atmosphere of Venus has not yet been obtained. Cloud patterns and convection on Venus have recently been described by Belton *et al.* (1976).

2.3. *The Case of Uniform Heating with Rotation*

An entire chapter in Chandrasekhar's (1961) book is concerned with the linear theory of convection on a uniformly heated plane under the influence of rotation, more precisely under the influence of the Coriolis force. The effect of rotation on convection becomes apparent through the dependence of the critical Rayleigh number on the Taylor number, which is a nondimensional measure of the rotation rate defined as

$$(2.6) \qquad T_N = \frac{4\Omega^2 d^4}{\nu^2}$$

where Ω is the rotation rate and d the depth of the fluid layer. Onset of stationary convection is, according to theory, inhibited by large rotation rates. This is due to the fact that the motions in the fluid are bent to a longer curved path by the Coriolis force and have hence to overcome more dissipation. The effect of rotation on the different convection patterns such as rolls, squares, and hexagons is discussed in Chandrasekhar (1961). Of importance here is the effect of rotation on the ring cell, which has been studied by Müller (1965). According to this investigation a rotating ring cell retains its form as a series of circular concentric rings, but the projection of the streamlines onto the bottom changes from straight radial lines in the resting case to logarithmic spirals in the rotating case. In formulas, the velocity field with free surfaces is then

$$(2.7a) \qquad v_r = A I_0(ar) F(z)$$

$$(2.7b) \qquad v_\phi = A \frac{\sqrt{\Omega}}{\pi^2 + a^2} I'_0(ar) F(z)$$

where $I_0'(ar)$ is the derivative of the Bessel function of order zero, and a the wave number. Müller's study is the only theoretical investigation of this topic.

An experimental study (Koschmieder, 1967) confirms the existence of such circular concentric rolls on a rotating plate heated uniformly from below at moderately high Rayleigh and Taylor numbers. With vanishing rotation the flow is then a pattern of circular concentric rings on a resting plate, as must be the case. Surprisingly, the concentric rings in rotating experiments were of alternately larger and smaller section, similar to those in the case of nonuniform heating. The appearance of rings of different sizes in the rotating case on a *uniformly* heated plate turned out to be due to the presence of an overall centrifugal circulation in the fluid layer. The centrifugal circulation moves warm light fluid inward and cold heavy fluid outward, just as a centrifuge separates the light and heavy components of a fluid. Any fluid with a vertical temperature stratification (whether stable or unstable) has such a centrifugal circulation, even when rotated at minimal rotation rates. The superposition of the centrifugal circulation upon the rolls caused by vertical instability creates the rolls of alternating section. Of only three rotating Bénard convection experiments that have been made, Koschmieder's (1967) experiment is the only one to result in an axisymmetric and reproducible flow.

The obvious presence of the centrifugal circulation in this experiment does not agree with the premise which all theoretical studies of convection with rotation make since Chandrasekhar's (1953b) paper. Namely, it is assumed that the fluid is in a state of rigid rotation (no motion relative to the plate) before the onset of convection. This is obviously not true, even in the stable case. The stabilizing effect that rotation theoretically has on the onset of convection is therefore unrealistic. In formal terms, the Boussinesq approximation made, which says that the density is supposed to be constant except when coupled with gravity, is not valid in this case. Previous theoretical treatments of convection with rotation have another unrealistic feature: they are concerned with convection on an infinite plane rotating about a normal axis. But a rotating fluid layer of infinite extent experiences, at its outside, infinite centrifugal accelerations which invalidate the solutions obtained. A theoretical investigation which drops the assumption of solid body rotation has recently been made by Torrest and Hudson (1974). The centrifugal circulation is certainly irrelevant when we discuss convection on a rotating plane in the context of the circulation of the atmosphere. But for all realistic theoretical studies which aspire to be verifiable in the laboratory, the centrifugal circulation is not negligible.

Chandrasekhar (1957, 1961) has also investigated the effect of rotation, meaning the Coriolis force, on *axisymmetric* convective motions in a fluid sphere. The motions are then governed by the equation

(2.8) $$\Delta_5^3 U + T_N \frac{\partial^2 U}{\partial z^2} = C \frac{\partial^2}{\partial \mu^2}[(1-\mu^2)U]$$

where Δ_5 is a five-dimensional Laplacian operator, U is a scalar function which describes motions which are entirely meridional and from which, through simple differentiation, v_r and v_θ is obtained. T_N is the Taylor number, C is a constant representing an eigenvalue, and $\mu = \cos \theta$, with the colatitude θ. The general solution of Eq. (2.8) is an infinite sum of products of Bessel functions and Gegenbauer polynomials. Simple special solutions of (2.8), corresponding to those in the case of a resting sphere, can unfortunately not be split off, so that one cannot make qualitative statements about the velocity field. In order to determine the velocity field in the rotating case the difficult eigenvalue problem has to be solved first, and this has not yet been done, although it is feasible.

There are, however, other theoretical studies based partly on Chandrasekhar's work that provide information concerning axisymmetric as well as nonaxisymmetric convective motions caused by heating from below on rotating spheres. Durney (1968) uses the so-called mean field approximation and develops the flow in poloidal and toroidal eigenfunctions. He obtains, for moderate Rayleigh and Taylor numbers, motions in the form of concentric rings around the axis of rotation. Extensive investigations of asymmetric motions in deep layers have been made by Gilman (1975), in context with the convection zone on the sun. In a numerical study, Williams and Robinson (1973) obtained also a flow in rings around the axis of rotation. This study, which aims at the circulation on Jupiter, will be discussed in more detail in Section 4.

To summarize the results for Bénard convection with rotation: There is theoretical as well as experimental evidence which indicates that convective motions in the form of *rings* around the axis of rotation can occur. There is neither unanimity of the results of the theories nor complete agreement of the experiments with the results of the theories. Many results can be considered only as tentative, and much more work has to be done to approach certainty.

2.4. Nonuniform Heating with Rotation

Because of obvious difficulties, an analytical study of axisymmetric convection in a fluid layer heated nonuniformly from below on either a rotating plane or a rotating sphere has not yet been made. One can, however, qualitatively predict the sort of solution to expect, using as a guide the solution of the subproblems which we have discussed before. As long as we stay within the domain of linear theory the solution should be the sum of a density circulation due to axisymmetric nonuniform heating, a centrifugal circulation due to rotation of a stratified fluid, and a cellular motion due to vertical instability, if the latter is present. This superposition of the solutions of the simpler subproblems is hypothetical, supported by the evidence of the laboratory experiments, but not demanded by mathematical theory. Indeed, model experiments made by Koschmieder (1968) confirm the qualitative prediction above. In the experiments, an axisymmetrically heated plane plate with a warm rim and a cool center covered by a fluid of uniform depth was rotated around its axis at moderate Taylor numbers. If the vertical temperature gradient was stable everywhere, a symmetric density circulation developed with an effective radial temperature difference $\Delta T_r = +40°C$ and 0.5 revolution per second (rps) rotation. This result is not mentioned in the paper by Koschmieder (1968), since it was taken for granted that such a symmetric density circulation had to occur. If the vertical temperature gradient became unstable over part of or over the entire plate, then the fluid moved in circular concentric rings as expected. The arrangement of rings, which had either large or small sections, depended strongly on the relation between the magnitude of the horizontal temperature gradient and the magnitude of the rotation rate. With rotation rates lower than 0.25 rps the flow was dominated by the density circulation and looked quite similar to the one shown in Fig. 10. The only characteristic difference was the appearance of two narrow counter rolls side by side at the location where the outward-turning centrifugal circulation ends and the inward-turning density circulation begins. At rotation rates ≥ 0.5 rps the centrifugal circulation became practically dominant, see Fig. 13. Figure 13a shows three strong rolls, turning outward on top with the centrifugal circulation, two narrow counter rolls, and around the center the density circulation turning inward on top. Figure 13b shows the flow at 0.75 rps. In this case all counter rolls are still suppressed by the centrifugal circulation. Only the density circulation at the center turns inward.

The question will be asked whether the flow was axisymmetric in all cases, regardless of Rayleigh number, radial temperature gradient, and Taylor number. This cannot be so since it is known from experiments of Bénard and Avsec (1938) that convective rolls orient themselves parallel to a shear, if the shear is sufficiently strong. The axes of the rolls in Fig. 13 are, however, oriented perpendicular to the shear in the density circulation, where the fluid moves inward at the top and outward at the bottom. Likewise, the axes of the circular rolls affected by the centrifugal circulation are perpendicular to the shear. If, in either of both circulations, the shear was too strong, the axisymmetry of the flow ceased. The transition to flow in rolls oriented parallel to the shear, i.e., in a radial direction, occurred in all cases first in the

(a)
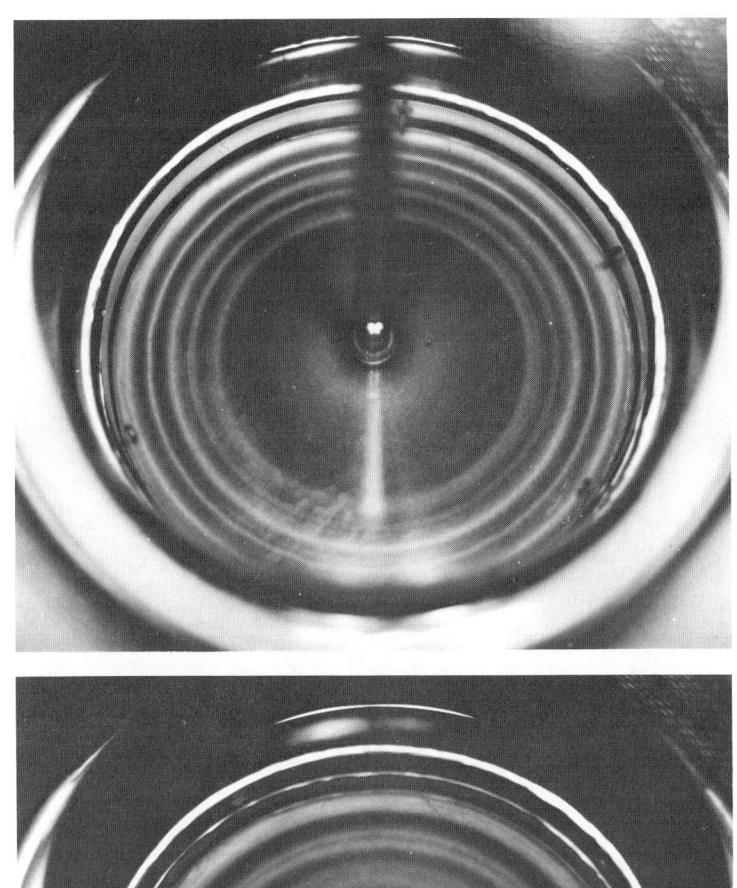

(b)

FIG. 13. Convection on a nonuniformly heated rotating plane with a positive radial temperature gradient. Top plate of glass, cooled uniformly. (a) Rotation rate 3.14 rad sec^{-1}, $\Delta T_r = 3.5°C$; (b) rotation rate 4.71 rad sec^{-1}, $\Delta T_r = 3°C$. (After Koschmieder, 1968.) (Copyright by Cambridge University Press. Used with permission.)

counter rolls. These are interestingly much more sensitive to azimuthal disturbances than the rolls which turn with either the centrifugal or the density circulation. A photograph showing the instability of a counter roll is shown in Fig. 14. A photograph of rolls pointing in radial direction can be found in Koschmieder (1968).

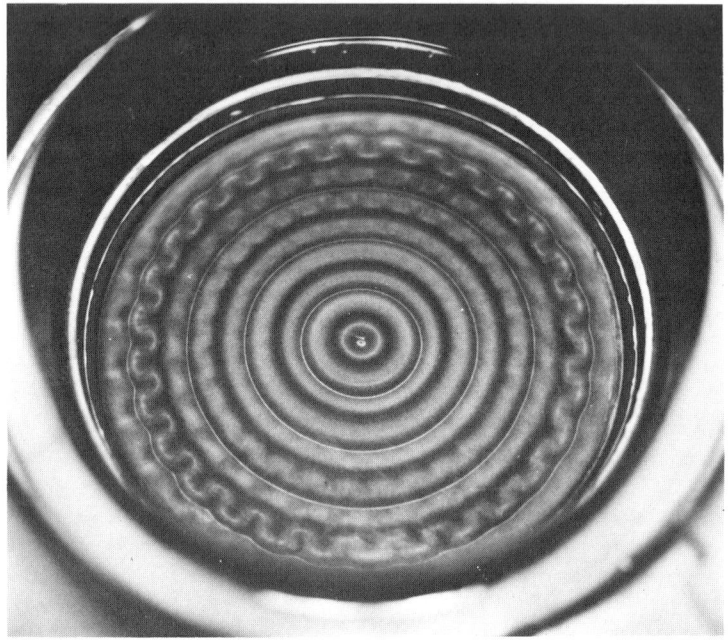

Fig. 14. Convection on a nonuniformly heated rotating plane with a positive radial temperature gradient. Transition to flow in rolls oriented in radial direction. Rotation rate 1.57 rad sec^{-1}. $\Delta T_r = 8°C$. (After Koschmieder, 1968.) (Copyright by Cambridge University Press. Used with permission.)

A different set of experiments dealing with the question of convection on a nonuniformly but axisymmetrically heated rotating plate are the dishpan experiments of Fultz *et al.* (1959). These experiments were made with the expressed purpose of modeling the general circulation of the atmosphere. The main result of these experiments is said to be that the flow in the dishpan was either axisymmetric (Hadley regime) when the Rossby number was large (≥ 1), or the flow was asymmetric, unsteady, and vaguely wavelike when the Rossby number was small ($10^{-3} \leq \text{Ro} \leq 0.1$). The Rossby number is defined as the ratio of a characteristic velocity c of the flow to the circumferential speed of the container of radius r and angular velocity Ω, so

$$\text{Ro} = \frac{c}{\Omega r} \tag{2.9}$$

It is, however, by no means certain that the different behavior of the fluid in both regimes is caused by a response of the fluid to variations of the Rossby number. Both regimes might as well be due to other dynamic factors. Although this is not stated by Fultz *et al.*, vertical instability (Bénard convection) must have occurred in the majority of their experiments. The critical vertical temperature difference for onset of Bénard convection in a resting layer of water 1 cm deep is about 0.1°C, if the upper surface is free. Taking the dependence of the critical Rayleigh number on the Taylor number into account and using $T_N = 10^7$ and $d = 4$ cm gives a critical vertical temperature difference of about 0.3°C. The mean water temperature of many dishpan experiments was, however, 10°C or more above the temperature of the air on top of the fluid. That means that at least part of the bottom of the fluid must have been 10°C warmer than the top of the fluid. Consequently the fluid must have been highly unstable. That concerns all Rossby regime flows discussed between pages 39 and 49 of Fultz *et al.* (1959). A combination of nonuniformities of the temperature on top of the fluid (caused by air motion and evaporation) with effects caused by torque on top of the fluid (caused by friction between the air and the rotating fluid) and with surface tension effects, then caused the irregular, not really reproducible, surface flows observed in the dishpan experiments. Note that according to Fultz *et al.* (1959, p. 13), " the top surface motions are representative of the top millimeter or so of depth."

Of the two reported cases of symmetric flow in the Hadley regime in the dishpan, the first (Fultz *et al.*, 1959, Fig. 10a) concerns an experiment with small rotation rate where the air temperature is only 0.5°C below the mean temperature of the fluid, and hence vertical instability is not important. Forced convection caused by the radial temperature differences in the bottom and in particular by the heated lateral wall might easily dominate the circulation. The second case of symmetric flow in the dishpan (Fultz *et al.*, 1959, Fig. 11) is again at low rotation rate but with a mean fluid temperature 14°C above air temperature. Strong vertical instability should be present. Apparently the likewise strong density circulation still dominates the flow. An extended theoretical and experimental investigation is needed to understand positively the result of this experiment. There are two more pictures of symmetric flow in Fultz *et al.* (1959, Figs. 15a and 16a), but they concern flow in a rotating annulus, a completely different matter, which will be discussed in Section 3.

We should now summarize the results that have been obtained from fluid dynamics studies of convection on a plane or a sphere. Theoretical and experimental investigations *pertaining to the linear aspects* of the flow of a viscous, heat-conducting fluid under the influence of heating from below as well as axisymmetric nonuniform heating, and of rotation, indicate that vertical instability is a *necessary condition* for the occurrence of "cells" (or in the axisymmetric case of rings). There is no evidence that under stable conditions a horizontal temperature gradient creates anything but a direct overall density circulation with or without rotation. Indirect cells come about by the superposition of a density circulation upon cells caused by vertical instability. Axisymmetric flows are clearly a possible solution, if the imposed conditions are axisymmetric. Whether or not the axisymmetric solutions are stable against azimuthal disturbances will be discussed in the next chapter. There are several aspects of convection on a nonuniformly heated plate or sphere where the present knowledge is incomplete and which are understood only partially. In particular, the problems associated with the nonlinear nature of the Navier–Stokes equations are mainly unsolved. Even in the most simple and most thoroughly studied case, the Bénard convection problem, no satisfactory theory for only slightly supercritical *i.e.*, nonlinear convection, let alone turbulent convection, has emerged; for detail see Koschmieder's (1974) review.

2.5. Symmetric Baroclinic Instability

We have finally to consider the possibility that a mechanism completely different from vertical instability might produce a fluid flow in the form of axisymmetric parallel rolls with opposite circulation in adjacent rolls. Such a different mechanism has indeed been suggested by Stone (1966), namely the so-called symmetric baroclinic instability. Stone studies the stability with regard to symmetric (*i.e.*, only y-dependent) disturbances of a shear flow on an infinite plane rotating with constant angular velocity about a vertical axis. The fluid is *stably* stratified in the vertical, the unperturbed (zonal) flow in x direction is a linear function of height z. For an inviscid fluid, symmetric instability is possible according to Stone, if the Richardson number $0 < \text{Ri} < 1$, where

$$(2.10) \qquad \text{Ri} = \alpha g \frac{dT}{dz} \frac{1}{(\partial u/\partial z)^2}$$

A picture of the expected flow, taken from Stone (1967), is shown in Fig. 15. The original assumptions of an inviscid, nonconducting fluid have been dropped in a recent study by Walton (1975). An attempt has been made to verify experimentally the existence of symmetric baroclinic instability (Stone

FIG. 15. Theoretical circulation of flow caused by symmetric baroclinic instability. (After Stone, 1967.) (Copyright by the American Meteorological Society. Used with permission.)

et al., 1969; Hadlock et al., 1972). For this purpose a fluid layer in a circular tank heated nonuniformly from below, hence *unstably* stratified, has been rotated about its axis. The resulting irregular, unsteady motion and a few temperature measurements have been interpreted as proof of the existence of this kind of instability. But, as pointed out by Koschmieder (1973), vertical instability as well as a density circulation must have been present in the tank. Not before the motions from these two causes are separated from the motion due to a possible baroclinic instability, has proof of the existence of symmetric baroclinic instability been provided. Anyway, the basic unperturbed state assumed in these studies, namely a motion with $T = T(y, z)$, $u = z$, and $v = w = 0$, does not satisfy the equation of thermal conduction for a fluid with a nonvanishing thermal conductivity.

2.6. *The Circulation of Jupiter's Atmosphere*

The circulation of Jupiter's atmosphere is a test case for the suggestion advanced here, that large-scale atmospheric motions are examples of convection on nonuniformly heated rotating spheres. The circulation on Jupiter takes place in rings, nearly symmetric about the axis of rotation (see Fig. 2). Close-up pictures taken from the Pioneer 10 and Pioneer 11 spaceprobes (Gehrels et al., 1975) reveal that the rings have a marked fine structure. Nevertheless, it is obvious that even short-time azimuthal averages of the circulation result in a meridional circulation in the form of rings. Observation of the rings covering a time span of 100 years are summarized in Peek's (1958) book. There are at least four bright zones in each hemisphere, separated by dark belts. Similar zones and belts occur on Saturn. The number of zones on Jupiter, which is not necessarily the same on each hemisphere, varies slowly over decades; there may be as many as six. The polar areas appear always to be dark and free of zones. The bright zones represent cloud fields, the clouds being presumably made of ammonia crystals. The dark belts are cloudless; one might be looking there at the surface of Jupiter. Zones indicate rising and belts indicate subsiding air. The air on Jupiter is composed mainly of hydrogen and helium. There are very large variations of the latitudinal extent of the zones, some of which are very narrow as compared to the others [see in particular the presentation of the latitudinal extent of belts and zones in Chapman (1969, Fig. 2a-e)].

The azimuthal velocity field of the flow has been studied in detail by Chapman (1969). A feature of particular interest are the very strong (~ 100 m sec^{-1}) westerly winds in the equatorial zone, which are in marked contrast to the easterly winds at the corresponding location (tropics) on Earth. It is, in this context, crucial to know for sure whether the so-called equatorial band, a very narrow dark belt a little south of the equator, is a permanent feature. The equatorial band often cannot be seen on photographs taken with visible light, but seems to be prominent on pictures taken in blue light (see Fig. 2; see also Fountain and Larson, 1973). The equatorial band is also indicated in Chapman's data as a small dent in the westerlies at the equator. The westerlies in the equatorial zone, combined with the presence of the equatorial band, show that the equator of Jupiter is the location of a *sink*, meaning a location of subsiding air. This is in startling contrast to the equator of Earth, which is the location of a source. It will be safe to assume that at the surface at the equator of Jupiter there is, as on Earth, the location of the highest surface temperature. This should be so, since one can assume that in a first approximation the internal energy release on Jupiter is uniform over the planet. The additional solar energy then makes the equator warmer than any other place on Jupiter. [The temperature of the cloud deck of the entire planet is however, fairly uniform, see Wildey *et al.* (1965), but there are substantial variations with time, see Keay *et al.* (1972).] Since one should expect that the equator of Jupiter is likewise a source, it requires a particular mechanism, referred to as the "equatorial acceleration," to explain the subsiding motion at this location. That will be discussed later on. There is one more important subject, namely the Great Red Spot, about which there has been ample speculation. We would like to point out only that the Great Red Spot is living proof that the circulation of the atmosphere of Jupiter in the form of rings is stable against a steady, finite, azimuthal disturbance.

Before we compare the convective motions on a plane with the circulation of Jupiter's atmosphere, we would like to emphasize that no artificial assumptions whatsoever are involved in the results of the convection studies discussed in Sections 2.1 to 2.4. These studies deal with the motion of an ordinary Newtonian fluid under the influence of some simple, axisymmetric heating and of rotation, under conditions when the linear approximation of the Navier-Stokes equation is valid. Comparing now the convective motions with the circulation on Jupiter we find good *qualitative* agreement in four points. (1) There is an obvious similarity between the rings in the convection experiments and the rings that occur in Jupiter's atmosphere. (2) There is subsidence at the center of the rotating experiments if the radial temperature gradient is positive. In an atmosphere, subsidence means the absence of clouds, and there are no clouds near the poles of Jupiter, which has a meridional temperature gradient due to solar radiation. The meridional temperature gradient corresponds to the radial temperature gradient in the convection experiments. (3) There are rings of different cross-section in the convection experiments, and there are rings of clearly different sections on Jupiter. (4) Finally, the circulation of the outermost ring in the rotating experiments is *downward* over the warmest location of the bottom plate, corresponding to the circulation in the equatorial zone of Jupiter, where we have likewise downward motion (a sink) over the warmest place at the surface.

We shall now discuss these points in detail. First, a comment concerning the point that the similarity of the convective motions to the planetary atmospheric motions is only qualitative. In order to show that the similarity is better than qualitative one would have to use similarity parameters, such as the Rayleigh, Reynold, Richardson, Rossby, Prandtl, and Taylor numbers. But the use of similarity parameters is ambiguous if many parameters are involved. Furthermore, many of the material constants involved in the similarity parameters are known only vaguely in the atmospheric case or vary by orders of magnitude since they are eddy coefficients. Most of all, none of the similarity parameters takes into account anisotropic eddy viscosity or anisotropic thermal diffusivity, both of which are possibly of great dynamic significance. Theoretical investigations of the circulation on Jupiter are likewise only qualitative; they are linear and actually neglect viscosity and thermal diffusivity. For a review of these studies see Maxworthy (1973). The numerical study of Williams and Robinson (1973) of a convectively unstable atmosphere on Jupiter does not neglect viscosity and thermal diffusivity. This study is in general in agreement with the concept of convection as discussed here. Details will be discussed in Section 4, together with the other numerical studies.

Now, concerning the similarity between the convective rings on a plane and the zones and belts in Jupiter's atmosphere, which represent motion in the form of rings, we note the following. It has been maintained that the rings in Jupiter's atmosphere cannot be of convective origin since their aspect ratio, namely the ratio of their depth to their latitudinal extent, is of order 1/1000, while the aspect ratio of rings of convective motion caused by vertical instability is of order 1. This objection to the similarity of both flows is erroneous. It is true that rings, caused by vertical instability, have a precisely defined wavelength of order 1, which makes them almost exactly as wide as they are deep. However, that is true only in the case of flow between two rigid, perfectly conducting plates. Linear theory says that cells with significantly different wavelengths, up to infinitely wide cells, occur if thermal conduction on top of the fluid is poor, or heat can be lost only by radiation. This has been shown by Hurle et al. (1967) and Nield (1968), has been applied by Sasaki (1970) to meteorological conditions, and has been verified experimentally by Koschmieder (1969). It is not really known whether the radiative boundary condition on top of Jupiter's atmosphere is the cause of the shallow rings, but it is likely to be a contributing factor.

Mesoscale polygonal convection cloud cells in the atmosphere of the Earth do not occur with an aspect ratio of order 1 either, but rather with a variable ratio which is, on the average, about 1/30, see Priestley (1962) and Hubert (1966). The aspect ratio of atmospheric convection cells can be explained theoretically using an anisotropic eddy viscosity (Ray, 1965), which is, however, a forced solution. There is, at best, only a remote chance that the polygonal convection cells in the atmosphere are caused by something else but vertical instability. Hence we conclude that aspect ratios as they occur in the rings of Jupiter's atmosphere are, by all means, compatible with the concept that these rings are caused by vertical instability. There is, as a matter of fact, no alternative. No other process is known in fluid dynamics which has produced circular concentric rings. The cloud cells serve to illustrate one other important point. They show that, although the motions in the cells are undoubtedly turbulent and nonlinear, the entire flow field is arranged in a pattern which is qualitatively the same as that of the linear convection regime. There is, as yet, no theoretical explanation based on the nonlinear Navier–Stokes equation, which would show that this has to be so or why that is so. We just note, as an empirical fact, that turbulent atmospheric convection is modeled qualitatively in a correct way by linear fluid dynamics.

With regard to the second aforementioned qualitative similarity between the convective motions and the circulation of Jupiter's atmosphere, we note the following. It is quite obvious that a meridional temperature gradient will cause a density circulation with subsidence at the poles. That means the absence of clouds there. A theoretical investigation of the density circulation on a rotating planet does not seem to exist, even in the linear approximation. Hence we have to content ourselves with the qualitative result. The third listed similarity, namely the similarity between the variable ring sizes in experiment and the variable size of the rings of Jupiter's atmosphere, is again qualitative. The fact that the sections of the rings of Jupiter's atmosphere differ substantially is certainly of dynamic significance. This matter has not yet been explored at all.

Finally, we consider the fourth qualitative similarity mentioned, namely the correspondence between the circulation of the outermost ring over the location of the warmest bottom temperature in the rotating convection experiment and the circulation in the equatorial zone of Jupiter. Following the experiments one is tempted to ascribe the westerly winds in the equatorial zone of Jupiter (and the much faster (~ 400 m sec^{-1}) westerly winds near the equator of Saturn) to the presence of the centrifugal force. Centrifugal effects are certainly negligible for large-scale atmospheric motions on Earth, since the ratio of centrifugal acceleration (at the equator) to gravity is $\Omega r^2/g = 0.0035$. However, this acceleration ratio has the value 0.089 on Jupiter and the value 0.159 on Saturn. Therefore, one cannot reject offhand the possibility that the centrifugal force is the cause for the occurrence of a sink at the equator of Jupiter and Saturn. As is well known, the centrifugal force deforms the surface of rotating planets into an ellipsoid. The effect of the centrifugal force may therefore become apparent only in a decrease of gravity in the equatorial region. Substantial theoretical work that has been done on the "equatorial acceleration" always neglects either the centrifugal force or the variation of gravity. A most elaborate nonlinear study of the equatorial acceleration has been made by Gilman (1972), who is mainly interested in the equatorial acceleration which occurs in the deep convection zone of the Sun. His studies look for the cause of the equatorial acceleration in nonaxisymmetric conditions. Durney (1974) obtains an equatorial acceleration from an axisymmetric flow with a pole–equator temperature difference; and Köhler (1970) obtains an equatorial acceleration from an axisymmetric flow through the introduction of an anisotropic viscosity. Williams and Robinson (1973) obtain likewise a westerly jet in the equatorial zone of Jupiter with a symmetric circulation. From the very different approaches taken by the theoretical studies it seems to follow that the final explanation of the equatorial acceleration has not yet emerged.

2.7. Convection in the Atmosphere on Earth

It is of great interest and importance to see whether the points discussed above apply also to the atmosphere of the Earth. We will first clarify whether the atmosphere is heated *from below*. That is a basic condition for the occurrence of vertical instability. According to Sellers (1965) the total energy intercepted per day by the Earth is 3.67×10^{21} cal day^{-1}. According to Hanson (1976), 52% of this amount is absorbed at the surface and hence heats the atmosphere from below; 29% of the incoming solar radiation is reflected and does not contribute to the heat balance; and 19% of the incoming solar energy is absorbed within the atmosphere and amounts to internal heating. According to Palmén and Newton (1969) about 10^{20} cal day^{-1} are crossing 40° latitude in a northward direction in winter (which is the season of maximal meridional heat transfer). That means that at most 1/20 of the total energy received is transferred by horizontal exchange to northern latitudes. Since very little heat absorbed at the surface is lost directly to space by infrared reradiation from the ground (about 6 %), the rest of the available energy must be transferred upward and radiated to space. This transfer occurs either through transport of sensible heat, or by transfer and release of latent heat, or by radiative flux. It has been known for a long time (London, 1957) that intensive radiative cooling takes place in the upper part of the tropical troposphere, at 8 to 9 km height in summer, at 12 km height in winter, with cooling rates of order of 2°C per day. Heat must therefore be transferred to this height to maintain the temperature. This heat transfer in the tropics is apparently mainly accomplished by cumulus convection.

According to fluid dynamics, a negative vertical temperature gradient or its equivalent, a net upward heat flux through the fluid layer, is a *necessary* condition for the occurrence of vertical instability. The discussion above confirms that this necessary condition is met at least in the tropics, if not also in midlatitudes. A summary of the empirical heat balance in winter for 10° latitude intervals is shown in Fig. 16. According to this figure the majority of

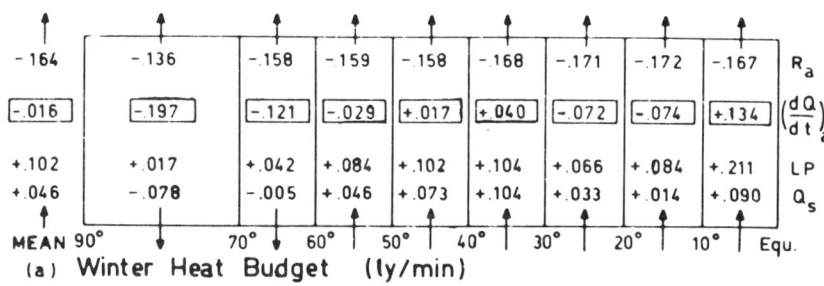

FIG. 16. The heat budget in latitude belts of the northern hemisphere in winter. R_a is the radiative loss from the atmosphere, dQ/dt is the divergence of the latitudinal heat flux in the atmosphere. LP is the release of latent heat, and Q_s is sensible heat transfer from earth to atmosphere; units langley per minute. (After Palmén and Newton, 1969.)

the radiative energy loss of the atmosphere in all latitudes up to 60° N is made up by vertical heat transfer. The meridional heat transfer becomes dominant for latitudes greater than 60° and actually supplies energy for a substantial downward heat flux to the ground, which implies stable conditions there, on the average. The stability criterion used in meteorological practice, namely whether the potential temperature increases with height (stable) or decreases with height (unstable) is only a *sufficient*, not a necessary condition. We repeat the statement made in Section 2.2 that the measured vertical temperature gradient in an already convecting fluid layer is not necessarily negative throughout the entire layer. The subsiding section of a convection cell can have a (potential) temperature increasing with height. Nevertheless this area is a part of an initially overall unstable fluid layer. Whether eddy viscosity or eddy thermal diffusivity in air, as viscosity and thermal diffusivity in fluids, postpone the onset of instability until a critical temperature gradient has been reached has apparently not yet been established by empirical data.

3. Azimuthal Disturbances

So far we have restricted the discussion to the case of axisymmetric flows. It is, however, questionable whether an axisymmetric circulation of the atmosphere of Earth would actually exist under idealized conditions with a homogeneous surface and an axisymmetric bottom temperature. Finite azimuthal disturbances obviously affect the circulation of the Earth's atmosphere. Otherwise we would not have stationary winter highs and summer lows over land, for example. But the finite disturbances do not pose the most intriguing question. The most important question is whether idealized axisymmetric circulations are stable or unstable with respect to infinitesimal azimuthal disturbances. If the atmospheric circulation were unstable to infinitesimal azimuthal disturbances that would, as Lorenz (1969, p. 10) put it, "mean that the instability renders it impossible for eddies not to be generally present."

Extensive investigations of the stability of atmospheric motions with respect to infinitesimal azimuthal waves have been made with the theory of baroclinic instability. This concept was introduced by Charney (1947) and Eady (1949). The problem is usually approached by considering the stability of an inviscid, nonconductive fluid layer that has a temperature gradient with uniform vertical and horizontal components, hence is baroclinic, but is *stably* stratified in vertical direction. The layer rotates with constant angular velocity about a normal axis. A baroclinic fluid can be unstable to infinitesimal disturbances in the form of azimuthal waves. The theory cannot specify the kind of flow that follows after onset of instability. The irregularity of the atmospheric motions on Earth seems to be in qualitative agreement with the results of theory of baroclinic instability. On the other hand, one wonders for what reasons the consequences of that instability do not appear in similar form on either Jupiter or Saturn or also on Venus. This is not the place to review the numerous theoretical studies concerned with baroclinic instability, for a summary the reader is referred to Charney (1973). What we will be concerned with here is the verification of the concept of baroclinic instability in fluid dynamics experiments and their relevance to atmospheric circulations.

3.1. The Rotating Annulus Experiment

The convective motions of a fluid in a rotating, laterally-heated annulus have been studied intensively since Hide's (1958) discovery that the motions in the annulus form, under certain circumstances, a pattern of azimuthal waves. A schematic sketch of an annulus apparatus is shown in Fig. 17. The fluid (water) is contained between two circular, concentric, vertical cylinders, which are made of a material with good thermal conductivity. The top and bottom plates of the fluid are insulating. Both cylinders are rigidly connected

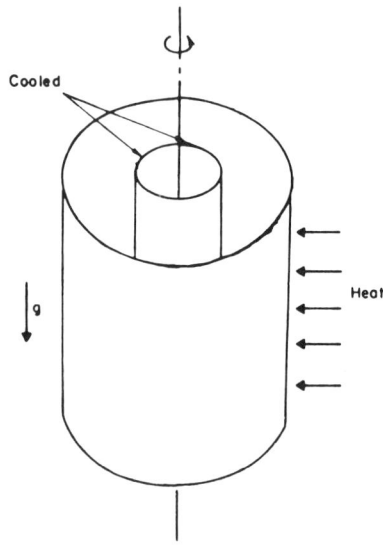

FIG. 17. Schematic apparatus of a rotating annulus experiment.

and rotate about their common axis. Different temperatures are applied to the outer and inner cylinder. Hence, forced convection must occur. Note that the impressed temperature gradient is radial. Heat is transferred by the fluid only from the warm cylinder to the cold cylinder, i.e., horizontally. There is no provision for a vertical heat transfer; the lid and bottom are insulated. If the annulus rotates and a sufficiently large temperature difference is applied, "waves" appear as shown in Fig. 18. The number of waves depends primarily on the temperature gradient, the rotation rate, and the gap width. If, at constant Ω, the temperature difference ΔT_r is increased, the number of waves decreases (see Fig. 18a–d), until at a large value of ΔT_r the flow changes into a symmetric density circulation. The waves must be caused by infinitesimal azimuthal disturbances, since the temperature gradient, as well as all other parameters, are made as axisymmetric as is feasible. Actually the more symmetric the conditions are, the more regular the waves become. The theoretical explanation for the appearance of the waves is that

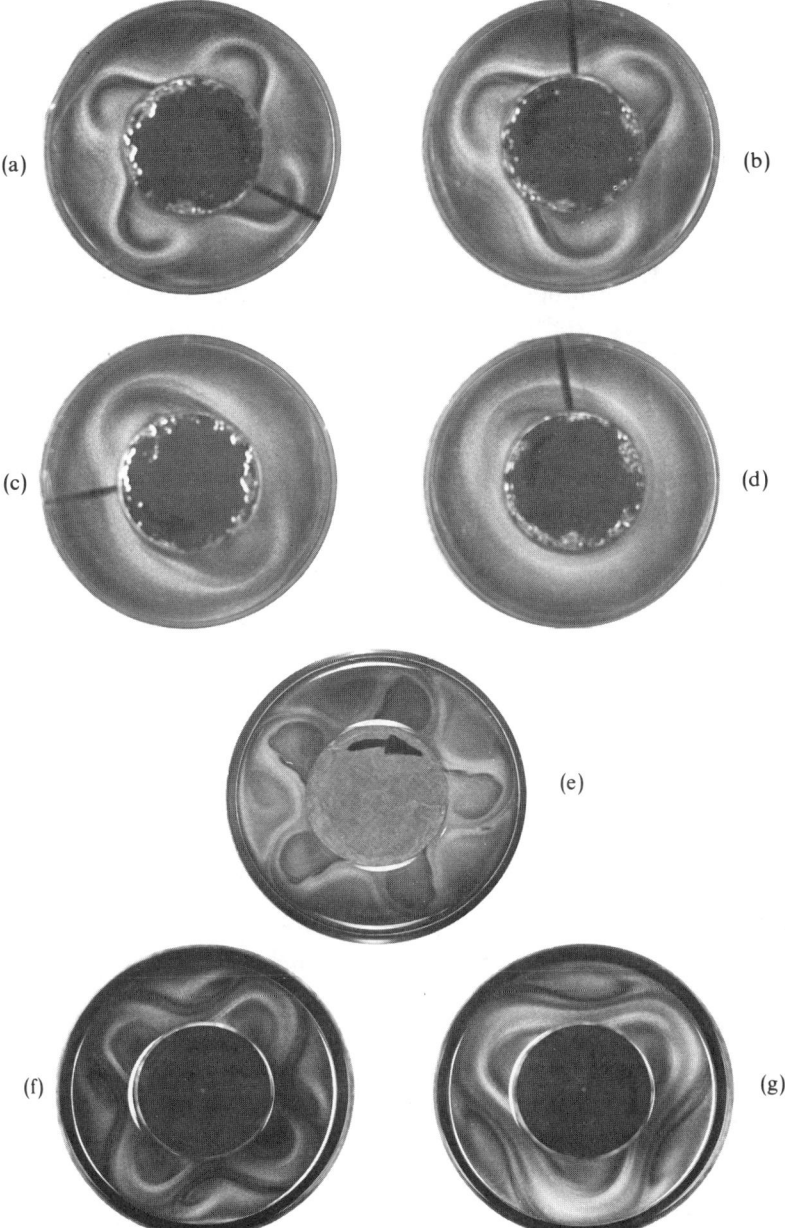

Fig. 18. Waves in a rotating, laterally heated annulus. Top view. Fluid water. Fluid in contact with a lucite lid. Visualization with aluminum powder. Dark lines indicate vertical motion, bright areas predominantly horizontal flow. Direction of rotation indicated by arrow on the inner cylinder. $\Omega = 2.5$ rad sec^{-1}. (a) $\Delta T_r = 14.6°C$, (b) $\Delta T_r = 17.1°C$, (c) $\Delta T_r = 22.0°C$, (d) $\Delta T_r = 30.7°C$, (e) under an air surface; $\Omega = 3.14$ rad sec^{-1}, $\Delta T_r = 9.4°C$, (f) with negative radial gradient, $\Omega = 2.5$ rad sec^{-1} and an air surface, $\Delta T_r = -15°C$, (g) as (f), with $\Delta T_r = -20°C$. (After Koschmieder, 1972.) (Copyright by Cambridge University Press. Used with permission.)

they originate from baroclinic instability, as was first suggested by Lorenz (1956). Details will be discussed later.

The rotating annulus experiment is frequently referred to, in a somewhat uncritical way, as a model of the general circulation of the atmosphere. That is certainly not correct. There is no significant horizontal heat flux through the lateral confines of a hemisphere of the atmosphere, namely through equator and pole, but there is a substantial heat flux from below (cf., Fig. 16). The fluid motions in the rotating annulus are, however, driven exclusively by a horizontal temperature difference. Furthermore, there is still little evidence that the meridional circulation in the midlatitudes and the radial circulation in the annulus are similar. According to the three-cellular model of the general circulation, and in agreement with the unquestionable fact that the surface winds in midlatitudes are westerlies, the meridional circulation over the midlatitudes is that of an indirect cell. Certainly in the symmetric regime the radial circulation in an annulus is that of a direct cell. There are, therefore, easterly fluid motions over the entire bottom of the annulus (if the temperature gradient is positive), while there are westerlies at the surface of the Earth in midlatitudes.

We proceed to the question of whether the waves in the wave regime cause westerly flow over part of the bottom of the annulus (with a positive temperature gradient), in other words, whether the waves set up a three-cellular circulation, as one would expect from the theoretical investigations of baroclinic wave motion mentioned before (Phillips, 1954; Kuo, 1956; Saltzman and Tang, 1972). There seems to be only one good piece of evidence pointing in this direction, namely the numerical investigation of a steady five-wave pattern under a free surface by Williams (1971). As is shown in his Fig. 4b, the zonally averaged flow has then a weak westerly component over the center of the annulus bottom. Note that the free surface has an extraordinarily strong influence on the zonal flow in the annulus. With a free surface only a vestige of the easterly flow in the lower half of the annulus remains in the symmetric solution, see Williams' Fig. 4a. Stronger thermal boundary layers along the side walls in the wave case make it likely that westerly flow, which nearly fills the entire annulus gap, penetrates to the bottom. There exists, as yet, no experimental verification of Williams' (1971) computations. His study makes two assumptions, constant coefficient of thermal expansion and negligible centrifugal effects, which as later experiments have shown do not really hold. Therefore it appears that the question of westerly flow caused by the wave motion has not yet been settled. It appears also that, in order to arrive at a convincing demonstration of the validity of such an important and delicate point, it will be necessary to study this question when the fluid in the annulus is in contact with a rigid lid and the flow is consequently much more symmetric in vertical direction.

If we summarize, we note that there does not seem to be a correct analogy between atmospheric motions and the fluid flow in the rotating annulus. Note that the vast majority of all annulus experiments were made with deep annuli, in which the depth of the fluid is a multiple of the gap width of the annulus. This does not match the shallow atmosphere at all. There are no annulus experiments in which the depth of the fluid was one-fifth of the gap width or less. This is for a good physical reason: It is just not possible to get sufficient thermal energy into the fluid through the walls of a shallow annulus. It would require a radial temperature difference of at least 1000°C in order to have the same amount of heat transferred through a fluid layer with the depth to width ratio of the troposphere, as the amount of heat (per unit area) that passes through the fluid in an annulus experiment. To put this in another way, shallow fluid layers as well as large-scale atmospheric motions depend on vertical heat exchange to maintain vigorous fluid motions. Finally, we should not fail to mention the feature that has perhaps fascinated atmospheric dynamicists most, namely the similarity between (idealized) weather maps [e.g., Fig. 19 taken from Palmén and Newton (1969)] and the corresponding four-wave patterns such as in Fig. 18a.

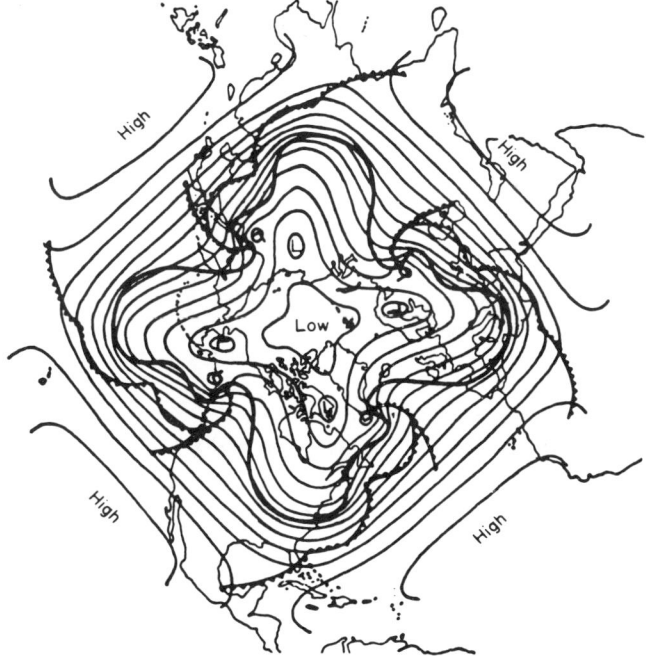

FIG. 19. Schematic circumpolar chart showing a simple atmospheric four-wave pattern. (After Palmén and Newton, 1969.)

The annulus is, however, a proper tool to investigate experimentally the existence of baroclinic instability. It is impossible to discuss here in detail the results of the numerous annulus experiments. A great deal of the evidence has recently been reviewed by Hide and Mason (1975). We discuss here only results which relate to the cause of the waves in the annulus. The first thing to do is to investigate the axisymmetric density circulation in the annulus. Due to complexities introduced by the viscous lateral and horizontal boundary layers a comprehensive analytical description of the symmetric flow has not yet been obtained. Numerical investigations (Williams, 1967; Dietrich, 1973) show that the symmetric meridional circulation takes place essentially in thin boundary layers along the walls, see Fig. 20, the warm fluid rising, of

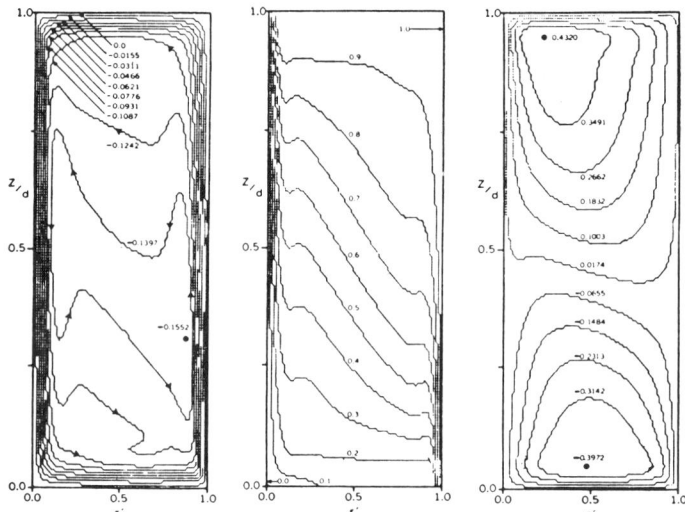

FIG. 20. Streamlines, isotherms, and zonal velocity (left to right) in a rotating, laterally heated annulus with a positive radial temperature gradient and a rigid upper surface. (After Williams, 1967.) (Copyright by the American Meteorological Society. Used with permission.)

course, along the warm wall and sinking at the cold wall. The bulk of the fluid in the interior of the gap is then in geostrophic balance, is stably stratified, has a temperature gradient with a radial and vertical component (Fig. 20), and a zonal velocity increasing with height.

Hence the flow in the interior, away from the boundaries, corresponds largely to the conditions set as the basic state in the theory of baroclinic instability. Disturbing the basic state with infinitesimal azimuthal waves yields then a theoretical stability criterion according to which specific wave numbers are unstable or not. Comparing the results of the stability calculations with the experimentally determined curve for transition from symmetric flows to waves should make it possible to decide the validity of the theoretical results. However, to quote Hide and Mason (1975), the "interpretation of the experimental transition curve on the basis of theoretical studies ... is not strongly supported by the laboratory work." We note that the transition curve is the most accurate result of all experimental studies with the annulus. Actually, the transition curve cannot coincide with the current stability criteria since it has been found experimentally that, contrary to long held firm convictions, the sign of the temperature gradient has a strong effect on the transition from symmetric flow to the wave regime (Koschmieder, 1972, 1978). All current theoretical results are independent of the sign of dT/dr.

Besides the uncertainty about the explanation of the transition curve, there are other features of the experiments which are perplexing from the point of view of baroclinic instability, e.g., the fact that the waves do not drift always as expected backward (with a negative temperature gradient) but sometimes in the opposite direction [provided both ends have a rigid boundary (see Koschmieder, 1978)]. Another question is: Why do the waves form, without exception, at the inner cylinder regardless of the initial condition or the sign of dT/dr? The way an instability forms is about the most characteristic information one can obtain about the nature of the instability. The fact that the waves do not form in the bulk of the fluid but rather at the inner wall points to the boundary layer, which is omitted in the stability analyses. It is, however, an ever reoccurring experience in fluid dynamics that instabilities form in regions of strong shear, as all four boundaries in the annulus are (see Fig. 20). We finally draw attention to the peculiar fact that waves with properties very similar to those of the annulus waves can be observed in a rotating tank filled with a fluid (water) of *uniform* temperature. In an article about detached shear layers in a rotating fluid, Hide and Titman (1967) describe waves driven by differential rotation of a homogeneous fluid. There is not only a remarkable visual similarity to the waves in the rotating annulus, but rather a strong functional similarity. The wave number of the waves observed by Hide and Titman decreased with increased differential rotation, so the transition curves are similar. The waves drift also relative to the frame of reference, as the annulus waves do.

To summarize, it appears that the waves in the annulus have, as yet, not been identified unambiguously as baroclinic waves. Essential features of the symmetric flow have been omitted in the theoretical investigations because of mathematical difficulties. Theoretical predictions based on simplified models of the basic state agree only qualitatively with the results of the experiments. There are some experimental observations concerning the wave regime in the annulus that have, as yet, not been understood at all. To return to the theme of this paper, we are, from the point of view of fluid dynamics, not certain whether a real fluid layer on a rotating planet heated axisymmetrically from below is necessarily unstable to infinitesimal azimuthal disturbances or not.

3.2. Finite Azimuthal Disturbances

Finite azimuthal disturbances which abound on the surface of the Earth must have a profound effect on the general circulation. What are the consequences of the presence of extended steady temperature anomalies on the flow over an otherwise homogeneous surface; in particular, what are the consequences of just one disturbance on an axisymmetric circulation? For example, what kind of effect has an area of, say, the size of Australia whose temperature is, say, 10°C above its surroundings, on the flow over the entire sphere? The theory of standing waves forced by nonuniform heating and topography on a continental scale has been worked on extensively, although many questions remain. See, for example, Saltzman (1968) and Manabe and Terpstra (1974) and references therein. The most relevant observational information concerning individual finite disturbances of atmospheric flow seems to come from the small-scale disturbances. Satellite or spacecraft pictures of insulated islands in the ocean, such as Guadelupe Island off Baja California or Madeira show a trail of vortices in lee of these islands. These vortices are strongly reminiscent of the famous v. Karman vortex street.

An investigation of the effects of large-scale disturbances is of great importance, in particular since the Earth is singled out as the planet with the most and probably strongest disturbances. This is so since no other planet has the ocean-continent contrast. Actually, Jupiter has a surface which must be nearly homogeneous. Its hydrogen-helium atmosphere liquefies at an (undetermined) depth; the basis of the atmosphere is thus similar to an ocean. Temperature variations there may account for some of the irregularities of the flow in the rings. One may, for example, speculate that the Great Red Spot is the consequence of a "hot spot" at the basis of Jupiter's atmosphere. It is well known that hot spots occur in the mantle of the Earth as well. Conditions similar to Jupiter exist on Saturn. Venus has, because of its high surface temperature, no water on its surface. The surface appears to be quite homogeneous although cratered, with surface elevations which do not exceed much more than a kilometer. Summarizing, we note that the planet with the largest surface inhomogeneities (Earth) is also the planet with the most irregular atmospheric motions.

4. Nonlinearity

In the preceding we have repeatedly stressed that the results of the theories concerning convection on a sphere as well as the results of the respective experiments are valid only for conditions which are properly described by linear theory. Linear theory and laminar flow characterize also the rotating annulus experiments. Therefore one can, at best, expect a *qualitative* correspondence between the results of such studies and the phenomena that occur in the atmosphere. Quantitative results, such as an answer to the simple question as to why are there three cells on one hemisphere of Earth and not, say, five or maybe one, cannot be provided from linear theory. It is common knowledge that at present nobody can solve analytically the fully nonlinear Navier–Stokes equation. Present theories of Bénard convection cannot even explain satisfactorily convection under just slightly supercritical conditions. Experiments have shown unambiguously that the wavelength, i.e., the size of supercritical convection cells, increases with increasing Rayleigh number. All theoretical studies predict just the opposite (for details, see Koschmieder, 1974).

Comparing the results of linear studies with turbulent atmospheric motions raises the following fundamental question: Is it at all likely that a fully turbulent flow reproduces in any way qualitatively the original solution of the problem under strictly linear conditions? It is a premise of many theoretical and practically all experimental models that one can infer from the linear case to the fully nonlinear case. In order to verify the validity of this premise it has at least to be shown that such an inference is justified in fluid dynamics experiments. Formidable experimental problems make it extremely difficult to prove the validity of said premise in the case of convection. There is apparently only one experiment (Deardorff and Willis, 1965) in which a two-dimensional convection roll existed at very high Rayleigh number, i.e., under turbulent conditions. Since the lateral walls played a very large role in this experiment, this result is not necessarily conclusive.

There is, on the other hand, a classic fluid dynamics problem where it can be shown convincingly that the solution of the linear problem returns qualitatively under turbulent conditions. That is the Taylor vortex instability, which has many theoretical similarities to the Bénard convection problem. Taylor vortices form at the critical Taylor number in the gap between two vertical concentric cylinders when the inner cylinder is rotated with sufficient speed. The vortices that appear under critical conditions are two-dimensional toroidal rings of perfect axisymmetry. At moderately high supercritical Taylor numbers the vortices become three-dimensional, forming large azimuthal waves. However, at extremely high Taylor numbers the azimuthal waves fade away, and fully turbulent flow in two-dimensional rings around the inner cylinder remains. In this case the linear solution is quite obviously qualitatively restored by fully turbulent flow (for details see Burkhalter and Koschmieder, 1973). This is, of course, only a qualitative statement. Current intensive investigations of two-dimensional turbulence will shed more light on this problem area [see, e.g., Batchelor (1946, 1969) and Lilly (1972), and concerning anisotropic two-dimensional turbulence, see Herring (1975)].

4.1. Numerical Studies

The computer is the only tool available at present to investigate the nonlinear aspects of convective motions on a rotating sphere. Since extensive studies of nonaxisymmetric atmospheric motions on the Earth have been made one would expect that the simple axisymmetric problem has been researched in all detail. However, that is not the case. There are only a very few papers either mentioning or studying the axisymmetric problem and a few more papers investigating the zonally averaged problem, i.e., the case when azimuthal variations are retained but the equations are averaged over all longitudes. In the first numerical study of the general circulation Phillips (1956) found a single, weak, direct meridional cell when the flow was independent of longitude. He did not elaborate on the case further but proceeded to the three-dimensional problem. Zonally averaged studies have been made by Williams and Davies (1965), Kurihara (1970), Saltzman and Vernekar (1971), Wiin-Nielsen and Fuenzalida (1975), and Egger (1975). These studies obtained a variety of results interesting from a meteorological point of view. But a tangible result with regard to the understanding of the basics of axisymmetric circulations has not emerged.

An investigation of the axisymmetric circulation on the Earth using the six-layer NCAR general circulation model has been made by Koschmieder and Walsh (1978). As is well known, the NCAR GCM reproduces satisfactorily the major characteristics of the general circulation. It uses the nonlinear Navier–Stokes equations applied to a compressible, moist medium. Koschmieder and Walsh's study reproduces exactly all features incorporated in the NCAR model, omitting only all terms depending on longitude and omitting humidity. The bottom temperature was $T = T_0 + \Delta T \cos \phi$, ΔT being the equator pole temperature difference. T is independent of time. The upper boundary condition of the NCAR model prohibits a sensible heat transfer through the boundary, and no radiative flux occurs through a dry atmosphere (which does not contain CO_2 and O_3, either). Hence this investigation studied the atmospheric analog to a density circulation of a fluid on a rotating sphere heated nonuniformly but axisymmetrically from below. Heating the atmosphere gradually from $\Delta T = 0$ to $\Delta T = 50°C$ over 50 days (that means increasing ΔT with each iteration by $2 \times 10^{-3}°C$) resulted in a nonsteady circulation as shown in Fig. 21. The circulation was necessarily nonsteady since the viscous and thermal relaxation times are so large that equilibrium is established only after many years. As expected, a direct density circulation was obtained, which reversed,

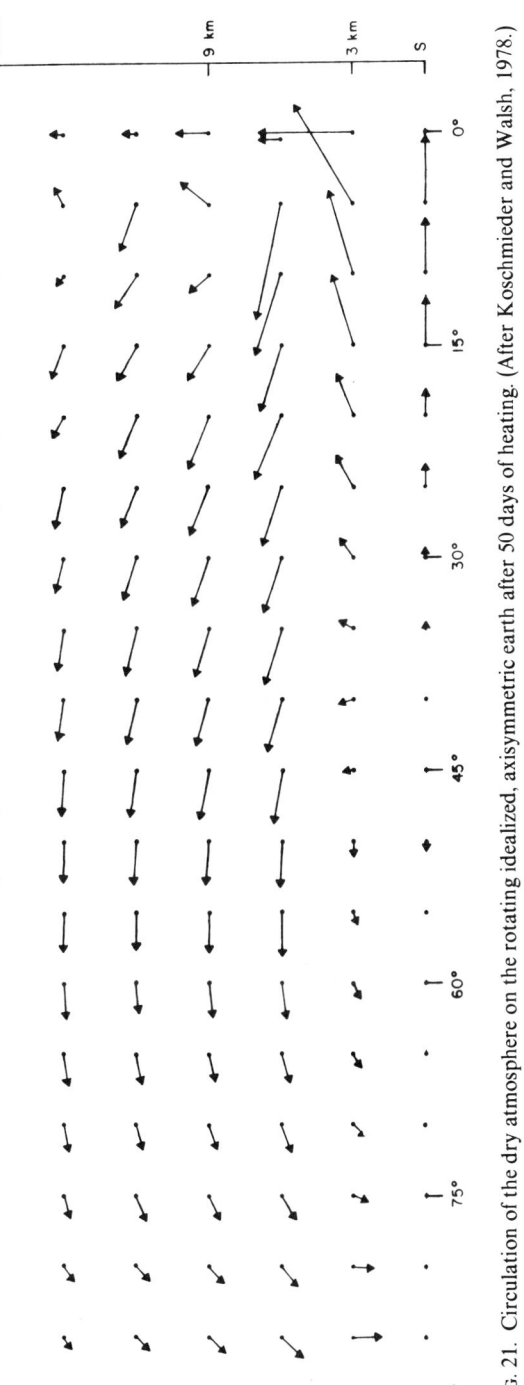

FIG. 21. Circulation of the dry atmosphere on the rotating idealized, axisymmetric earth after 50 days of heating. (After Koschmieder and Walsh, 1978.)

however, into a single indirect cell after 195 days, in order to change again into a direct circulation afterward. No cellular motion occurred up to that time, since the necessary condition for the occurrence of vertical instability was not fulfilled, i.e., the upper boundary condition did not permit an upward heat transfer through the entire depth of the layer. Figure 21 seems to support the notion of the qualitative similarity between large-scale nonlinear atmospheric motions and the laminar motions in fluid dynamics model experiments. It also indicates that the circulation of the atmosphere on the *rotating* Earth is not necessarily three-cellular, if axial symmetry is imposed.

A very thorough numerical investigation of the dynamics of a convectively unstable atmosphere has been made by Williams and Robinson (1973; see also Williams, 1975). This study aims in particular at the understanding of the circulation on Jupiter. These authors study first the classic Bénard problem for a spherical rotating shell in which the fluid layer is heated uniformly from below, is incompressible, and obeys the Boussinesq approximation. The centrifugal force is hence neglected. A number of cases have been solved for laboratory scale parameters, which means for gravitating spheres of radius 10 cm, fluid depth 1.5 cm, and for fluid water, for example. According to these studies the sphere is then covered by different systems of *circular rings*, concentric about the axis of rotation. Onset of convection is suppressed at the poles, due to a higher local Taylor number there; a result consistent with linear theories of convection with rotation but, as we believe, unrealistic because of the omission of the centrifugal force. The paper then proceeds with a corresponding investigation of planetary-scale convection. Parameters are chosen to fit conditions on Jupiter, namely the radius of the sphere, the rotation rate, the depth of the layer, etc. Heating is assumed to be *uniform from below*, meaning that the contribution of the incoming solar light (one-third of total energy) is neglected. Centrifugal force is likewise not taken into account. A basic assumption made concerns the eddy diffusivities, which are assumed to be anisotropic. That means that the horizontal eddy viscosity is between 10^3 to 10^5 times larger than the vertical eddy viscosity. This is done in order to arrive at realistic aspect ratios of the convective cells. The assumption of anisotropy of the eddy diffusivities is consistent with the practice used in the models of the general circulation on Earth, where the horizontal and vertical diffusivities differ also. For a number of parameter combinations Williams and Robinson obtain again axisymmetric circular concentric rings, the details of the rings varying with the parameters. The most realistic theoretical solution had five zones, four belts, and an area of suppressed convection poleward from 45° latitude on. This circulation is shown in Fig. 22 compared with an empirical circulation.

The study of Williams and Robinson is by all means the most thorough

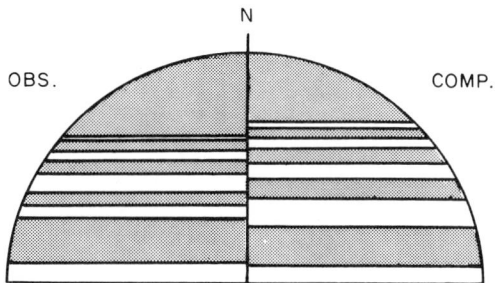

FIG. 22. Comparison of observed (left) and computed bands (right) on Jupiter. (After Williams and Robinson, 1973.) (Copyright by the American Meteorological Society. Used with permission.)

and most advanced investigation of the circulation of Jupiter's atmosphere. The results of their study are in good agreement with our contention that atmospheric motions in the form of axisymmetric rings on a sphere are the consequence of vertical instability.

5. Summary and Conclusions

It is astonishing that after several decades of investigation of the general circulation and, in particular, in view of the extraordinary efforts with the numerical models, we still do not possess a satisfactory mathematical description of the comparatively simple problem of the axisymmetric circulations of a fluid layer on a rotating planet. We seem to have made progress with the physical understanding of the problem. In particular, an explanation of the cells (or rings) of the circulation has emerged. One hears, sometimes, the objection that the understanding of the fluid dynamics aspects of the general circulation is irrelevant for atmospheric problems. However, this objection contradicts the very old experience that science proceeds from the simple to the complex.

There are a substantial number of problems where progress can be made presently with routine methods of fluid dynamics, for example, the question of the density circulation on a rotating planet, or Bénard convection on a rotating sphere. There are a number of problems where basic insight can rapidly be gained if only the numerical models are asked questions of a more physical than meteorological nature. Since further progress with the numerical models might depend very much on a better understanding of the basic physics, such studies will be highly beneficial for progress in the numerical modeling.

There are finally a number of problems where no definite answer is in sight, as long as no better understanding of turbulence has been obtained.

The persistent demand for something better than a mere qualitative similarity will not be met, as long as we cannot solve the nonlinear Navier–Stokes equation.

Acknowledgments

The first version of this manuscript was written while the author enjoyed the hospitality of the Advanced Study Program of the National Center for Atmospheric Research, Boulder, Colorado. It is a pleasure to acknowledge suggestions and comments made by Professor B. Haurwitz, Professor E. N. Lorenz, and Dr. P. A. Gilman. The author is also obliged to Professor R. M. Goody, Dr. C. W. Newton, Dr. G. P. Williams, and Professor J. Zierep for permission to reproduce some of their original figures. Suggestions by Professor J. Van Mieghem and a painstaking review by Professor B. Saltzman are also gratefully acknowledged. Thanks are also due to the Astrophysical Journal, Beiträge zur Physik der Atmosphäre, Journal of Fluid Mechanics, Academic Press, John Wiley and Sons, Oxford University Press, University of Chicago Press, and the American Meteorological Society for permission to reproduce figures for which they hold the copyright.

List of Symbols

- a Wave number
- d Fluid layer depth
- g Gravitational acceleration
- M   Marangoni number, Eq. (2.5)
- P_n^m Legendre polynomial
- R   Rayleigh number, Eq. (2.4a)
- Ri   Richardson number, Eq. (2.10)
- Ro   Rossby number, Eq. (2.9)
- S Surface tension coefficient
- T_N Taylor number, Eq. (2.6)
- u Velocity component in x direction
- v_r Radial velocity component
- v_θ Meridional velocity component
- v_ϕ Azimuthal velocity component
- w Vertical velocity component
- α Volume expansion coefficient
- Δ Laplace operator
- ΔT Vertical temperature difference
- ΔT_r Radial temperature difference
- θ Colatitude
- κ Thermal diffusivity
- ν Kinematic viscosity
- λ Wavelength $\lambda = 2\pi/a$
- ρ Density
- ϕ Longitude
- Ω Rotation rate

References

Alexander, A. F. (1962). "The Planet Saturn." Faber & Faber, London.
Batchelor, G. K. (1946). The theory of axisymmetric turbulence. *Proc. R. Soc., Ser. A* **186**, 480–503.
Batchelor, G. K. (1969). Computation of the energy spectrum in homogeneous two-dimensional turbulence. *Phys. Fluids* **12**, Suppl. II, 223–239.
Belton, M. J. S., Smith, G. R., Schubert, G., and Del Genio, A. D. (1976). Cloud patterns, waves and convection on Venus. *J. Atmos. Sci.* **33**, 1394–1417.
Bénard, H., and Avsec, D. (1938). Travaux récents sur les tourbillons cellulaires et les tourbillons en bandes. *J. Phys. Radium* **9**, 486–500.
Burkhalter, J. E., and Koschmieder, E. L. (1973). Steady supercritical Taylor vortex flow. *J. Fluid Mech.* **58**, 547–560.
Chandrasekhar, S. (1952). The thermal instability of a fluid sphere heated within. *Philos. Mag.* **43**, 1317–1329.
Chandrasekhar, S. (1953a). The onset of convection by thermal instability in spherical shells. *Philos. Mag.* **44**, 233–241.
Chandrasekhar, S. (1953b). The instability of a layer of fluid heated below and subject to Coriolis forces. *Proc. R. Soc., Ser. A* **217**, 306–327.
Chandrasekhar, S. (1957). The thermal instability of a rotating fluid sphere heated within. *Philos. Mag.* **2**, 845–858.
Chandrasekhar, S. (1961). "Hydrodynamic and Hydromagnetic Stability." Oxford Univ. Press (Clarendon), London and New York.
Chapman, C. R. (1969). Jupiters zonal winds: Variation with latitude. *J. Atmos. Sci.* **26**, 986–990.
Charlson, G. S., and Sani, R. L. (1970). Thermoconvective instability in a bounded cylindrical fluid layer. *Int. J. Heat Mass Transfer* **13**, 1479–1496.
Charney, J. G. (1947). The dynamics of long waves in a baroclinic westerly current. *J. Meteorol.* **4**, 135–163.
Charney, J. G. (1973). Planetary fluid dynamics. *In* "Dynamic Meteorology" (P. Morel, ed.), pp. 97–351. *Reidel Publ.* Dordrecht, Netherlands.
Deardorff, J. W., and Willis, G. E. (1965). The effect of two-dimensionality on the suppression of thermal turbulence. *J. Fluid Mech.* **23**, 337–353.
de Rivas, E. K. (1973). Numerical models of the circulation of the atmosphere of Venus. *J. Atmos. Sci.* **30**, 763–779.
de Rivas, E. K. (1975). Further numerical calculations of the circulation of the atmosphere of Venus. *J. Atmos. Sci.* **32**, 1017–1024.
Dickinson, R. E. (1969). The steady circulation of a nonrotating, viscous, heat-conducting atmosphere. *J. Atmos. Sci.* **26**, 1199–1215.
Dietrich, D. E. (1973). A numerical study of rotating annulus flows using a modified Galerkin method. *Pure Appl. Geophys.* **109**, 1826–1861.
Durney, B. (1968). Convective spherical shell: II. With rotation. *J. Atmos. Sci.* **25**, 771–778.
Durney, B. R. (1974). On the sun's differential rotation: Its maintenance by large-scale meridional motions in the convection zone. *Astrophys. J.* **190**, 211–221.
Eady, E. T. (1949). Long waves and cyclone waves. *Tellus* **13**, 33–52.
Egger, J. (1975). A statistical-dynamical model of the zonally averaged steady-state of the general circulation of the atmosphere. *Tellus* **27**, 325–350.
Ferrel, W. (1859). The motions of fluids and solids relative to the Earth's surface. *Math. Mon.* **1**, 140.
Fountain, J. W., and Larson, S. M. (1973). Multicolor photography of Jupiter. *Lun. Planet. Lab., Commun.* No. 175, p. 9.

Fultz, D., Long, R. R., Owens, G. V., Bohan, W., Kaylor, R., and Weil, J. (1959). Studies of thermal convection in a rotating cylinder with some implications for large-scale atmospheric motions. *Meteorol. Monogr.* **4**, No. 21.

Gehrels, T., *et al.* (1975). The imaging photopolarimeter experiment on Pioneer 10. *Science* **183**, 318–320.

Gilman, P. A. (1972). Nonlinear Boussinesq convective model for large-scale solar circulations. *Sol. Phys.* **27**, 3–26.

Gilman, P. A. (1975). Linear simulation of Boussinesq convection in a deep rotating spherical shell. *J. Atmos. Sci.* **32**, 1331–1352.

Goody, R. M. (1964). "Atmospheric Radiation." Oxford Univ. Press, London and New York.

Goody, R. M. (1969). Motions of planetary atmospheres. *Annu. Rev. Astron. Astrophys.* **7**, 303–352.

Goody, R. M., and Robinson, A. R. (1966). A discussion of the deep circulation of the atmosphere of Venus. *Astrophys. J.* **146**, 339–355.

Hadley, G. (1735). Concerning the cause of the general trade winds. *Philos. Trans.* **39**, 58–62.

Hadlock, R. K., Na, J. Y., and Stone, P. H. (1972). Direct thermal verification of symmetric baroclinic instability. *J. Atmos. Sci.* **29**, 1391–1393.

Hanson, K. J. (1976). A new estimate of solar irradiance at the earth's surface on zonal and global scales. *J. Geophys. Res.* **81**, 4435–4443.

Herring, J. R. (1975). Theory of two-dimensional anisotropic turbulence. *J. Atmos. Sci.* **32**, 2254–2271.

Hide, R. (1958). An experimental study of thermal convection in a rotating liquid. *Philos. Trans. R. Soc. London, Ser. A* **250**, 443–478.

Hide, R., and Mason, P. J. (1975). Sloping convection in a rotating fluid. *Adv. Phys.* **24**, 47–100.

Hide, R., and Titman, C. W. (1967). Detached shear layers in a rotating fluid. *J. Fluid Mech.* **29**, 39–60.

Hubert, L. F. (1966). Mesoscale cellular convection. *Meteorol. Satellite Lab. Rep.* No. 37.

Hurle, D. T., Jakeman, E., and Pike, E. R. (1967). On the solution of the Bénard problem with boundaries of finite conductivity. *Proc. R. Soc., Ser. A* **296**, 469–475.

Jeffreys, H. (1926). On the dynamics of geostrophic winds. *Q. J. R. Meteorol. Soc.* **52**, 85–104.

Jeffreys, H., and Bland, M. E. M. (1951). The instability of a fluid sphere heated within. *Mon. Not. R. Astron. Soc., Geophys. Suppl.* **6**, 148–158.

Jeffreys, H., and Bland, M. E. M. (1952). Problems of thermal instability in a sphere. *Mon. Not. R. Astron. Soc., Geophys. Suppl.* **6**, 272–277.

Jones, C. A., Moore, D. R., and Weiss, N. O. (1976). Axisymmetric convection in a cylinder. *J. Fluid Mech.* **73**, 353–388.

Keay, C. S. L., Low, F. J., and Rieke, G. H. (1972). Infrared maps of Jupiter. *Sky Telescope* **44**, 296–297.

Keldysh, M. V. (1977). Venus exploration with the Venera 9 and Venera 10 spacecraft. *Icarus* **30**, 605–625.

Köhler, H. (1970). Differential rotation caused by anisotropic turbulent viscosity. *Sol. Phys.* **13**, 3–18.

Koschmieder, E. L. (1959). Über Konvektionsströmungen auf einer Kugel. *Beitr. Phys. Atmos.* **32**, 34–42.

Koschmieder, E. L. (1965). Convection on a sphere II. *Beitr. Phys. Atmos.* **38**, 23–27.

Koschmieder, E. L. (1966a). Convection on a uniformly heated plane. *Beitr. Phys. Atmos.* **39**, 1–11.

Koschmieder, E. L. (1966b). Convection on a nonuniformly heated plane. *Beitr. Phys. Atmos.* **39**, 208–216.

Koschmieder, E. L. (1967). Convection on a uniformly heated, rotating plane. *Beitr. Phys. Atmos.* **40**, 216–225.

Koschmieder, E. L. (1968). Convection on a nonuniformly heated, rotating plane. *J. Fluid Mech.* **33**, 515–527.
Koschmieder, E. L. (1969). On the wavelength of convective motions. *J. Fluid Mech.* **35**, 527–530.
Koschmieder, E. L. (1972). Convection in a rotating, laterally heated annulus. *J. Fluid Mech.* **51**, 637–656.
Koschmieder, E. L. (1973). Comments on " Direct thermal verification of symmetric baroclinic instability." *J. Atmos. Sci.* **30**, 1705–1706.
Koschmieder, E. L. (1974). Bénard convection. *Adv. Chem. Phys.* **26**, 177–212.
Koschmieder, E. L. (1975). Supercritical Bénard convection and Taylor vortex flow. *Adv. Chem. Phys.* **32**, 109–133.
Koschmieder, E. L. (1978). Convection in a rotating annulus with a negative radial temperature gradient. *Geophys. Astrophys. Fluid Dyn.* **10** No. 3.
Koschmieder, E. L., and Walsh, J. F. (1978). Numerical investigation of the circulation on an idealized axisymmetric earth.
Krishnamurti, R. (1975). On cellular cloud patterns. Part 2: Laboratory model. *J. Atmos. Sci.* **32**, 1364–1372.
Kulacki, F. A., and Goldstein, R. J. (1972). Thermal convection in a horizontal fluid layer with uniform volumetric energy sources. *J. Fluid Mech.* **55**, 271–287.
Kuo, H. L. (1956). Forced and free meridional circulations in the atmosphere. *J. Meteorol.* **13**, 561–568.
Kurihara, Y. (1970). A statistical-dynamical model of the general circulation of the atmosphere. *J. Atmos. Sci.* **27**, 847–870.
Lilly, D. K. (1972). Numerical simulation studies of two-dimensional turbulence. I. Models of statistically steady turbulence. *Geophys. Fluid Dyn.* **3**, 289–319.
London, J. (1957). "A Study of the Atmospheric Heat Balance," AFCRC-TR-57-28. U.S. Air Force Cambridge Res. Lab., Cambridge, Massachusetts.
Lorenz, E. N. (1956). A proposed explanation for the existence of two regions of flow in a rotating symmetrically heated cylindrical vessel. *In* "Fluid Models in Geophysics," pp. 73–80. U.S. Gov. Print. Off., Washington, D.C.
Lorenz, E. N. (1967). The nature and theory of the general circulation of the atmosphere. *W.M.O.* No. 218, TP. 115.
Lorenz, E. N. (1969). The nature of the global circulation of the atmosphere: A present view. *In* "The Global Circulation of the Atmosphere" (G. A. Corby, ed.), pp. 3–23. R. Meteorol. Soc., London.
Manabe, S., and Terpstra, T. B. (1974). The effects of mountains on the general circulation of the atmosphere as identified by numerical experiments. *J. Atmos. Sci.* **31**, 3–42.
Maxworthy, T. (1973). A review of Jovian atmospheric dynamics. *Planet. Space Sci.* **21**, 623–641.
Mintz, Y. (1961). Temperature and circulation of the Venus atmosphere. *Planet. Space Sci.* **5**, 141–152.
Müller, U. (1965). Untersuchungen an rotationssymmetrischen Zellularkonvektionsströmungen. *Beitr. Phys. Atmos.* **38**, 1–8.
Müller, U. (1966). Über Zellularkonvektionsströmungen in horizontalen Flüssigkeitsschichten mit ungleichmässig erwärmter Bodenfläche. *Beitr. Phys. Atmos.* **39**, 217–234.
Murray, B. C., *et al.* (1974). Venus: Atmospheric motion and structure from Mariner 10 pictures. *Science* **183**, 1307–1315.
Newell, R. E., Vincent, D. G., Dopplick, T. G., Ferruzza, D., and Kidson, J. W. (1969). The energy balance of the global atmosphere. *In* "The Global Circulation of the Atmosphere" (G. A. Corby, ed.), pp. 42–90. R. Meteorol. Soc., London.
Nield, D. A. (1968). The Rayleigh-Jeffreys problem with boundary slab of finite conductivity. *J. Fluid Mech.* **32**, 393–398.

Oort, A. H., and Rasmusson, E. M. (1971). Atmospheric circulation statistics. *NOAA* (*Natl. Oceanic Atmos. Adm.*) *Prof. Pap.* No. 5.
Palmén, E., and Newton, C. W. (1969). "Atmospheric Circulation Systems." Academic Press, New York.
Peek, B. M. (1958). "The Planet Jupiter." Faber & Faber, London.
Phillips, N. A. (1954). Energy transformations and meridional circulations associated with simple baroclinic waves in a two-level, quasi-geostrophic model. *Tellus* **6**, 272–286.
Phillips, N. A. (1956). The general circulation of the atmosphere: A numerical experiment. *Q. J. R. Meteorol. Soc.* **82**, 123–164.
Priestley, C. H. B. (1962). Width–height ratio of large convection cells. *Tellus* **14**, 123–124.
Ray, D. (1965). Cellular convection with non-isotropic eddies. *Tellus* **17**, 434–439.
Roberts, P. H. (1967). Convection in a horizontal layer with internal heat generation. Theory. *J. Fluid Mech.* **30**, 33–49.
Saltzman, B. (1968). Surface boundary effects on the general circulation and macroclimate: A review of the theory of the quasi-stationary perturbations in the atmosphere. *Meteorol. Monogr.* No. 30, 4–19.
Saltzman, B., and Tang, C.-M. (1972). Analytical study of the evolution of an amplifying baroclinic wave. *J. Atmos. Sci.* **29**, 427–444.
Saltzman, B., and Vernekar, A. D. (1971). An equilibrium solution for the axially symmetric component of the earth's macroclimate. *J. Geophys. Res.* **76**, 1498–1524.
Sasaki, Y. (1970). Influences of thermal boundary layer on atmospheric convection. *J. Meteorol. Soc. Jpn.* **48**, 492–502.
Sasamori, T. (1971). A numerical study of the atmospheric circulation on Venus. *J. Atmos. Sci.* **28**, 1045–1057.
Schwiderski, E. W., and Schwab, H. J. A. (1971). Convection experiments with electrolytically heated fluid layers. *J. Fluid Mech.* **48**, 703–719.
Sellers, D. W. (1965). "Physical Climatology." Univ. of Chicago Press, Chicago, Illinois.
Stone, P. H. (1966). On non-geostrophic baroclinic stability. *J. Atmos. Sci.* **23**, 390–400.
Stone, P. H. (1967). An application of baroclinic stability theory to the dynamics of the Jovian atmosphere. *J. Atmos. Sci.* **24**, 642–652.
Stone, P. H. (1968). Some properties of Hadley regimes on rotating and nonrotating planets. *J. Atmos. Sci.* **25**, 644–657.
Stone, P. H., Hess, S., Hadlock, R., and Ray, P. (1969). Preliminary results of experiments with symmetric baroclinic instabilities. *J. Atmos. Sci.* **26**, 991–996.
Thirlby, R. (1970). Convection in an internally heated layer. *J. Fluid Mech.* **44**, 673–693.
Torrest, M. A., and Hudson, J. L. (1974). The effect of centrifugal convection on the stability of a rotating fluid heated from below. *Appl. Sci. Res.* **29**, 273–289.
Tritton, D. J., and Zarraga, M. N. (1967). Convection in horizontal layers with internal heat generation. Experiments. *J. Fluid Mech.* **30**, 21–31.
von Tippelskirch, H. (1959). Weitere Konvektionsversuche: Der Nachweis der Ringzellen und ihrer Verallgemeinerung. *Beitr. Phys. Atmos.* **32**, 2–22.
Walton, I. C. (1975). The viscous nonlinear symmetric baroclinic instability of a zonal shear flow. *J. Fluid Mech.* **68**, 757–769.
Wasiutinsky, J. (1946). Studies in hydrodynamics and structure of stars and planets. *Astrophys. Norv.* **4**, 1–497.
Weber, J. E. (1973). On thermal convection between non-uniformly heated planes. *Int. J. Heat Mass Transfer* **16**, 961–970.
Wiin-Nielsen, A., and Fuenzalida, H. (1975). On the simulation of the axisymmetric circulations of the atmosphere. *Tellus* **27**, 199–214.
Wildey, R. L., Murray, B. C., and Westphal, J. A. (1965). Thermal infrared emission of the Jovian disk. *J. Geophys. Res.* **70**, 3711–3719.

Williams, G. P. (1967). Thermal convection in a rotating fluid annulus: Part I and Part II. *J. Atmos. Sci.* **24**, 144–161, 162–174.
Williams, G. P. (1971). Baroclinic annulus waves. *J. Fluid Mech.* **49**, 417–449.
Williams, G. P. (1975). Jupiter's atmospheric circulation. *Nature (London)* **257**, 778.
Williams, G. P., and Davies, D. R. (1965). A mean motion model of the general circulation. *Q. J. R. Meteorol. Soc.* **91**, 471–489.
Williams, G. P., and Robinson, J. B. (1973). Dynamics of a convectively unstable atmosphere: Jupiter? *J. Atmos. Sci.* **30**, 684–717.
Zierep, J. (1958). Eine rotationsymmetrische Zellularkonvektionsströmung. *Beitr. Phys. Atmos.* **30**, 215–222.
Zierep, J. (1959). Zur Theorie der Zellularkonvektion III. *Beitr. Phys. Atmos.* **32**, 23–33.

Note Added in Proof

A study of a symmetric circulation of the atmosphere driven by internal heating can be found in:

Schneider, E. K., and Lindzen, R. S. (1977). Axially symmetric steady-state models of the basic state for instability and climate studies: Part I and Part II. *J. Atmos. Sci.* **34**, 263–279, 280–296.

A SURVEY OF STATISTICAL-DYNAMICAL MODELS OF THE TERRESTRIAL CLIMATE

BARRY SALTZMAN

Department of Geology and Geophysics
Yale University
New Haven, Connecticut

1. Introduction 184
2. Fundamental Equations Governing the Terrestrial Climatic System 185
   2.1 Continuity Equations 186
   2.2 The Equations of Motion 188
   2.3 The Thermodynamical Energy Equation 191
   2.4 Constitutive Equations 193
3. Climatic Averaging 193
   3.1 The Ensemble Monthly Average 193
   3.2 The Climatological-Mean Equations 194
   3.3 Spatial Averaging 202
4. Climate Models—An Overview 206
5. The Vertically Integrated, Thermodynamic Models 225
   5.1 Foundations 225
   5.2 Applications 237
   5.3 Critique 248
6. Momentum Models 250
   6.1 Equations for the Axially Symmetric and Asymmetric Mean States 250
   6.2 Symmetric (Zonal) Models 252
   6.3 Asymmetric (Nonzonal) Models 264
   6.4 The Complete Time Average State 265
7. Modeling the Evolution of Climate 267
   7.1 General Concepts 267
   7.2 Studies of Forced Climatic Change 271
   7.3 Studies of Free Climatic Change 277
   7.4 Climatic Prediction 279
Appendix. A Simple Sea Ice–Ocean Temperature Oscillator Model 281
   A.1 Introduction 281
   A.2 The Ice Limit Equation 282
   A.3 The Surface Heat Balance 283
   A.4 The Mean Ocean Temperature Equation 285
   A.5 Equilibrium Conditions and Constants 286
   A.6 Linear Analysis 288
   A.7 Concluding Remarks 290
List of Symbols 290
References 295

1. Introduction

The last decade has seen an accelerated growth of interest in the Earth's climate, in its variations, and in the possibilities for its management. Included in this growth has been the development of numerous theoretical models aimed at accounting for the present climate and elucidating the mechanisms for climatic change.

It is our goal here to discuss the physical foundation and review the development of an important group of these models known variously as quasi-steady state, equilibrium, mean motion, or, energy balance models, or more inclusively, as statistical–dynamical models. These models are characterized by the fact that they govern directly, as dependent variables, the atmospheric climatic statistics (i.e., monthly or longer normals, normal variances, and higher moments of the probability distributions of the climatic variables, and their spatial averages). On the other hand, the so-called explicit–dynamical numerical models (Egger, 1975), which include the atmospheric general circulation models (i.e., GCMs), govern the "synoptic" variables (i.e., quantities averaged over roughly a half-hour time period). The asymptotic, near-equilibrium solutions for the day-to-day weather variations obtained from these explicit–dynamical models are postprocessed to yield the climate statistics in much the same manner as they are compiled from real weather observations. Thus, the essential difference between the two main groups of models, the statistical– and explicit–dynamical models, lies in the point of truncation of the frequency spectrum of atmospheric variations. In the explicit–dynamical group the equations are designed to govern all phenomena of frequencies lower than and *including* the synoptic cyclone, the effects of all higher frequencies associated with smaller scale motions being parameterized. In the statistical–dynamical group, the equations govern much lower frequency variations (say, of seasonal period and longer) thus requiring that the effects of synoptic frequencies also be parameterized in terms of the macroclimatic fields. The crux of the difficulty of this statistical–dynamical approach is largely this need to formulate physically sound parameterizations for the effects of the synoptic-scale weather phenomena (and even of the seasonal cycle if *annual* means are considered) in terms of the average conditions.

In spite of the difficulties, it is now widely accepted (though this was not the case 10 years ago) that the development of statistical–dynamical models is of great value, not only for the practical reason that they consume much less computer time than the high-resolution models (Smagorinsky, 1974), and hence may represent the only feasible means for integrating over geologic time, but also because the atmosphere may indeed behave in a statistically determinant way, equilibrating quasi-statically to a much more slowly varying ocean and cryosphere that are the main "carriers" of the time dependence of climatic change.

In the next section we shall present a review of the fundamental equations that are the basis of both the explicit- and statistical-dynamical climate models, to be followed in later sections by a discussion of the specializations (e.g., modes of averaging and parameterizations) leading to the statistical-dynamical models. The symbols used throughout the discussion are listed at the end of the review.

Our aim will be to show how these models stem from the most general statements of the hydrothermodynamic laws and fit into a hierarchy of increasing complexity and simulation capability. It is hoped that the levels of approximation inherent in these models will thus become more clear, that one will therefore be better able to judge the validity of the climatic consequences predicted from the models, and that the most promising avenues of approach to improving and extending these models will become more evident.

2. Fundamental Equations Governing the Terrestrial Climatic System

We now set down the general statements of conservation of mass, momentum, and energy, together with the constitutive relations appropriate for all components of the complete climatic system (the atmosphere, hydrosphere, cryosphere, and climatically relevant portions of the lithosphere and biosphere that comprise the land surface). A schematic pictorialization of these components is shown in Fig. 1 (which includes some symbols representing vertical distances, to be discussed later).

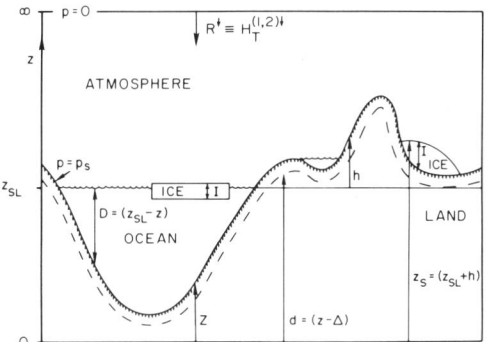

Fig. 1. Schematic representation of the land–ocean–ice–atmosphere climatic system, showing the symbols used to denote height above a subsurface reference level $z = 0$.

We note first that synoptic meteorological and oceanographic measurements of the climatic state variables are usually in the nature of averages over finite time intervals ΔT (e.g., 15–30 minutes) that smooth out small-scale turbulent fluctuations of periods less than ΔT. Accordingly, we shall write the equation in forms applicable to such representative synoptic values, with the effects of eddy stresses due to subsynoptic variations included as complements to the purely molecular stresses.

For this purpose we shall denote the *synoptic average* by $\bar{\psi}^{(s)}$, representing the mean of the instantaneous values ψ over ΔT, the departures from which are denoted by ψ^* (i.e., $\psi = \bar{\psi}^{(s)} + \psi^*$).

Moreover, we shall assume that the Reynolds condition $\overline{\partial \psi^{(s)}/\partial t} \approx \partial \bar{\psi}^{(s)}/\partial t$ is valid. This implies that the dominant periods of variation of the mean $\bar{\psi}^{(s)}$ are much longer than the sampling interval ΔT, which in turn is much longer than the dominant periods of the departures ψ^*, i.e., ΔT falls within a spectral gap in the period spectrum of ψ (Monin and Yaglom, 1971).

2.1. Continuity Equations

Let us consider an arbitrary volume of the climatic system that can, in principle, contain a major carrier constituent (e.g., dry air, liquid water, or continental ice) whose mass concentration is χ_c, plus j minor constituents or phases of matter whose respective mass concentrations are χ_j. The total density of the volume is thus,

$$(2.1) \qquad \rho = \chi_c + \sum_j \chi_j$$

The continuity equations expressing conservation of mass for each constituent can be written in the forms,

$$(2.2a) \qquad \partial \chi_c/\partial t = -\nabla_3 \cdot \chi_c \mathbf{V}_c + \mathscr{S}_c$$

$$(2.2b) \qquad \partial \chi_j/\partial t = -\nabla_3 \cdot \chi_j \mathbf{V}_j + \mathscr{S}_j$$

where $\mathbf{V}_c$ is the instantaneous three-dimensional velocity of the carrier constituent, $\mathbf{V}_j$ is the velocity of the jth constituent, and $\mathscr{S}_{c,j}$ are the source functions giving the rate of production of the constituent in the unit volume due to phase changes and chemical reactions. In general, we shall use the notation $\mathbf{V} = \mathbf{v} + w\mathbf{k}$, where $\mathbf{v} = u\mathbf{i} + v\mathbf{j}$ is the *horizontal* component of velocity.

If we sum (2.2) over all constituents, noting that $\mathscr{S}_c + \sum_j \mathscr{S}_j = 0$ (since the source function represents only a transfer of mass from one constituent form to another) and define the velocity of the center of mass (i.e., the "barycentric" velocity, see deGroot and Mazur, 1962) by

(2.3) $$\mathbf{V}_B = \left(\chi_c \mathbf{V}_c + \sum_j \chi_j \mathbf{V}_j\right)/\rho$$

we obtain

(2.4) $$\partial \rho/\partial t = -\nabla_3 \cdot \rho \mathbf{V}_B$$

Alternatively, we can decompose the velocity of a component substance as follows:

(2.5) $$\mathbf{V}_j = \mathbf{V}_c + \mathscr{V}_j - W_j \mathbf{k}$$

where $\mathbf{V}_c$ is the velocity of the main carrier fluid (e.g., dry air, or liquid water), $\mathscr{V}_j$ the diffusion velocity of the component relative to the carrier current, and W_j the fall or sedimentation speed of a particulate.

In terms of (2.5) the continuity equation for a minor constituent can be written in the form

(2.6) $$\partial \chi_j/\partial t = -\nabla_3 \cdot \chi_j \mathbf{V}_c - \nabla_3 \cdot (\mathscr{D}_j + \mathscr{W}_j) + \mathscr{S}_j$$

where $\mathscr{D}_j = \chi_j \mathscr{V}_j$ is the diffusive flux of the jth constituent and $\mathscr{W}_j = -\chi_j W_j \mathbf{k}$ is the downward sedimentation flux of the jth particulate, and the continuity equation for total density takes the form

(2.7) $$\partial \rho/\partial t = -\nabla_3 \cdot \rho \mathbf{V}_c - \nabla_3 \cdot \sum_j (\mathscr{D}_j + \mathscr{W}_j)$$

We shall now synoptic average (2.2a), (2.6), and (2.7) over ΔT, assuming here and henceforth that the carrier fluids of the climatic system are "Boussinesq," i.e., we may neglect variations of density χ_c^* except insofar as they introduce buoyancy forces in the equations of motion (e.g., Van Mieghem, 1973). We obtain

(2.8) $$\partial \bar{\chi}_c^{(s)}/\partial t = -\nabla_3 \cdot \bar{\chi}_c^{(s)} \bar{\mathbf{V}}_c^{(s)} + \bar{\mathscr{S}}_c^{(s)}$$

(2.9) $$\partial \bar{\chi}_j^{(s)}/\partial t = -\nabla_3 \cdot (\bar{\chi}_j^{(s)} \bar{\mathbf{V}}_c^{(s)} + \mathscr{F}_j) + \bar{\mathscr{S}}_j^{(s)}$$

(2.10) $$\partial \bar{\rho}^{(s)}/\partial t = -\nabla_3 \cdot \left(\bar{\rho}^{(s)} \bar{\mathbf{V}}_c^{(s)} + \sum_j \mathscr{F}_j\right)$$

in which the mass flux of constituent j due to all subsynoptic motions is given by

$$\mathscr{F}_j = \mathscr{E}_j + \bar{\mathscr{D}}_j^{(s)} + \bar{\mathscr{W}}_j^{(s)}$$

where $\mathscr{E}_j = \overline{\chi_j^* \mathbf{V}_c^{*(s)}}$ is the eddy flux of constituent j; $\bar{\mathscr{D}}_j^{(s)} = \overline{\chi_j \mathscr{V}_j}^{(s)}$ is the mean diffusion flux, and $\bar{\mathscr{W}}_j^{(s)} = -\overline{\chi_j W_j \mathbf{k}}^{(s)}$ is the mean sedimentation flux.

Henceforth, we shall simplify the notation by replacing $\bar{\psi}^{(s)}$ by ψ, proceeding with the understanding that all quantities are synoptic averages. Thus (2.8)–(2.10) can be written in the simpler forms

(2.11) $$\partial \chi_c / \partial t = -\nabla_3 \cdot \chi_c \mathbf{V}_c + \mathscr{S}_c$$

(2.12) $$\partial \chi_j / \partial t = -\nabla_3 \cdot (\chi_j \mathbf{V}_c + \mathscr{F}_j) + \mathscr{S}_j$$

and

(2.13) $$\partial \rho / \partial t = -\nabla_3 \cdot \rho \mathbf{V}_c - \nabla_3 \cdot \left(\sum_j \mathscr{F}_j \right)$$

or

(2.14) $$d\rho / dt = -\rho \nabla_3 \cdot \mathbf{V}_c - \nabla_3 \cdot \left(\sum_j \mathscr{F}_j \right)$$

where

$$(\chi_j, \rho, \mathscr{S}_j, \mathbf{V}) \equiv \overline{(\chi_j, \rho, \mathscr{S}_j, \mathbf{V})}^{(s)},$$

and

(2.15) $$d/dt = (\partial/\partial t) + \mathbf{V}_c \cdot \nabla_3$$

The forms for χ_j, $\xi_j = \chi_j \rho^{-1}$ (the mass fraction or mixing ratio of the jth constituent), and $\mathscr{S}_j$ for various constituents of the climatic system, are given in Table I. In the following we shall assume that for nearly all volumes $\chi_c \approx \rho \gg \sum_j \chi_j$ and $\sum_j \mathscr{F}_j \ll \rho \mathbf{V}_c$ so that the continuity equation for total density is simply,

(2.16) $$\partial \rho / \partial t \approx \nabla_3 \cdot \rho \mathbf{V}_c$$

2.2. The Equations of Motion

Applying the same averaging procedures as in Section 2.1, we can write the generalized equation expressing conservation of momentum for the barycentric, *synoptic* motions within any volume of the climatic system (mainly due to the carrier constituent, i.e., $\mathbf{V} \approx \mathbf{V}_c$) in the form,

(2.17) $$\rho (d\mathbf{V}/dt) = \partial \rho \mathbf{V} / \partial t + \nabla_3 \cdot \rho \mathbf{V} \mathbf{V}$$
$$= -\nabla_3 p - \rho g \mathbf{k} - 2\rho \boldsymbol{\Omega} \times \mathbf{V}$$
$$+ \rho \mathbf{G} + \rho \mathbf{E}_c + \rho \sum_j \mathbf{N}_j + \rho \mathbf{F}$$

where we have again assumed $\chi_c \approx \rho$, $p = \bar{p}^{(s)}$ is pressure, g is the acceleration of gravity (assumed constant with height in the climatic system), $\boldsymbol{\Omega}$ is the angular velocity of the earth, $\mathbf{G}$ is the tide-producing external gravitational force per unit mass, $\mathbf{E}_c$ is an arbitrary external force per unit mass acting on the carrier constituents (e.g., the Lorentz electromagnetic force

TABLE I. Forms for $\chi_{c,j}$, $\xi_{c,j}$, and $\mathcal{S}_j$ Appearing in the Continuity Equations (2.11) and (2.12) and Constitutive Relations for Internal Energy (e), and for the Density for Various Components of the Climatic System[a]

Component	c, j	Mass concentration (density) $\chi_{c,j}$	Mass fraction (mixing ratio) $\xi_{c,j}$	Source function $\mathcal{S}_j$	Internal energy e	Equation of state
Atmosphere						
dry air	d	χ_d	—	—	$c_V T$	$\rho_d = p/RT$
water vapor	v	χ_v	$\xi_v\,(=\varepsilon)$	$\mathcal{E}_n$	$c_{vV}(T-273) - 273 R_v + L_v$	$\rho_v = p_v/R_v T$
moist air	A	$\chi_A = \chi_d + \chi_v$	—	$\mathcal{E}_n$	$(1-\varepsilon)c_V T + \varepsilon[L_v - c_{vV} \\ \times(T-273) - 273 R_v]$ \newline $\approx c_V T + \varepsilon L_v$	$\rho_A = p/RT_{\text{virt}} \approx p/RT$
cloud particle	n	χ_n	ξ_n	$-\mathcal{E}_n$	—	—
trace substance	j	χ_j	ξ_j	$\mathcal{S}_j$	—	—
Hydrosphere						
fresh water	w	χ_w	—	$\mathcal{M} - \mathcal{E}$	$c_w(T-273)$	$\rho_w \approx \rho_{w0}(1 - \mu_T \delta T + \mu_p \delta p + \mu_j \delta\xi_j)$
salinity	s	χ_s	$\xi_s\,(=S)$	$\mathcal{S}_s$	—	—
trace substance	j	χ_j	ξ_j	$\mathcal{S}_j$	—	—
sea water	M	χ_M	—	$\mathcal{M} - \mathcal{E}$	$c_{wM}(T-273)$	$\rho_M = \chi_w + \chi_s + \sum \chi_j$ \newline $\approx \rho_{M0}(1 - \mu_T \delta T + \mu_p \delta p$ \newline $+ \mu_s \delta s + \mu_j \delta\xi_j)$
Cryosphere						
ice (including snow)	i	χ_i	ξ_i	$-\mathcal{M}$	$c_i(T-273) - L_f$	$\rho_i \approx \rho_{i0}(1 - \mu_{iT}\delta T$ \newline $+ \mu_{ip}\delta p + \mu_{ij}\delta\xi_j)$
Lithosphere						
land (soil, sand lava, biomass, etc.)	l	χ_l	ξ_l	$\mathcal{S}_l$	$c_l T$	$\rho_l \approx \rho_{l0}(1 - \mu_{lT}\delta T$ \newline $+ \mu_{lp}\delta p + \mu_{lj}\delta\xi_j)$

[a] In the equations of state $\delta\psi$ represents the departure of ψ from the "standard" values embodied in ρ_0.

acting on a charged carrier current in the ionosphere), $\mathbf{N}_j$ is the force per unit mass exerted on the jth constituent, and $\rho\mathbf{F} = \rho\mathbf{F}_{(m)} + \rho\mathbf{F}_{(E)}$ is the frictional force per unit volume that can be subdivided into a part due to molecular viscosity

$$\rho\mathbf{F}_{(m)} = \nabla_3 \cdot \mathbf{P} = \mu[\nabla_3^2 \mathbf{V} + \tfrac{1}{3}\nabla_3(\nabla_3 \cdot \mathbf{V})]$$

($\mathbf{P}$ is the Navier–Stokes tensor, μ the coefficient of molecular viscosity), and a part due to eddy viscosity (i.e., Reynolds stresses),

$$\rho\mathbf{F}_{(E)} = \nabla_3 \cdot \mathbf{R} = -\nabla_3 \cdot \overline{\rho\mathbf{V}^*\mathbf{V}^{*\,(s)}}$$

($\mathbf{R}$ is the Reynolds stress tensor). For the slow movements of solid earth and glacier we may set $\mathbf{R} = 0$ and identify $\mathbf{P}$ with the elastic stress tensor.

For future reference, the scalar forms of (2.17), for motions in the λ, ϕ, and z directions, respectively, are

(2.17a) $\quad \dfrac{\partial u}{\partial t} = -\mathbf{V} \cdot \nabla_3 u + fv - (2\Omega \cos\phi)w + \dfrac{uv \tan\phi}{r}$

$\qquad\qquad - \dfrac{uw}{r} - \dfrac{1}{\rho r \cos\phi}\dfrac{\partial p}{\partial \lambda} + G_\lambda + E_\lambda + \sum_j N_{j\lambda} + F_\lambda$

(2.17b) $\quad \dfrac{\partial v}{\partial t} = -\mathbf{V} \cdot \nabla_3 v - fu - \dfrac{u^2 \tan\phi}{r} - \dfrac{vw}{r}$

$\qquad\qquad - \dfrac{1}{\rho r}\dfrac{\partial p}{\partial \phi} + G_\phi + E_\phi + \sum_j N_{j\phi} + F_\phi$

(2.17c) $\quad \dfrac{\partial w}{\partial t} = -\mathbf{V} \cdot \nabla_3 w + (2\Omega \cos\phi)u + \dfrac{u^2 + v^2}{r} - g$

$\qquad\qquad - \dfrac{1}{\rho}\dfrac{\partial p}{\partial z} + G_z + E_z + \sum_j N_{jz} + F_z$

where $f = 2\Omega \sin\phi$.

In principle, the air-, water-, and ice-borne movements of snow, soil, sand, rock, sediment, and other particulates, are governed by equations similar to (2.17), though in practice these movements are usually treated separately by empirical, kinematical, relationships of a less fundamental nature than (2.17) [for examples, see Bagnold (1954) and Kessler (1969) for formulas relating the movements of sand and raindrops, respectively, to wind speed].

For most considerations of the terrestrial climate we can neglect the external forces ($\mathbf{G}$, $\mathbf{E}$, and $\mathbf{N}_j$) compared to the others, and, for simplicity, we shall usually set these quantities equal to zero.

As shown by Van Mieghem (1973), the mechanical energy equation governing the rate of change of kinetic energy of the *synoptic* motions,

$k = \mathbf{V}^2/2 \equiv \overline{\mathbf{V}^{(s)2}}/2$ (obtained by scalar multiplying (2.17) by $\mathbf{V}$), can be written in the form,

(2.18) $$\frac{\partial \rho k}{\partial t} = -\nabla_3 \cdot [(\rho k + p)\mathbf{V} - (\mathbf{P} + \mathbf{R}) \cdot \mathbf{V}] - g\rho w$$
$$+ p\nabla_3 \cdot \mathbf{V} - d_k - \mathbf{R} \cdot \nabla_3 \mathbf{V}$$

where $d_k = \mathbf{P} \cdot \nabla_3 \mathbf{V}$. Similarly, the equation for the rate of change of the mean kinetic energy of the *subsynoptic* motions, $\kappa = \overline{\rho \mathbf{V}^{*2\,(s)}}/2\rho$ can be written in the form

(2.19) $$\frac{\partial \rho \kappa}{\partial t} = -\nabla_3 \cdot (\rho \kappa \mathbf{V} + \mathbf{A} + \mathbf{C}) + \mathbf{R} \cdot \nabla_3 \mathbf{V} - d_\kappa$$

where

$$\mathbf{A} = \overline{\rho \mathbf{V}^{*3\,(s)}}/2, \qquad \mathbf{C} = (-\overline{\mathbf{V}^* \cdot \mathbf{P}^{*\,(s)}} + \overline{p^* \mathbf{V}^{*\,(s)}})$$

and

$$d_\kappa = (\overline{\mathbf{P}^* \cdot \nabla_3 \mathbf{V}^{*\,(s)}} - \overline{p^* \nabla_3 \cdot \mathbf{V}^{*\,(s)}} + \overline{g\rho^* w^{*\,(s)}})$$

To a good first approximation, we can assume the total kinetic energy in any volume of the climatic system $(k + \kappa)$ is very nearly that of the carrier fluids [i.e., $(k + \kappa) \approx (k_c + \kappa_c)$].

In addition, we have from the definition of the geopotential energy per unit volume of the climatic system, $\rho\Phi = g\rho z = \chi_c \Phi_c + \sum_j \chi_j \Phi_j$, the balance equation,

(2.20) $$\frac{\partial \rho \Phi}{\partial t} = -\nabla_3 \cdot \rho \Phi \mathbf{V} + g\rho w + \sum_j \overline{\Phi_j \nabla_3 \cdot (\mathscr{L}_j + \mathscr{W}_j)}^{(s)}$$

Assuming we can neglect the last term, we can combine (2.20) with (2.18) and (2.19), to obtain the following equation for the rate of change of total mechanical (i.e., kinetic plus potential) energy,

(2.21) $$\frac{\partial}{\partial t}[\rho(\mathscr{K} + \Phi)] = -\nabla_3 \cdot \{[\rho(\mathscr{K} + \Phi) + p]\mathbf{V}$$
$$- (\mathbf{P} + \mathbf{R}) \cdot \mathbf{V} + \mathbf{A} + \mathbf{C}\} + p\nabla_3 \cdot \mathbf{V} - d$$

where

$$\mathscr{K} = (k + \kappa) \quad \text{and} \quad d = (d_k + d_\kappa)$$

2.3. The Thermodynamical Energy Equation

The generalized first law of thermodynamics, expressing conservation of energy for an arbitrary multiphase or multiconstituent volume of the climatic system can be written in the form,

(2.22) $\rho(de/dt) = (\partial \rho e / \partial t) + \nabla_3 \cdot \rho e \mathbf{V}$
$= -p\nabla_3 \cdot \mathbf{V} - \nabla_3 \cdot (\mathbf{H}_{\text{rad}} + \mathbf{H}_{\text{cond}} + \mathbf{H}_{\text{conv}} + \mathbf{H}_{\text{lat}})$
$+ \rho Q + d$

where the internal energy per unit volume is $\rho e \equiv \overline{\rho \bar{e}^{(s)}} \approx \overline{\chi_c e_c} + \sum_j \overline{\chi_j e_j}$. The functional forms of e for various significant constituents of the climatic system are listed in Table I.

In (2.22), $\mathbf{H}_{\text{rad}}$ is the *radiative* flux [which may be decomposed into a short-wave (solar) flux $\mathbf{H}^{(1)}$ and a long-wave (terrestrial) flux $\mathbf{H}^{(2)}$, i.e., $\mathbf{H}_{\text{rad}} = \mathbf{H}^{(1)} + \mathbf{H}^{(2)}$]; $\mathbf{H}_{\text{cond}}$ is the heat flux due to molecular conduction; $\mathbf{H}_{\text{conv}}$ is the heat flux due to all subsynoptic eddy and sedimentation processes given by

$$\mathbf{H}_{\text{conv}} \approx [c_c \overline{(\chi_c \tau_c)^* \mathbf{V}_c^{*\,(s)}} + \sum_j c_j \overline{(\chi_j \tau_j)^* \mathbf{V}_c^{*\,(s)}} - \sum_j c_j \overline{(X_j T_j) W} \mathbf{k}^{(s)}]$$

where $\tau = T - 273$; $\mathbf{H}_{\text{lat}}$ is the flux of latent energy due to all subsynoptic motions, given by

$$\mathbf{H}_{\text{lat}} = L_v(\overline{\chi_v \mathscr{V}_v}^{(s)} + \overline{\chi_v^* \mathbf{V}_c^{*\,(s)}}) - L_f(\overline{\chi_i \mathscr{V}_i}^{(s)} + \overline{\chi_i^* \mathbf{V}_c^{*\,(s)}} - \overline{\chi_i W} \mathbf{k}^{(s)})$$
$= L_v \mathscr{F}_v - L_f \mathscr{F}_i$

and ρQ is the rate of heat addition due to internal (e.g., radiochemical) sources. For convenience we shall define $\mathbf{H}^{(3)} = \mathbf{H}_{\text{cond}} + \mathbf{H}_{\text{conv}}$, representing the combined sensible heat flux due to all subsynoptic motions.

For all modes of mass and heat transfer we shall use the following notation to resolve horizontal and vertical fluxes;

$$\mathscr{F}_j = \mathbf{f}_j + f_j^\uparrow \mathbf{k}$$
$$\mathbf{H} = \mathbf{h} + \mathscr{H}^\uparrow \mathbf{k}$$

where $(\mathbf{h}, \mathbf{f}) = (h, f)_\lambda \mathbf{i} + (h, f)_\phi \mathbf{j}$, and $(f, \mathscr{H})^\uparrow = -(f, \mathscr{H})^\downarrow$.

The heat added per unit volume due to friction is given by

(2.23) $d = d_k + d_\kappa > 0$

where d_k is the rate of dissipation of the synoptic kinetic energy into thermal energy by molecular viscosity, and d_κ is the rate of dissipation of the high-frequency eddy kinetic energy into thermal energy per unit volume.

The energy equation (2.22) can be written in a convenient form by combining it with (2.21) thereby eliminating $(-p\nabla_3 \cdot \mathbf{V} + d)$ to yield the equation of balance for total energy per unit volume,

$$\rho \Upsilon = \rho(\mathscr{K} + \Phi + e) = \chi_c \Upsilon_c + \sum_j \chi_j \Upsilon_j$$

i.e.,

(2.24) $\quad \partial \rho \Upsilon/\partial t = -\nabla_3 \cdot [(\rho \Upsilon + p)\mathbf{V} + \mathbf{D} + \mathbf{H}^{(1)} + \mathbf{H}^{(2)} + \mathbf{H}^{(3)} + \mathbf{H}_{\text{lat}}] + \rho Q$

where $\mathbf{D} = [\mathbf{A} + \mathbf{C} - (\mathbf{P} + \mathbf{R}) \cdot \mathbf{V}] \equiv \mathbf{d} + d^\dagger \mathbf{k}$ represents the flux of mechanical energy due to all subsynoptic processes.

2.4. Constitutive Equations

In order to close the system represented by (2.11)–(2.13), (2.17), and (2.22) and express the internal energy in terms of a measurable quantity—temperature—we have at our disposal two sets of constitutive relationships that can be applied to arbitrary, possibly heterogeneous, masses within the climatic system. These relationships are (1) the expressions for internal energy per unit mass referred to above, of the form,

(2.25) $\quad e = \xi_c e_c(p, T) + \sum_j \xi_j e_j(p, T) = e(p, T, \xi_j)$

and (2) the "equations of state" relating the thermodynamic variables, of the form,

(2.26) $\quad \rho = \chi_c(p, T) + \sum_j \chi_j(p, T) = \rho(p, T, \chi_j)$

Table I contains a list of approximate functional forms for $e(p, T, \xi_j)$ and $\rho(p, T, \chi_j)$ for various pure and heterogeneous components of the climatic system.

In addition, we must either neglect, prescribe, or specify the functional forms (i.e., parameterizations) for $\mathscr{F}_j$, $\mathscr{S}_j$, $\mathbf{F} = \mathbf{F}_{(m)} + \mathbf{F}_{(E)}$, $\mathbf{H}^{(1)}$, $\mathbf{H}^{(2)}$, $\mathbf{H}^{(3)}$, Q, and d, examples of which are the Navier–Stokes relation for $\mathbf{F}_{(m)}$, the associated Stokes dissipation function for the production of heat by molecular viscous processes embodied in d, and the Fourier law of conduction,

$$\mathbf{H}_{\text{cond}} = -k_T \nabla T$$

where k_T is the thermal conductivity. Further relationships of this sort will be necessary for the representation of the other quantities, particularly those representing radiative heating and small-scale eddy processes.

3. Climatic Averaging

3.1. The Ensemble Monthly Average

As we have said, equations (2.11)–(2.13), (2.17), (2.22), (2.25), and (2.26) govern the synoptic, point, values of the climatic variables. In order to specialize these equations to apply to climatological-mean values of the variables, we shall average them over 1 month periods, centered at a running time t for an ensemble sample consisting of several consecutive years (say, 5 years). Such ensemble monthly averages are the basis for the so-called normal charts that represent such a large part of the classical climatic description of the atmosphere.

We define this average by

$$(3.1) \qquad \bar{\psi}(\lambda, \phi, z, t) = \frac{1}{5} \sum_{n=-2}^{2} \left[\frac{1}{T} \int_{(t+nP)-T/2}^{(t+nP)+T/2} \psi(\lambda, \phi, z, t') \, dt' \right]$$

where ψ is an arbitrary climatic variable, $P = 1$ year, $T = 1$ month, t' is a "dummy" time coordinate, and $n = 0, \pm 1, \pm 2$.

With this definition, the departure from the mean,

$$(3.2) \qquad \psi' = \psi - \bar{\psi}$$

is almost completely due to the transient, weather bearing, waves, and vortices of frequencies close to a week, but also includes small effects of seasonal variations within a month as well as of interannual differences in the monthly means.

The choice of a month (or ensemble of months) as a basic sampling interval defining climate is dictated to a large extent by our innate desire to portray the mean atmospheric environmental condition that would be experienced if we average over several "swings" of weather associated with storm passages, at the same time we allow for a portrayal of the pronounced seasonal variation of these mean environmental conditions. In choosing a 5-year ensemble mean we have removed the details of specific month-to-month variations for a given year from consideration, but we have made it more likely that we can use the approximation,

$$(3.2a) \qquad \overline{\partial \psi / \partial t} \approx \partial \bar{\psi} / \partial t$$

in dealing with the longer term (i.e., 10-year and longer) climatic changes in which we are principally interested. In summary, then, equations governing $\bar{\psi}$ clearly will not describe the day-to-day variations of weather and their important transport properties, and they will also have little predictive value for month-to-month forecasts within a given year. In this sense, statistical–dynamical models based on these equations can be classified as low temporal resolution models as distinct from the high resolution explicit–dynamic models.

3.2. *The Climatological-Mean Equations*

When (3.1) and (3.2) are applied to the fundamental synoptic equations developed in Section 2 and we again assume the validity of the Boussinesq approximation applied now to departures of density from monthly mean conditions, we obtain the following system of equations governing the ensemble monthly averages (Van Mieghem, 1973):

Continuity equations:

(3.3) $$\partial \bar{\chi}_c/\partial t = -\nabla_3 \cdot \bar{\chi}_c \bar{V}_c + \bar{\mathcal{S}}_c$$

(3.4) $$\partial \bar{\chi}_j/\partial t = -\nabla_3 \cdot (\bar{\chi}_j \bar{V}_c + M_j + \bar{\mathcal{F}}_j) + \bar{\mathcal{S}}_j$$

(3.5) $$\partial \bar{\rho}/\partial t = -\nabla_3 \cdot \bar{\rho} \bar{V}_c$$

where $M_j = \overline{\chi'_j V'}$ represents the mass transport of constituent j due to transient synoptic eddies in the carrier fluid.

Equation of motion:

(3.6) $$\frac{\partial \bar{\rho} \bar{V}}{\partial t} = -\nabla_3 \cdot \bar{\rho} \bar{V}\bar{V} - \nabla_3 \bar{p} - \bar{\rho}g k - 2\bar{\rho}\Omega \times \bar{V} + \rho \bar{F} - \nabla_3 \cdot J$$

where

$$J = \bar{\rho}\overline{V'V'} = \begin{pmatrix} J^{(\lambda)} \\ J^{(\phi)} \\ J^{(z)} \end{pmatrix} = \bar{\rho} \begin{pmatrix} \overline{u'u'} & \overline{u'v'} & \overline{u'w'} \\ \overline{v'u'} & \overline{v'v'} & \overline{v'w'} \\ \overline{w'u'} & \overline{w'v'} & \overline{w'w'} \end{pmatrix}$$

is the synoptic eddy momentum transport (i.e., the "Jeffreys stress") tensor.

Kinetic energy equations:

(3.7) $$\frac{\partial \bar{\rho} k_m}{\partial t} = -\nabla_3 \cdot [(\bar{\rho} k_m + \bar{p})\bar{V} - (\bar{P} + \bar{R} + J) \cdot \bar{V}]$$
$$- g\bar{\rho}\bar{w} + \bar{p}\nabla_3 \cdot \bar{V} - \bar{P} \cdot \nabla_3 \bar{V} - \bar{R} \cdot \nabla_3 \bar{V} + J \cdot \nabla_3 \bar{V}$$

(3.8) $$\frac{\partial \bar{\rho} \overline{k_E}}{\partial t} = -\nabla_3 \cdot [\bar{\rho}\overline{k_E}\bar{V} + \overline{p'V'} - \overline{(P'+R') \cdot V'}]$$
$$- J \cdot \nabla_3 \bar{V} - \overline{(P'+R') \cdot \nabla_3 V'} + \overline{p'\nabla_3 \cdot V'} - g\overline{\rho'w'}$$

(3.9) $$\frac{\partial \bar{\rho}\bar{\kappa}}{\partial t} = -\nabla_3 \cdot (\bar{\rho}\bar{\kappa}\bar{V} + \bar{A} + \bar{C}) + \bar{R} \cdot \nabla_3 \bar{V} + \overline{R' \cdot \nabla_3 V'} - \bar{d}_\kappa$$

(3.10)
$$\frac{\partial \bar{\rho}\bar{\mathcal{K}}}{\partial t} = -\nabla_3 \cdot [\bar{\rho}(k_m\bar{V} + \overline{k_E V} + \overline{\kappa V}) + \bar{p}\bar{V} + \bar{A} + \bar{C} - \overline{(P+R) \cdot V} + J \cdot \bar{V}]$$
$$- g\overline{\rho w} + \overline{p\nabla_3 \cdot V} - \bar{d}$$

where $\mathcal{K} = k_m + k_E + \kappa$, $k_m = \bar{V}^2/2$, $k_E = \overline{V'^2}/2$, $\kappa = \overline{V^{*2\,(s)}}/2$, and as before, $d = d_k + d_\kappa$.

Potential energy equation:

(3.11) $\quad \partial \bar{\rho}\bar{\Phi}/\partial t = -\nabla_3 \cdot [\bar{\rho}(\bar{\Phi}\bar{\mathbf{V}} + \overline{\Phi'\mathbf{V}'})] + g\bar{\rho}\bar{w} + g\overline{\rho'w'}$

Thermodynamical energy equation:

(3.12) $\quad \dfrac{\partial \bar{\rho}\bar{e}}{\partial t} = -\nabla_3 \cdot (\bar{\rho}\bar{e}\bar{\mathbf{V}} + \mathbf{M}_{\text{conv}} + \mathbf{M}_{\text{lat}}) - \overline{p\nabla_3 \cdot \mathbf{V}} + \nabla_3 \cdot \overline{p'\mathbf{V}'}$

$\quad\quad\quad -\nabla_3 \cdot (\bar{\mathbf{H}}^{(1)} + \bar{\mathbf{H}}^{(2)} + \bar{\mathbf{H}}^{(3)} + \mathbf{H}_{\text{lat}}) + \overline{\rho Q} + \bar{d}$

where

$$\mathbf{M}_{\text{conv}} = \mathbf{m}_{(c)} + m_{(c)}^{\uparrow}\mathbf{k}$$
$$= [c_{cp}\overline{(\chi_c T_c)'\mathbf{V}'} + \sum_j c_{jp}\overline{(\chi_j T_j)'\mathbf{V}'_j}]$$

is the sensible heat (i.e., enthalpy) flux due to the transient synoptic eddies, and

$$\mathbf{M}_{\text{lat}} = \mathbf{m}_{\text{lat}} + m_{\text{lat}}^{\uparrow}\mathbf{k}$$
$$= (L_v \overline{\chi'_v \mathbf{V}'} - L_f \overline{\chi'_i \mathbf{V}'_i})$$
$$= L_v \mathbf{M}_v - L_f \mathbf{M}_i$$

is the latent heat flux due to the transient synoptic eddies.

Total energy equation:

(3.13) $\quad \dfrac{\partial \bar{\rho}\Upsilon_m}{\partial t} = -\nabla_3 \cdot [\bar{\rho}(\bar{e}\bar{\mathbf{V}} + \overline{\mathcal{K}\mathbf{V}} + \overline{\Phi\mathbf{V}}) + \mathbf{M}_{\text{conv}} + \mathbf{M}_{\text{lat}} + \bar{p}\bar{\mathbf{V}} + \mathbf{B}]$

$\quad\quad\quad -\nabla_3 \cdot (\bar{\mathbf{H}}^{(1)} + \bar{\mathbf{H}}^{(2)} + \bar{\mathbf{H}}^{(3)} + \bar{\mathbf{H}}_{\text{lat}}) + \overline{\rho Q}$

where $\Upsilon_m = \bar{\rho}(\bar{\mathcal{K}} + \bar{\Phi} + \bar{e})$ is the total energy of the monthly mean climatic state, and $\mathbf{B} = (\bar{\mathbf{D}} + \mathbf{J} \cdot \bar{\mathbf{V}})$.

Internal energy equations (see Table I):

(3.14) $\quad\quad\quad\quad\quad\quad \bar{e} = f(\bar{T}, \bar{\xi}_j)$

Equations of state (see Table I):
Assuming Boussinesq behavior ($\rho' = 0$),

(3.15) $\quad\quad\quad\quad\quad\quad \bar{\rho} \approx f(\bar{p}, \bar{T}, \bar{\xi}_j)$

Equations (3.3)–(3.15) provide the basic relationships connecting the climatological mean variables and are the basis for all the statistical–dynamical models. In particular Eqs. (3.3)–(3.6), (3.12), (3.14), and (3.15) will comprise a "closed" system providing we either deduce, prescribe, parameterize, or neglect the new synoptic transient eddy stress terms that arise (e.g., $\mathbf{M}_{\text{conv}}$) along with the previously mentioned subsynoptic stress terms.

As written, these equations are quite general, applying to all carrier fluid domains of the climatic system—the atmosphere, ocean, and cryosphere. In addition, they are expressed in an extremely compact vector form that must be rewritten in a scalar form if they are to constitute a model capable of solution by standard methods. The expanded systems, applicable separately to the atmosphere, ocean, and ice sheets are given in Table II. For both atmosphere and oceans we have neglected the gravitational-tidal forces $\mathbf{G}$ and forces $\mathbf{N}_j$ on tracer constituents within these carrier fluids.

Note that to a high degree of accuracy all the carrier fluids are Boussinesq and hydrostatic. Because the atmosphere is compressible and hence density is highly stratified in the vertical [i.e., $\rho_0 = \rho_0(z)$], it is desirable to use the hydrostatic relation to rewrite the equations with pressure p as a vertical coordinate rather than z. For the ocean and ice sheets, on the other hand, ρ_0 is nearly a constant and there is little advantage in using p as a vertical coordinate even though the hydrostatic equation is generally valid in these domains too.

As remarked in the Introduction, much of the difficulty of the statistical-dynamical models lies in the necessity to develop physically sound closure parameterizations for all the stresses appearing in these equations. We shall summarize some of the attempts along this line in Sections 5 and 6. We note, also, that to deduce rather than parameterize these stresses we need to formulate equations for them by first subtracting the averaged equations, e.g. (3.6), from the unaveraged equations, e.g. (2.17), to yield equations for the departures, and then forming the appropriate products to yield prognostic equations for the relevant quadratic stress terms. These new equations will then involve triple products of the eddy departures, which in turn must be deduced, prescribed, parameterized, or neglected to effect closure, and so forth to a higher order if desirable.

A further reduction in temporal resolution is possible by averaging the monthly means over several months (e.g., 3-month "seasons" or 6-month "winter or summer half-years"), and an even greater reduction in resolution is effected by averaging over a year, thus eliminating the seasonal cycle. We define this annual average by

$$(3.16) \qquad \bar{\bar{\psi}}(t) = \frac{1}{P} \int_{t-P/2}^{t+P/2} \bar{\psi}(t') \, dt'$$

and the seasonal departure by

$$(3.17) \qquad \psi'' = \bar{\psi} - \bar{\bar{\psi}}$$

TABLE II. Fundamental Equations Governing the Time-Average State of the Main Carrier Fluids of the Climatic System

i. General equations applying to all domains (atmosphere, oceans, ice sheets and glaciers)

Boussinesq equations of motion

(3.6)
$$\frac{\partial \bar{u}}{\partial t} = -\left(\bar{\mathbf{v}} \cdot \nabla \bar{u} + \bar{w}\frac{\partial \bar{u}}{\partial z}\right) + f\bar{v} - (2\Omega \cos\phi)\bar{w} + \frac{\tan\phi}{r}\bar{u}\bar{v} + \overline{uw}\frac{1}{r} - \frac{1}{\rho_0 r \cos\phi}\frac{\partial \bar{p}}{\partial \lambda} + \bar{E}_\lambda + \bar{F}_\lambda - \left(\overline{\mathbf{v}' \cdot \nabla u'} + \overline{w'\frac{\partial u'}{\partial z}} - \overline{u'v'}\frac{\tan\phi}{r} - \frac{\overline{u'w'}}{r}\right)$$

$$\frac{\partial \bar{v}}{\partial t} = -\left(\bar{\mathbf{v}} \cdot \nabla \bar{v} + \bar{w}\frac{\partial \bar{v}}{\partial z}\right) - f\bar{u} - \frac{\bar{u}^2 \tan\phi}{r} - \frac{\bar{v}\bar{w}}{r} - \frac{1}{\rho_0 r}\frac{\partial \bar{p}}{\partial \phi} + \bar{E}_\phi + \bar{F}_\phi - \left(\overline{\mathbf{v}' \cdot \nabla v'} + \overline{w'\frac{\partial v'}{\partial z}} - \overline{u'^2}\tan\phi\right)$$

$$\frac{\partial \bar{w}}{\partial t} = -\left(\bar{\mathbf{v}} \cdot \nabla \bar{w} + \bar{w}\frac{\partial \bar{w}}{\partial z}\right) + (2\Omega\cos\phi)\bar{u} + \frac{\bar{u}^2+\bar{v}^2}{r} - g\frac{\bar{\rho}}{\rho_0} - \frac{1}{\rho_0}\frac{\partial \bar{p}}{\partial z} + \bar{E}_z + \bar{F}_z - \left(\overline{\mathbf{v}' \cdot \nabla w'} + \overline{w'\frac{\partial w'}{\partial z}} - \frac{\overline{(u'^2+v'^2)}}{r}\right)$$

Continuity equations for trace substances

(3.4) (a) $\quad \frac{\partial \overline{\chi}_j}{\partial t} = -\left(\nabla \cdot \overline{\chi_j \mathbf{v}}_c + \frac{\partial \overline{\chi_j w_c}}{\partial z}\right) - \left[\nabla \cdot \overline{(\chi_j \mathbf{v}_c' + \mathbf{f}_{\lambda j})} + \frac{\partial}{\partial z}\overline{(\chi_j w_c' + f\uparrow_j)}\right] + \mathcal{G}_j$

(b) $\quad \overline{\chi}_j = \overline{\chi}_c \overline{\xi}_j : \frac{\partial \overline{\xi}_j}{\partial t} = -\left(\nabla \cdot \overline{\xi}_j \bar{\mathbf{v}}_c + \frac{\partial}{\partial z}\overline{\xi}_j \bar{w}_c\right) - \nabla \cdot \left(\overline{\xi'_j \mathbf{v}'_c} + \frac{\mathbf{f}_{\lambda j}}{\overline{\chi}_c}\right) - \frac{\partial}{\partial z}\left(\overline{\xi'_j w'_c} + \frac{f\uparrow_j}{\overline{\chi}_c}\right) + \frac{\xi_j}{\overline{\chi}_c}\mathcal{G}_c + \frac{\mathcal{G}_j}{\overline{\chi}_c}$

$(\mathbf{f}_{\lambda j} \approx k^{(i)}\nabla \overline{\chi}_j, f\uparrow_j \approx k^{(i)} \partial \overline{\chi}_j/\partial z)$

Specialized equations for separate domains

Atmosphere (hydrostatic, p system)

(3.6)
$$\frac{\partial \bar{u}}{\partial t} \approx -\left(\bar{\mathbf{v}} \cdot \nabla \bar{u} + \bar{\omega}\frac{\partial \bar{u}}{\partial p}\right) + \left(f + \frac{\tan\phi}{a}\bar{u}\right)\bar{v} - \frac{\partial \bar{\Phi}}{a\cos\phi\,\partial\lambda} + \bar{F}_\lambda - \left(\nabla_3 \cdot \overline{\mathbf{V}'u'} - \frac{\tan\phi}{a}\overline{u'v'}\right)$$

$$\frac{\partial \bar{v}}{\partial t} \approx -\left(\bar{\mathbf{v}} \cdot \nabla \bar{v} + \bar{\omega}\frac{\partial \bar{v}}{\partial p}\right) - \left(f + \frac{\tan\phi}{a}\bar{u}\right)\bar{u} - \frac{\partial \bar{\Phi}}{a\,\partial\phi} + \bar{F}_\phi - \left(\nabla_3 \cdot \overline{\mathbf{V}'v'} + \frac{\tan\phi}{a}\overline{u'^2}\right)$$

$\partial \bar{\Phi}/\partial p = -\bar{\alpha}$

(3.15) $\bar{\alpha} = \dfrac{R\bar{T}}{p}$

(3.12) $\dfrac{\partial \bar{T}}{\partial t} = -\left(\bar{\mathbf{v}} \cdot \nabla \bar{T} + \bar{\omega}\dfrac{\partial \bar{T}}{\partial p}\right) - \dfrac{R}{c_p p}\overline{\omega T} + \dfrac{\bar{q}_A^{1-4}}{c_p} + \dfrac{d_A}{\rho c_{p_A}} - \left(\nabla_3 \cdot \overline{\mathbf{V}'T'} - \dfrac{R}{c_p p}\overline{\omega'T'}\right)$

$q^{(4)} = -L_v\left(\dfrac{d\varepsilon}{dt}\right) + L_f\left(\dfrac{d\xi_{ni}}{dt}\right) = \dfrac{L_v}{\rho}\mathscr{C}_w + \dfrac{L_f}{\rho}\mathscr{C}_i$

(3.3) $\partial\bar{\omega}/\partial p = -\nabla \cdot \bar{\mathbf{v}}$

(3.4) (a) $\dfrac{\partial \bar{\chi}_v}{\partial t} = -\left[\nabla \cdot \left(\bar{\chi}_v \bar{\mathbf{v}} + \overline{\chi_v' \mathbf{v}'} + \mathbf{f}_v\right) + \dfrac{\partial}{\partial z}(\bar{\chi}_v \bar{w} + \overline{\chi_v' w'} + f_v^{\uparrow})\right] - \mathscr{C}$

(b) $\bar{\chi}_v = \bar{\rho}\bar{\varepsilon}$: $\dfrac{\partial \bar{\varepsilon}}{\partial t} = -\left(\bar{\mathbf{v}} \cdot \nabla\bar{\varepsilon} + \bar{\omega}\dfrac{\partial \bar{\varepsilon}}{\partial p}\right) - \left[\nabla \cdot \left(\overline{\varepsilon' \mathbf{v}'} + \dfrac{\bar{\mathbf{f}}_v}{\bar{\rho}}\right) + g\dfrac{\partial}{\partial p}\left(\overline{\bar{\rho}\varepsilon' w'} + f_v^{\uparrow}\right)\right] - \dfrac{\mathscr{C}}{\bar{\rho}}$

Ocean (hydrostatic, z system)

(3.6) $\dfrac{\partial \bar{u}}{\partial t} \approx -\left(\bar{\mathbf{v}} \cdot \nabla \bar{u} + \bar{w}\dfrac{\partial \bar{u}}{\partial z}\right) + f\bar{v} - \dfrac{1}{\rho_{MO} a \cos\phi}\dfrac{\partial \bar{p}}{\partial \lambda} + \bar{E}_{M\lambda} + \bar{F}_{M\lambda} - \nabla_3 \cdot \overline{\mathbf{V}'u'}$

$\dfrac{\partial \bar{v}}{\partial t} \approx -\left(\bar{\mathbf{v}} \cdot \nabla \bar{v} + \bar{w}\dfrac{\partial \bar{v}}{\partial z}\right) - f\bar{u} - \dfrac{1}{\rho_{MO} a}\dfrac{\partial \bar{p}}{\partial \phi} + \bar{E}_{M\phi} + \bar{F}_{M\phi} - \nabla_3 \cdot \overline{\mathbf{V}'v'}$

$\dfrac{\partial \bar{p}}{\partial z} = -\bar{\rho}g, \quad (F_{M\lambda} \approx \nu_M \nabla^2 \bar{u}, F_{M\phi} \approx \nu_M \nabla^2 \bar{v})$

(3.15) $\bar{\rho}_M \approx \rho_{MO}[1 - \mu_T(\bar{T} - T_0) + \mu_S(\bar{S} - S_0)]$

(3.12) $\dfrac{\partial \bar{T}}{\partial t} = -\left(\bar{\mathbf{v}} \cdot \nabla \bar{T} + \bar{w}\dfrac{\partial \bar{T}}{\partial z}\right) + \dfrac{\bar{q}_M^{(1-4)}}{c_M} + \dfrac{d_M}{\rho_w c_M} - \nabla_3 \cdot \overline{\mathbf{V}'T'}$

$q_M^{(3)} \approx \left(\dfrac{k_T}{\rho}\right)_w \nabla_3^2 \bar{T}, \quad q^{(4)} \approx -\rho_M^{-1}(L_v \mathscr{E} + L_f \mathscr{M}_M)$

(Continued)

TABLE II—*Continued*

Specialized equations for separate domains

(3.3) (a) fixed volume

$$\frac{\partial \bar{\chi}_w}{\partial t} = -\left[\nabla \cdot (\bar{\chi}_w \bar{\mathbf{v}}_w + \overline{\chi'_w \mathbf{v}'_w} + \mathbf{f}_w) + \frac{\partial}{\partial z}(\bar{\chi}_w \bar{w}_w + \overline{\chi'_w w'_w} + f^{\dagger}_w)\right] + \overline{\mathcal{M}}_M - \overline{\mathcal{E}}_M$$

(b) interior, incompressible, homogeneous mass

$$\nabla \cdot \bar{\mathbf{v}}_M + \frac{\partial \bar{w}}{\partial z} = 0$$

(3.4) (a) $\dfrac{\partial \bar{\chi}_S}{\partial t} = -\left[\nabla \cdot (\bar{\chi}_S \bar{\mathbf{v}}_M + \overline{\chi'_S \mathbf{v}'_M} + \mathbf{f}_S) + \dfrac{\partial}{\partial z}(\bar{\chi}_S \bar{w}_M + \overline{\chi'_S w'_M} + f^{\dagger}_S)\right] + \bar{\mathcal{G}}_S$

(b) $\bar{\chi}_S = \bar{\chi}_w \bar{S}; \dfrac{\partial \bar{S}}{\partial t} = -\left(\bar{\mathbf{v}} \cdot \nabla \bar{S} + \bar{w} \dfrac{\partial \bar{S}}{\partial z}\right) - \left[\nabla \cdot \left(\overline{S'\mathbf{v}'} + \dfrac{\mathbf{f}_S}{\bar{\chi}_w}\right) + \dfrac{\partial}{\partial z}\left(\overline{S'w'} + \dfrac{f^{\dagger}_S}{\bar{\chi}_w}\right)\right] + \dfrac{\bar{S}}{\bar{\chi}_w}\left[\bar{\mathcal{E}}_M + \nabla_3 \cdot \overline{\mathcal{F}}_w - \overline{\mathcal{M}}_M\right] + \dfrac{\mathcal{G}_S}{\bar{\chi}_M}$

$(\bar{\chi}_M^{-1} \nabla_3 \cdot \overline{\mathcal{F}}_S \approx -\kappa^{(s)} \nabla_3^2 \bar{S})$

Ice sheets and glaciers (hydrostatic, highly "viscous" fluid)

(3.6) $\bar{E}_{i\lambda} + \bar{F}_{i\lambda} - \dfrac{1}{\rho_i a \cos\phi} \dfrac{\partial \bar{p}}{\partial \lambda} \approx 0 \quad (F_{i\lambda} \approx \rho_i^{-1} \nabla \cdot \bar{\mathbf{P}}^{(\lambda)} \approx \nu_i \nabla^2 \bar{u})$

$\bar{E}_{i\phi} + \bar{F}_{i\phi} - \dfrac{1}{\rho_i a}\dfrac{\partial \bar{p}}{\partial \phi} \approx 0 \quad (F_{i\phi} \approx \rho_i^{-1} \nabla \cdot \bar{\mathbf{P}}^{(\phi)} \approx \nu_i \nabla^2 \bar{v})$

$-g - \dfrac{1}{\rho_i}\dfrac{\partial \bar{p}}{\partial z} \approx 0$

(3.15) $\rho_i \approx \rho_{i0}[1 - \mu_{iT}(\bar{T} - T_0) + \mu_{ij}\bar{\xi}_j]$

(3.12) $\dfrac{\partial \bar{T}_i}{\partial t} \approx \dfrac{\bar{q}_i^{(1,4)}}{c_i} + \dfrac{d_i}{\rho_i c_i}, \quad \left[q_i^{(3)} \approx \left(\dfrac{k_T}{\rho}\right)_i \nabla_3^2 T, q_i^{(4)} \approx -\dfrac{L_f}{\rho_i}\mathcal{M}, d_i = \nabla \cdot (\mathbf{P} \cdot \mathbf{V}_i)\right]$

(3.3) (a) fixed volume

$\dfrac{\partial \bar{\chi}_i}{\partial t} = -\left|\nabla \cdot (\bar{v}_i \bar{v}_i + \overline{\chi_i' \mathbf{v}_i'} + \mathbf{f}_i) + \dfrac{\partial}{\partial z}(\bar{\chi}_i \bar{w}_i + \overline{\chi_i' w_i'} + J_i^l)\right| - \mathcal{M}$

(b) interior, incompressible, homogeneous mass

$\nabla \cdot \bar{\mathbf{v}}_i + \dfrac{\partial \bar{w}_i}{\partial z} = 0$

As we shall see, many of the simpler models are designed to govern only such annual quantities, particularly the surface temperature $\overline{\overline{T}}_s$. The equations governing these annual means are obtained by applying (3.16) and (3.17) to the above equations (3.3)–(3.15), and they will be identical in form to these ensemble monthly mean equations, with the following substitutions:

$$\overline{(\ )} \to \overline{\overline{(\ )}}$$
$$\overline{a'b'} \to \overline{\overline{(a'b' + a''b'')}}$$
$$\mathbf{J} \to \overline{\mathbf{J}} + \mathbf{S}$$

where $\mathbf{S} = \overline{\overline{\rho \mathbf{V''V''}}}$ is the seasonal-eddy momentum transport tensor,

$$\mathbf{M}_{\text{conv}} \to \overline{\overline{\mathbf{M}}}_{\text{conv}} + \mathbf{N}_{\text{conv}}$$
$$\mathbf{M}_{\text{lat}} \to \overline{\overline{\mathbf{M}}}_{\text{lat}} + \mathbf{N}_{\text{lat}}$$

where

$$\mathbf{N}_{\text{conv}} = \left[c_{cp}\overline{\overline{(\chi_c T_c)'' \mathbf{V''}}} + \sum_j c_{jp}\overline{\overline{(\chi_j T_j)'' \mathbf{V''}_j}} \right]$$

and

$$\mathbf{N}_{\text{lat}} = L_v \overline{\overline{\chi''_v \mathbf{V''}}} - L_f \overline{\overline{\chi''_i \mathbf{V''}_i}}$$

Note that these annual mean equations will now involve seasonal eddy stresses as well as the synoptic eddy stresses.

3.3. Spatial Averaging

The above equations apply to the values of the climatic variables at all locations in the climatic system. However, we are often most interested in the large spatial scale aspects of climate (i.e., the macroclimate), as distinct from the local variabilities, and to study these aspects the resolution can be still further reduced by applying space averaging in addition to the temporal averaging discussed above. Thus, by suitably defined spatial averaging operators, it is possible to formally resolve the complete field of climatic means into an arbitrary number of components, each of which is identifiable with a prominent observed feature of the total macroclimatic variability.

A major resolution of this kind consists of two components: (1) the zonally averaged (i.e., axially symmetric) mean field representing the North–South climatic variations, and (2) the axially asymmetric departure of the mean field from the zonal average values, representing the East–West variations. To express this resolution in mathematical form (cf. List of Symbols, Modes of Averaging), we first define a zonal average,

(3.18) $$\langle \psi \rangle = \frac{1}{2\pi} \int_0^{2\pi} \psi \, d\lambda$$

and the departure from the zonal average,

(3.19) $$\psi^* = \psi - \langle \psi \rangle$$

which, for the atmosphere, includes the effects of all the large scale, weather-producing, waves and vortices. Applying (3.18) to a climatic mean variable, say $\bar{\psi}$, we have,

(3.20) $$\bar{\psi}(\lambda, \phi, z, t) = \psi_0(\phi, z, t) + \psi_1(\lambda, \phi, z, t)$$

where

(3.21) $$\psi_0 = \langle \bar{\psi} \rangle = \frac{1}{2\pi} \int_0^{2\pi} \bar{\psi} \, d\lambda$$

and

(3.22) $$\psi_1 = \bar{\psi} - \psi_0 = \bar{\psi}^*$$

The zonal average component ψ_0 isolates the primary variation of climate from equator to pole that tends to be forced by the basic axial symmetry of rotation and monthly mean solar radiation. The departure field ψ_1, often called the stationary or standing-wave field in atmospheric studies, isolates all the nonzonal features arising from the nonhomogeneity of the earth's surface (e.g., continent–ocean structure and orography). For a purely homogeneous earth's surface, ψ_1 would vanish.

Just as it is possible to further resolve the time-eddy transients, represented by ψ', into a spectrum giving the contributions of different frequencies to the total time variability, it is also possible to further resolve ψ^* or ψ_1 into one-dimensional Fourier components around latitude circles, or into spherical harmonics. This is especially relevant for the atmosphere, where the instantaneous departures ψ^*, and normal departures ψ_1, have a wavelike appearance, and where, unlike the oceans, a cyclic continuity around latitude circles prevails.

We can further average the climatic variables over all latitudes to obtain a mean value over a complete spherical surface, as a function of height only,

(3.23) $$\tilde{\psi} = \{\{\psi_0\}\} = \{\{\langle \bar{\psi} \rangle\}\} = \frac{1}{4\pi} \int_{-\pi/2}^{\pi/2} \int_0^{2\pi} \bar{\psi} \cos \phi \, d\lambda \, d\phi$$

$$= \frac{1}{\sigma} \int_{-\pi/2}^{\pi/2} \int_0^{2\pi} \bar{\psi} \, d\sigma$$

where $\sigma = 4\pi a^2$ (surface area of the globe) and $d\sigma = a^2 \cos \phi \, d\lambda \, d\phi$ (an element of area). Moreover, if we let σ_j represent the area covered by surficial component j, we can define separate averages for the different component areas as follows,

$$\tilde{\psi}^{(j)} = \frac{1}{\sigma_j} \int \psi \, d\sigma_j \tag{3.24}$$

Thus, for example, we can consider separate averages over the surface area covered by oceans, including sea ice σ_M, and over the surface area covered by land σ_L. Since $\sigma_M + \sigma_L = \sigma$, it follows that,

$$\tilde{\psi} = \left(\frac{\sigma_M}{\sigma}\right)\tilde{\psi}^{(M)} + \left(\frac{\sigma_L}{\sigma}\right)\tilde{\psi}^{(L)} \tag{3.25}$$

Other area averages can be defined based on such further resolutions as

$$\sigma_L = \sigma_{Li} + \sigma_{Lw} + \sigma_{Ll} \tag{3.26}$$

$$\sigma_M = \sigma_{Mi} + \sigma_{Mw} \tag{3.27}$$

where σ_{Li}, σ_{Mi}, σ_{Lw}, σ_{Mw} and σ_{Ll} are the areas of the globe covered by continental ice, sea ice, continental water, ocean water, and unglaciated land, respectively.

To complete the possibilities for spatial averaging, we define a vertical, mass-weighted, average representing a mean value through the depth of the atmosphere, ocean, ice sheet, or land surface,

$$\overline{\overline{\psi}} = \frac{1}{m} \int \psi \, dm \quad \text{or} \quad \frac{1}{\overline{m}} \int \psi \, d\overline{m} \tag{3.28}$$

where dm is an element of mass per unit area given by

$$dm = \begin{cases} g^{-1} \, dp & \text{(hydrostatic atmosphere)} \\ \rho \, dz & \text{(arbitrary subsurface medium)}, \end{cases}$$

m is the mass of the medium per unit area, and the departures are denoted by ψ_* (see List of Symbols). The limits are $p = 0$ to p_s for the atmosphere, and $z = z_1$ to z_2 for the subsurface medium.

In Fig. 1 we show various vertical distances, above an arbitrary subsurface reference level ($z = 0$) located below the deepest part of the ocean, which can serve as the limits for the subsurface vertical averages, z_1 and z_2. These are as follows:

z_{sL} height of mean sea level above the arbitrary subsurface reference level.
Z height of the biolithosphere above the reference level.
d base of the climatic system ($Z - \Delta$), where Δ is an arbitrary fixed depth below which (1) there is no available water in any form, and (2) there is negligible temperature change due to conduction from the lithosphere surface Z.

h height of the earth's surface, including ice and continental water bodies, above sea level.
z_s $(z_{sL} + h)$, height of the lower boundary of the atmosphere above the reference level.
D $(z_{sL} - Z)$, depth of ocean.
I thickness of ice and/or snow.

This type of averaging gives a vertically representative value of the climatic variables as a function of the horizontal coordinates, as well as providing a means of separating purely inertial dynamical processes in the atmosphere and oceans (the vertically averaged barotropic mode) from the thermally driven convective processes related to the departures (baroclinic mode).

When the mass-weighted vertical average is combined with the horizontal area average we obtain the complete mass average, defined by

$$(3.29) \qquad \hat{\psi} = \frac{1}{M}\int_M \psi \, dM = \frac{\sigma}{M}\widetilde{\overline{m\psi}} = \frac{1}{\tilde{m}}\widetilde{\overline{m\psi}}$$

where $M = \sigma\tilde{m}$ is the total mass of the climatic system above $z = d$ (see Fig. 1), and $dM = \rho \, dV = d\sigma \, dm$, where $dV = r^2 \cos\phi \, d\lambda \, d\phi \, dz$ (an element of volume).

A mass average over a specific component of the complete climatic system is denoted by

$$(3.30) \qquad \hat{\psi}^{(j)} = \frac{1}{M_j}\int_{M_j} \psi \, dM$$

where, again, j can be M(ocean), L(continent), i(ice), l(lithosphere), or other subcategories (e.g., sea ice, M_i).

In Fig. 2 we show two hierarchies of possible spatial resolutions of the climatic system, starting with the instantaneous and synoptic fields of climatic variables. These are followed by the time-averaged state representing the three-dimensional field of the complete set of climate variables, $\{\bar{\psi}(\lambda, \phi, z, t)\}$, which may be taken to include the temporal variances and covariances as well as the means. One path (scheme A) then involves first averaging this three-dimensional climatic field *vertically* to yield the sets $\{\widetilde{\bar{\psi}}(\lambda, \phi, t) + \bar{\psi}_*(\lambda, \phi, z, t)\}$, while the other (scheme B) involves first averaging *zonally* to yield the sets $\{\psi_0 + \psi_1\}$. By further successive spatial averaging we can proceed to the lowest resolution description of a planetary climate consisting of the set of characteristic time and mass-averaged values for the entire planet $\{\hat{\bar{\psi}}\}$. The spatial dimensionality of the averages are indicated in the boxes (e.g., $\langle\bar{\psi}\rangle$ is a function of the two space coordinates, ϕ, z). As indicated above, any or all of the entries in this chart can be further

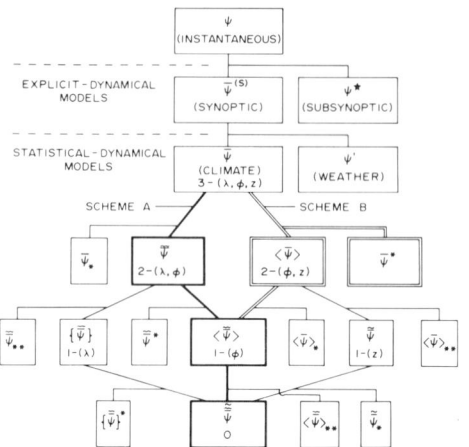

FIG. 2. Averaging hierarchies for the resolution of climate.

averaged over a year or ensemble of years, which in each case would yield an additional component term representing the seasonal departure.

4. CLIMATE MODELS—AN OVERVIEW

Let us begin by defining a climate model as some simplified form of the set of fundamental equations [(2.11)–(2.13), (2.17), (2.22), (2.25), and (2.26)], necessarily including the thermodynamic energy equation, from which the mean temperature and other climatic mean variables can be deduced for a given domain or spatial component of the complete climatic system, subject only to prescriptions of (*i*) the physical and planetary constants, (*ii*) parameterization constants for internal fluxes of heat, momentum, and matter, (*iii*) arbitrary initial conditions, (*iv*) boundary conditions at the limits of the domain under consideration, and, hopefully to a minimal extent (*v*) some observed distribution of climatic variables that are not being deduced (e.g., clouds) or that represent conditions in other domains or components of the complete climatic system. By this working definition we arbitrarily exclude all models in which internal sources and sinks of heat and momentum are prescribed as forcing functions and therefore are not self-determining, although a similar restrictiveness is also inherent to some degree in models using prescribed climatic mean values of quantities not being deduced (item *v* above). We now proceed to discuss some of the possibilities for classifying the climatic models that fall within the framework of the above working definition.

One way of classifying climatic models is in terms of the type and number of averaging operations that are applied. In general, the greater the number of averaging operations, the lower the resolution and the more departures there will be whose effects must be parameterized. Thus, Fig. 2 is in fact a chart showing the hierarchy of climate models, starting at the top with the highest resolution model requiring the least amount of parameterization (i.e., the GCMs) proceeding through intermediate resolution models requiring moderate amounts of parameterization, and ending at the bottom with the lowest resolution model for characteristic planetary values $\tilde{\bar{\psi}}$ requiring the most severe parameterization.

As we have noted, Fig. 2 actually portrays two alternate hierarchal schemes (labeled A and B). Scheme A, on the left, is based on an initial averaging of the mean state in the vertical, followed by zonal and/or meridional averaging. Scheme B, on the right, is based on an initial zonal averaging that separates the three-dimensional state into an axially symmetric (zonal) component ψ_0 and an axially asymmetric (standing eddy) component ψ_1 (as discussed in Section 3; Saltzman, 1961, 1968a) followed by vertical and/or meridional averaging. In this review we discuss the foundation for models embraced by both branches of the hierarchy (Sections 5 and 6).

From one point of view, the various levels of modeling, proceeding upward from the bottom rows to the top in Fig. 2, may be regarded as successive approximations to the solution of the complete climate problem posed by the explicit–dynamical equations, the hope being that from such successive approximations one may cumulatively build up an understanding of the complex climatic mechanisms (e.g., feedbacks and interactions) that are involved in the full problem. From another point of view, it can be argued that some of the more sophisticated lower resolution statistical-dynamical models may already contain the maximum amount of information that is possible to deduce or verify in a very-long-term integration over geologic eras and hence may be optimum simulation models for this purpose.

A complete theory of climate would have to account not only for the averages along any row of Fig. 2, but also for the departures (indicated by primes or asterisks) along that row and along every row above it. Moreover, as seen in Section 3, equations for the averages in any row will involve stresses due to all the departures connected to the lines leading upward from that average to the uppermost box representing the instantaneous values. In practice, however, not all of these departures (and the stresses associated with them) are of equal importance, just as not all the averages are of equal importance (some representing more of the total climate variance than others). In fact, a rather coarse description based on just a few of the mean statistics shown in Fig. 2 may capture a significant part of the total macroclimate variability. Those quantities that have been the subject of most active modeling interest are enclosed in double-lined boxes and connected by double lines.

Perhaps an even more meaningful classification of climate models than one based on the modes of averaging portrayed in Fig. 2 is one based on the physical content of the models—e.g., on the degree to which the full set of conservation statements are used explicitly to deduce the climatic variables and the degree to which all modes of transport and heating are taken into account. To a large extent, however, such physical content is strongly related to the modes of averaging used. For example, vertical averaging ($\tilde{\psi}$), and horizontal averaging over the globe ($\bar{\psi}$) both tend to eliminate the applicability of the equations of motion for deducing the climatic-mean motions [i.e., the mean poloidal (or meridional), toroidal (or zonal), and monsoonal circulations] and their important transport properties, but the energy equation governing the horizontally or vertically averaged temperature is still applicable.

In view of the desirability of taking into account both the physical and formal averaging considerations, as well as the ideas discussed in the Introduction, we propose the following first-order classification of climate models (cf. GARP, 1975; Egger, 1975).

I. *Explicit–Dynamical Models*

These models are based on the synoptic equations, by means of which the day-to-day evolution of the large-scale transient atmospheric eddies are treated in nearly the same detail as in weather forecast models. These can be further divided into two groups.

A. *Truncated Wave Models*, based on a simplified spatial resolution of the evolving synoptic eddies by means of a limited number of harmonics—i.e., a truncated Fourier representation (cf. Lorenz, 1960; Kraus and Lorenz, 1966). In principle, by progressively adding Fourier components this method can be made to approach the full resolution GCMs to be mentioned next. A program of this sort is now being developed at the Geophysical Fluid Dynamics Laboratory in Princeton.

B. *General Circulation Models (GCMs)*, requiring the long-term integration of the three-dimensional synoptic equations that govern the complete spectrum of atmospheric wave activity resolved on daily synoptic weather maps, with the spatial distributions represented either by a full Fourier description or by a gridpoint array.

In both these cases the climate is deduced by statistically operating on the quasi-steady state day-to-day variations obtained as a time-dependent solution, in much the same manner as climatic statistics are obtained from observed day-to-day variations. As we have said, we shall not be concerned with these groups in this review. Excellent recent reviews of these models are given by Smagorinsky (1974) and by Chang (1977).

II. *Statistical–Dynamical Models*

These models are based on the time averaged equations, wherein the effects of large-scale atmospheric and oceanic transient eddies are parameterized in terms of the climatic means (or neglected entirely as a degenerate null form of parameterization). These may be further divided into two main subgroups.

A. *Thermodynamic, or Energy Balance, Models,* based on the continuity equations and thermodynamic energy equation, with varying levels of parameterization of mass and heat fluxes due to all causes including large-scale eddy motions. These models exclude reference to the atmospheric and oceanic equations of motion (except possibly the geostrophic equation) and in particular exclude consideration of momentum fluxes and/or calculation of the mean atmospheric poloidal motions. This group generally embraces all globally and vertically averaged models ($\bar{\tilde{\bar{\psi}}}, \tilde{\bar{\psi}}, \langle \tilde{\bar{\psi}} \rangle, \tilde{\bar{\psi}}$) as well as some models dealing only with zonal averages but containing vertical resolution $\langle \bar{\psi} \rangle$.

B. *Momentum Models* that include, along with the thermodynamic energy equation, the mean equation of motion (3.6), with parameterized momentum transports (Jeffreys stresses) at least in the zonally averaged models. As shown in the diagnostic theoretical study of Kuo (1956), for example, such inclusion of momentum transport is necessary for a reasonably accurate deduction of the atmospheric mean poloidal wind field and its influence on the thermal field. These momentum models generally fall under scheme B of Fig. 2 governing the sets of mean variables $\langle \bar{\psi} \rangle$, $\bar{\psi}*$, and $\bar{\psi}$.

All of the statistical–dynamical models can additionally be resolved according to the following.

(a) The extent and manner of inclusion of horizontal heat fluxes by large-scale eddies.

(b) The spatial coordinates involved as independent variables (i.e., spatial resolution or degrees of freedom), which are closely related to the modes of averaging used.

(c) The climatic domains studied, emphasizing the degree to which separate energy balances are considered for the atmosphere and the subsurface domains (hydrosphere, biolithosphere, and cryosphere).

(d) The main dependent variables and significant feedbacks of the model.

(e) The vertical resolution of the atmospheric variables.

(f) The degree to which the hydrologic cycle and latent heat release are included.

(g) The particular averaging period studied.

(h) The nature of the time dependence, emphasizing whether the solution is of the steady-state (i.e., equilibrium), boundary-value type, or of the time-dependent, initial-value type.

(i) The main application of the model (e.g., simulation, sensitivity, prediction).

(j) The main prescribed parameters that are varied for sensitivity study, and, in addition, for the momentum models only,

(k) The manner of inclusion of horizontal momentum fluxes by large-scale eddies.

In Table III, we provide an index system for categorizing models according to the above features (a)–(k).

In Table IV we attempt to list in chronological order all studies made thus far based on statistical–dynamical models in group IIA (thermodynamic models), and in the zonal-average IIB group (momentum models), respectively, using the indices listed in Table III as a means for classifying and summarizing them. Within the thermodynamic group we shall include the models focusing on the cryospheric domain [sea ice, glaciers, ice sheets, i.e., (c) = 1e in Table III]. These models generally involve some consideration of ice motions, but usually rely on prescribed or highly parameterized atmospheric and ocean effects emphasizing surface mass and thermodynamic balances.

A few further remarks concerning the thermodynamic group (IIA) listed in Table IV, with particular regard to index (b) describing the dimensionality and mode of averaging of the models, are in order. We note that the first category of (b) listed in Table III, of dimension 0 corresponding to spatial averaging over the whole climatic system $\tilde{\tilde{\psi}}$, represents those simplest models by means of which one aims merely to determine the gross, globally averaged, values of variables characteristic of the planetary climatic system (e.g., the effective planetary temperature). For this purpose one can invoke global integral conservation constraints that involve the crudest possible parameterizations of physical processes (e.g., the statement of near equilibrium of long- and short-wave radiation for the planet), or one can bypass physical mechanism altogether by invoking so-called "similarity" theory (Golitsyn, 1970). In this latter approach, characteristic global mean values of the climatic variables are expressed in terms of basic planetary parameters that are assumed to be relevant from purely dimensional arguments (e.g., planetary rotation rate and radius, specific heat). In this review we shall not elaborate, or list in Table IV, the contributions based on this similarity methodology.

The second entry under (b) labeled 1-(z), corresponding to averaging of the type $\tilde{\psi}$, represents the so-called "one-dimensional, vertical coordinate" models in the terminology of Schneider and Dickinson (1974). These models have proved most useful in perfecting atmospheric radiation and vertical-convection modules for ultimate use in more elaborate models, and for preliminary testing of the sensitivity of climate to changes in atmospheric

TABLE III. An index system for the classification of climate models

Index	Description
I	Explicit-dynamical models
A	Truncated wave representation of large-scale atmospheric dynamics
B	Full general circulation models (GCMs)
II	Statistical-dynamical models
A	Thermodynamic or "energy balance" models
B	Momentum models
(a) Horizontal eddy heat flux	
0	No heat flux, $\overline{v'T'} = 0$. "Vertical column" models (Coakley, 1977), either involving no interaction between adjacent latitudes or dealing only with the complete, horizontally averaged earth
1	Newtonian-type heating due to the convergence of the horizontal heat flux, of the form $\nabla \cdot \overline{v'T'} = \beta(\overline{T} - \tilde{T})$, used by Budyko (1969; see also Leovy, 1964)
2	Linear diffusive heat flux, $\overline{v'T'} = -KV\overline{T}$, with a prescribed constant eddy diffusivity K (Defant, 1921) A simple form of this approximation is embodied in the Newtonian-type heating function listed above as (a) = 1
3	Linear diffusive heat flux, with K prescribed as a function of latitude (e.g., Sellers, 1969)
4	Diffusive heat flux, with K deduced as a dependent variable, from baroclinic wave theory, either (a) as a constant leading to a linear formula (cf. Saltzman, 1968; Saltzman and Vernekar, 1971a), or (b) as a function of the temperature field (e.g., the temperature gradient) leading to a nonlinear formula (Saltzman, 1968b; Green, 1970; Stone, 1972, 1973, 1974, see also Clapp, 1970)
5	Diagnostic second-order perturbation method for single amplifying baroclinic wave (Kuo, 1952; Charney, 1959)
6	Prognostic equation for heat flux, with closure at higher order

(*Continued*)

TABLE III—*Continued*

Index	Spatial averaging Usually implied	Description Comments
(b) Spatial resolution of solutions		
0	$\hat{\psi}$	Characteristic "planetary" values, e.g., the "standard atmosphere"
1–(z)	$\widetilde{\psi}$	
1–(λ)	$\{\{\psi\}\}$	
1–(φ) or 1½–(φ)	$\langle\psi\rangle$	½ indicates that continent and ocean are differentiated along latitude circles
2–(λ, φ)	$\widetilde{\psi}$	
2–(λ, z)	$\{\{\psi\}\}$	
2–(φ, z) or 2½–(φ, z)	$\langle\psi\rangle$	
3–(λ, φ, z)	ψ^*, ψ	
(c) Number of *climatic domains* for which separate energy or mass balances are considered		
1a		Complete climatic system
b		Atmosphere only
c		Ocean only
d		Complete subsurface (land and ocean) only, not necessarily in local equilibrium
e		Cryosphere only (sea ice, glaciers, ice sheets)
2a		Separate atmosphere, and subsurface in local equilibrium (equivalent to zero heat capacity subsurface)
b		Separate atmosphere, and subsurface not necessarily in local equilibrium
3		Separate atmosphere, subsurface mixing layer, and deep ocean

(d) Main dependent variables and feedbacks

T	Temperature
$\mathbf{V} = \mathbf{v} + w\mathbf{k}\ (\mathbf{v} = u\mathbf{i} + v\mathbf{j})$	Velocity
$\xi_v\ (\equiv \varepsilon)$	Water vapor mixing ratio
$\xi(x)$	Trace substance mixing ratio [e.g., $x = CO_2$ (carbon dioxide), O_3 (ozone), $\mathscr{A}$ (aerosol) cfc (chlorofluorocarbons)]
r_v	Relative humidity
$\mathscr{C}$	Cloud parameters (e.g., liquid water concentration, fraction of sky covered, depth, type)
P	Precipitation rate
E	Evaporation rate
σ_i	Area covered by ice and snow
m_i	Mass of ice per unit area
$v(\xi_v)$	Water vapor → long-wave emissivity feedback
$r_a(\mathscr{C})$	Cloud → albedo feedback
$r_s(\sigma_i)$	Ice → albedo feedback

Note: the subscripts A, M, L, and S may be applied to the above to signify atmosphere, ocean, land, and surface values, respectively

(e) Vertical resolution of atmospheric variables

1	One level only (e.g., surface, or mid-tropospheric level)
2	Two levels
3	Multilevel
4	Continuous analytical

(f) Hydrology

0	No explicit hydrologic cycle or latent heat release
1	No hydrologic cycle, but evaporation (E) and associated latent heat of condensation calculated for subsurface
2	No hydrologic cycle, but net latent heat release of condensation calculated for complete column (E–P), using water vapor continuity equation for atmosphere
3	Hydrologic cycle and latent heat release calculated for atmosphere and surface (E–P): (a) for vapor phase only, (b) for vapor and ice phases

(Continued)

TABLE III—*Continued*

Index	Description
(g) Averaging period	
1	Annual mean
2	Winter–summer half-year mean, or seasons
3	Monthly mean
(h) Time-dependence	
1	Steady-state solution (emphasizing the study of comparative equilibria under various boundary and parameter prescriptions) obtained (a) as a pure boundary value problem, or (b) as an asymptotic mixed boundary and initial value problem
2	Time-dependent seasonal evolution of monthly means deduced (annual oscillation)
3	Long-term evolution deduced (a) as a sequence of quasi-equilibrium solutions forced externally or (b) as a time-dependent solution usually carried mainly by the ocean and ice domains, the thermal inertia of the atmosphere being negligible on climatic time scales
(i) Application of model	
1	Simulation of presently observed state
2	Idealized study of special mechanisms and sensitivity, holding many factors constant, as problem in (a) comparative equilibria, or (b) time-dependent behavior
3	Deduction of past, or prediction of future, climatic change

(j) Main prescribed parameters varied
    for sensitivity study

S_o — Solar constant
(e, ε, Π) — Milankovitch orbital elements measuring eccentricity, obliquity, and precession, respectively
$\xi(x)$ — Trace substance composition of the atmosphere [see (c)]
r_v — Relative humidity of atmosphere
     — Cloud parameters
σ_i — Area covered by ice and snow
k — Conductive capacity of subsurface medium
$h(\lambda, \phi)$ — Height of surface topography
$L/M(\lambda, \phi)$ — Continent–ocean distribution
K — Large-scale eddy diffusivity

(k) Meridional momentum flux

0 — No momentum flux, $\overline{u'v'} = 0$
1 — Diffusive approximation, $\overline{u'v'} = -K_u \, \partial \langle \bar{u} \rangle / \partial \phi$ (essentially equivalent to a Rayleigh viscosity; this is unacceptable for large-scale atmospheric behavior (Starr, 1968)
2 — $\overline{u'v'} = c \langle u \rangle$ (Lahiff, 1975)
3 — $\overline{u'v'} = -K \, \partial \langle T \rangle / \partial \phi$ (Davies and Oakes, 1962)
4 — Diagnostic relations to surface zonal wind stress (Smagorinsky, 1964)
5 — $\overline{u'v'} = T_c \langle v'^2 \rangle \cos \phi \, d\mu / d\phi$, μ = angular phase speed of baroclinic eddies (Saltzman and Vernekar, 1968)
6 — Relation based on combined diffusive approximation for potential vorticity and sensible heat (Green, 1970; see also Wiin-Nielsen and Sela, 1971)
7 — Diagnostic second-order perturbation method for single amplifying baroclinic wave (Charney, 1959)
8 — Prognostic equation for momentum flux, with closure at higher order

TABLE IV. SURVEY OF CONTRIBUTIONS TO THE THEORY OF GLOBAL CLIMATE BASED ON STATISTICAL–DYNAMICAL MODELS[a]

A. Thermodynamic (Energy-Balance) Models

Year	Authors	(a)	(b)	(c)	(d)	(e)	(f)	(g)	(h)	(i)	(j)	Comments
1921	A. Defant	2	1–(ϕ)	1b	T_A	1	0	1	1a	1	—	First use of Austausch coefficient K for large-scale horizontal heat flux.
1926	Ångström	2	1–(ϕ)	1b	T_s	1	0	1	1a	1	($\epsilon, \epsilon_c, \Pi$)	Radiative equilibrium
1930	Milankovitch	0	1–(ϕ)	1b	T_s	1	0	1	3a	2a	—	Latent heat prescribed
1950	F. Defant	2	2–(ϕ, z)	1b	T_A	4	0	1	1a	1	—	
1953	Hess and Frank	2	1–(ϕ)	1b	T_s	1	0	3	2	1	$\epsilon, \xi(CO_2)$	
1956	Plass (a,b)	0	1–(z)	1b	T_s	1	0	1	1a	2a	$\epsilon, \xi(CO_2)$	
1960	Fritz	2	1–(ϕ)	1b	T_A	1	0	3	2	2a	S_0	Effect of variations of eddy diffusivity K as function of radiation gradient discussed
1961	Manabe and Möller	0	2–(ϕ, z)	1b	$T_{A,s}$	3	0	3	1b	2a	$\epsilon, \xi(CO_2, O_3)$	See references for earlier work (e.g., Emden, 1913)
1962	Adem	2	1–(ϕ)	1b	T_A	1	0	3	1a	1	K	
1963	Adem	2	2–(ϕ, z)	2b	$T_{A,s}$	2	2	3	2	1, 3	K	
1963	Möller	0	0	1b	T_s	1	0	1	1a	2a	$\epsilon, \xi(CO_2), \mathscr{C}$	Noted positive feedback of water vapor and temperature
1963	Nye	—	2–(λ, ϕ, z)	1c	m_i	—	0	1	3b	3	—	Model of *glacier* variations
1964	Adem (a,b)	2	3–(λ, ϕ, z)	2b	$T_{A,s}$	2	2	3	2	1, 3	$\xi(O_3)$	
1964	Manabe and Strickler	0	2–(ϕ, z)	1b	$T_{A,s}$	3	0	3	1b	1, 2a	—	Convective adjustment introduced
1964	Ohring and Mariano	0	0	1b	T_s	1	0	1	1a	2a	$\mathscr{C}$	
1964	Weertman	—	2–(ϕ, z)	1c	m_i	—	0	1	3b	3	—	Model of ice sheet variation (see also Nye, 1960)
1965	Adem	2	3–(λ, ϕ, z)	2b	$T_{A,s}$	2	2	3	2	3	—	
1967	Manabe and Wetherald	0	1–(z)	1b	$T_{A,s}$	3	0	1	1b	2a	$r_r, \xi(CO_2, O_3)\mathscr{C}, r_s$	
1967	Saltzman	0	1–(λ)	1d	T_s	1	1	3	1	1, 2a	$k, L/M$	Use of surface heat balance condition to determine T_s

Year	Author											Notes
1968	Bryson	0	0		T_e	1	0	1	1a	2a	$\xi(\mathscr{A}, CO_2), r_s$	
1968	Saltzman	4a, b	2-(ϕ, z)	2b	$T_{A,s}(v^{-2})_b, (T_s^2)_b$	2	2	2	1a	1, 2a	r_s, k, w	K deduced, not prescribed; thermally active subsurface introduced
1968	Shaw and Donn	2			Adem (1965) model							Earth-orbital parameters at 22,000 BP
1969	Budyko (also 1970, 1972)	1	1-(ϕ)	1a	$T_s, (\sigma_i, r_s)$	1	0	1	1a, b	2a	$S_0, (e, \varepsilon, \Pi)$	Ice-albedo feedback introduced
1969	Sellers (also 1970a,b)	3	1-(ϕ)	1a	$T_s, (\sigma_i, r_s)$	1	2	1	1a	2a	$S_0, r_s, \xi(\mathscr{A}), \varepsilon$	Ice-albedo feedback (see Bowling, 1971)
1970	Adem (a,b, also 1971, 1973, 1975)	2	2-(λ, ϕ)	1c	T_M	1	2	3	2	3	—	Mixed ocean layer model
1971	Barrett	0	0	1b	T_s	3	0	1	1a	2a	$\xi(\mathscr{A})$	Latitudinal distribution of seasonal surface irradiance discussed
1971	Kurihara	0	1-(ϕ)	1b	T_A	2	0	3	2	1, 2a	—	Survey of $\xi(CO_2)$ sensitivity studies
1971	Manabe	0										
1971	Maykut and Untersteiner (also 1969)	—	1-(z)	1e	m_i	—	1	3	2	1, 2	Snowfall	Model of sea ice variation
1971	Rasool and Schneider	0	0	1b	T_s	3	0	1	1a	2a	$\xi(CO_2, \mathscr{A})$	For discussion of relative heating or cooling effects of aerosols, see also Charlson and Pilat (1969, 1971); Atwater (1970); Mitchell (1971); Schneider (1971)
1971	Shaw and Donn	2			Adem (1965) model							Response to ice-free arctic
1972	Faegre	2	1-(ϕ)	1a	$T_s, (\sigma_i, r_s)$	1	2	1	1a	2a	S_0	r_s continuous function of T_s
1972	Kubota	*	2½-(ϕ, z)	2b	$T_{A,s}$	2	3a	3	2	1, 2a	$\mathscr{C}, r_s$	*Divergence of total energy flux proportional to latent heat release
1972	Schneider	0	0, 1-(ϕ)	1b	T_s	1	0	1, 3	1a	2a	$\mathscr{C}$	Rasool and Schneider (1971) model

(Continued)

TABLE IV—Continued

Year	Authors	(a)	(b)	(c)	(d)	(e)	(f)	(g)	(h)	(i)	(j)	Comments
1972	Stone	4b	0	1b	$T_A, \frac{\partial T}{\partial z}, \frac{\partial T}{\partial \phi}, R_i$	1	0	1	1a	2a	S_0	Parameterization of large-scale eddy stresses presented
1972	Yamamoto and Tanaka	0	1–(z)	1b	T_s, ε	3	0	1	1a	1, 2a	$\xi(\mathscr{A})$	
1973	Dwyer and Petersen	3	1–(φ)	1a	T_s	1	2	1	1b	2a	S_0, D	
1973	Petukhov and Feygel'son	0	1–(z)	2b	$T_M, T_{As}, \varepsilon, \xi_n$	1	3a	1, 2	1, 3b	2a, b	$|V_A|$	Long period ocean-atmosphere interactions with clouds as feed-back regulator
1973	Schneider and Gal-Chen	1, 2	1–(φ)	1a	$T_s, (\sigma_i, r_s)$	1	2	1	1b	2a	S_0, init. condit.	Comparison with Budyko, Sellers, and Faegre parameterizations
1973	Sellers (also 1974)	4b	1½–(φ)	1a	$T_s, u_s, v_s, (E-P), (\sigma_i, r_s)$	1	2	3	1b	1, 2a	$\xi(CO_2, \mathscr{A}), S_0$	Extension of Stone (1972) paper—improved parameterizations
1973	Stone	4b	0	1b	$T_A, \frac{\partial T}{\partial z}, \frac{\partial T}{\partial \phi}, R_i$	1	0	1	1a	1, 2a	R_0, K	
1974	Cess	0	0	1b	T_s	4	0	1	1a	2a	$\mathscr{C}, T_e$	Modification of Budyko (1970) parameterizations
1974	Gordon and Davies	1	2–(φ, z)	1b	$T_{A,s}, (\sigma_i, r_s)$	2	0	1	1a	2a	S_0	
1974	Held and Suarez	1, 4b	1–(φ)	1a	$T_s, (\sigma_i, r_s)$	1	0	1	1b	2a	S_0, K	Analytical study of stability and sensitivity of Budyko (1969) and Sellers (1969) models
1974	Paltridge	0	0	1b	$T_{Ms}, \mathscr{C}, P$	1	3a	1	1a	2a	S_0	
1974	Petukhov	0	0	2b	$T_{Ms}, T_{As}, \varepsilon, \xi_n^T, \mathscr{C}$	1	3a	1	1, 3b	2a, b	$|V_A|$	Extension of Petukhov and Feygel'son (1973)
1974	Reck	0	1–(z)	1b	T_s	3	0	3	1b	2a	$\xi(\mathscr{A}), r_s$	Extension of Manabe-Stricker/Wetherald models
1974	Wang and Domoto	0	1–(z)	1b	$T_{s,A}$	3	0	1	1a	2a	$\xi(\mathscr{A})$	

Year	Author											Remarks
1974	Weare and Snell	0	0	1b	$T_s, \ell, (\sigma_s, r_s)$	4	0	1	1a	2a	$\xi(\mathscr{A}, CO_2), S_0$	Cloud feedback included; see also Henderson–Sellers (1975) and Henderson–Sellers and Weare and Snell (1975)
1975	Cess	0	0	1b	T_s	3	0	1	1a	2a	$\dfrac{\partial T}{\partial z}$, $\mathscr{C}$-height	
1975	Chýlek and Coakley	1	$1-(\phi)$	1a	$T_s, (\sigma_s, r_s)$	1	0	1	1a	2a	S_0	Analytical analysis of stability of Budyko (1969) model
1975	North (a,b)	2	$1-(\phi)$	1a	$T_s, (\sigma_s, r_s)$	1	0	1	1a	2a	S_0, K	Analytical analysis of stability of Budyko (1969) and Sellers (1969) models
1975	Paltridge	2	$1-(\phi)$	2b	$T_s, \mathscr{C}$	1	3a	1	1a	1, 2a	S_0	Closure by a "principle of minimum entropy exchange"
1975	Ramanathan	0	0	1a	T_s	3	0	1	1a	2a	$\xi(c|c)$	
1975	Reck (a,b)	0	0	1b	T_s	3	0	3	1a	2a	$\xi(\mathscr{A})$	Rasool and Schneider (1971) model applied to polar conditions
1975	Schneider											Survey of $\xi(CO_2)$ sensitivity studies
1975	Schneider and Mass	0		1a	$T_{s,e}$	1	0	1	3b	3	$S_0, \xi(\mathscr{A})$, sunspots	
1975	Sellers and Meadows	0	0	1a	T_s	1	0	1	1a	2a	L/M	
1975	Temkin et al.	0	0	1b	T_s	3	0	1	1a	2a	$\mathscr{C}, r_{cs}$, solar zenith angle	
1975	Vernekar	0	$2-(\lambda, \phi)$	1d	T_s	1	1	3	1	1, 2a		Two-dimensional extension of Saltzman (1967)
1976	Bryson and Dittberner (also 1977)	0	0	1d	T_s	1	0	1	3b	3	$\xi(\mathscr{A}, CO_2)$	Variations over last 80 years studied (see comments by Woronko, 1977)
1976	Coakley and Grams	0	0	1b	T_s	1	0	1	1a	2a	$\xi(\mathscr{A}$-stratosphere)	
1976	Frederiksen	1	$1-(\phi)$	1a	$T_s, (\sigma_s, r_s)$	1	0	1	1a, b	2a, b	r_s	Analytical study of Budyko (1969) model
1976	Gal-Chen and Schneider	1, 4b	$1-(\phi)$	1a	$T_s, (\sigma_s, r_s)$	1	2	1	1b	2a	S_0, K	
1976	Ghil	2	$1-(\phi)$	1a	$T_s, (\sigma_s, r_s)$	1	0	1	1b	2a, b	S_0	Analytical study of Sellers (1969) model

(Continued)

TABLE IV—Continued

Year	Authors	(a)	(b)	(c)	(d)	(e)	(f)	(g)	(h)	(i)	(j)	Comments
1976	Harshvardhan and Cess	0	0	1b	T_s	1	0	1	1	2a	$\xi(\mathscr{A}$-stratosphere)	
1976	Pollack et al.	0	0	1b	T_s	1	0	1	1a	2a	$\xi(\mathscr{A})$	
1976	Ramanathan	0	$1-(z)$	1b	T_s	3	0	1	1a	2a	$(\partial T/\partial z)$, $\mathscr{C}$, r_{ss}	
1976	Ramanathan et al.	0	$1-(z)$	1b	$T_{s,A}$	3	0	3	1a	2a	$\xi(O_3, NO_2)$	
1976	Reck	0		1b	T_s	3	0		1b	2a	$\xi(O_3)$	Manabe and Wetherald (1967) model
1976	Sellers	4b	$2-(\lambda, \phi)$	1a	$T_s, p_a, u_a, P, \mathscr{C}, (\sigma_i, r_s)$	1	3b	3	1b	1, 2a	S_0, init. condit.	
1976	Sergin and Sergin	2	$2\frac{1}{2}-(\phi, z)$	2b	$T_{s,A}, u, \mathscr{C}, \sigma_i, \sigma_{M}, M_1, M_M$	1	3b	1	3b	2b		Angström (1926) atmospheric model
1976	Su and Hsieh	1	$1-(\phi)$	1a	$T_s, (\sigma_i, r_s)$	1	0	1	1a, b	2a, b	Time consts. for ocean and ice	Analytical study of stability of Budyko (1969) model
1976	Suarez and Held	2	$2\frac{1}{2}-(\phi, z)$	2b	$T_{s,A}, (m_i, r_s)$	2	3b	3	3a	3	(e, ϵ, Π)	Earth-orbital variations over 150,000 years
1976	Taylor	0	$2-(\phi, z)$	2b	$T_{s,A}$	2	0	2	2	2a, b	k	
1976	Temkin and Snell	3	$1-(\phi)$	1a	$T_s, (\sigma_i, r_s), \mathscr{C}$	1	0	1	1	2a	$\xi(\mathscr{A}, CO_2), S_0$	
1976	Weertman		$2-(\phi, z)$	1e	m_i		0	1	3b	3	(e, ϵ, Π)	Application of Weertman (1964)
1977	Augustsson and Ramanathan	0	0	1b	T_s	3	0	1	1a	2a	$\xi(CO_2)$	
1977	Berger (a)				Sellers (1969) model				1, 3b	2, 3	(e, ϵ, Π)	T_s response to varying earth-orbital parameters
1977	Birchfield		$2-(\phi, z)$	1e	m_i		0	1	3b	3	(e, ϵ, Π)	Extension of Weertman (1964, 1976) studies
1977	Coakley	0	0	1b	T_s	4	0	1	1a	2a	$\partial T/\partial z, r_{ss}, \partial r_s/\partial z$	Review of feedback mechanisms in "vertical column" models, Cess (1974) radiative model used
1977	Donn and Shaw				Adem (1965) model						$L/M(\lambda, \phi)$	T_s response to continental drift

Year	Authors		(c)	(d)	(e)	(f)	(g)	(h)	(i)	(j)	(k)		Comments
1977	Drazin and Griffel	2	1-(ϕ)	1a	$T_s, (\sigma_1, r_s)$	1	0	1	1	1a, b	2a	S_0	Stability analysis of Sellers (1969)–North (1975) models, showing existence of stable asymmetric equilibrium
1977	Lee and Snell	2	1½-(ϕ)	1a	$T_s, (\sigma_1, r_s), \mathscr{C}$	1	0	1	1	1a	2a	$\xi(\mathscr{A}, CO_2, cfc), S_0$	Extension of Temkin and Snell (1976)
1977	Lian and Cess	1	1-(ϕ)	1a	$T_s, (\sigma_1, r_s)$	1	0	1	1	1a	2a	$S_0, \bar{T}_s$	Comparison of ice albedo feedback parameterization
1977	Linden and Farrell	1, 2	1-(ϕ)	1a	$T_s, (\sigma_1, r_s)$	1	0	1	1	1a	2a	S_0	Role of tropical and ocean heat transports considered
1977	Luther et al.	0	1-(z)	1b*	$T_4, \xi(O_3)$	3	0	1	1	1a	2a	$\xi_i, \xi(NO_x, CO_2, cfc)$	* Stratosphere only

B. Momentum Models [Zonally Averaged, (b) = 2-(ϕ, z)]

Year	Authors	(a)	(c)	(d)	(e)	(f)	(g)	(h)	(i)	(j)	(k)	Comments
1959	Charney	5	1b	v, T	2	0	1	1b	1	—	7	Newtonian diabatic heating
1964	Leovy	1	1b	v, T	4	0	2	1b	1	—	1	Mesospheric climate only
1964	Saltzman											Generalized development of the steadystate, statistical–dynamical problem including both thermodynamic and momentum equations
1965	Williams and Davies	2	1b	v, T	3	0	1	1a	1	—	3	Extension of Davies and Oakes (1962) including thermodynamics
1969	Dolzhanskiy	2	1b	v, T	3	0	1	1b	1	—	3	Williams and Davies (1965) parameterizations
1970	Kurihara	6	1b	v, T, p	2	0	1	1b	1	—	4	
1970	Wiin-Nielsen	2	1b	v, T	2	0	3	2	1	—	0	Newtonian diabatic heating
1971	Dolzhanskiy	2	1b	v, T	3	0	1	1b	1	K	3	Same as Dolzhanskiy (1969) with improved radiation treatment

(*Continued*)

TABLE IV—Continued

Year	Authors	(a)	(c)	(d)	(e)	(f)	(g)	(h)	(i)	(j)	(k)	Comments
1971	Pike	2	2b	$\mathbf{V}, T$	3	3a	3	1b	1	—	1	Tropical ocean–atmosphere system
1971	Saltzman and Vernekar (a,b)	4a	2b	$\mathbf{V}, T, \varepsilon, P, E, \overline{v'^2}$	2	3a	2	1a	1, 2	(e, ε, Π)	5	No ice feedback
1971	Sela and Wiin-Nielsen (see also Wiin-Nielsen, 1972)	2	1b	$\mathbf{V}, T$	2	0	3	2	1	—	6	No feedback effects of mean poloidal motions
1972	Pike	2	2b	$\mathbf{V}, T$	3	3a	3	2	1	—	1	Tropical ocean–atmosphere system (seasonal variation deduced)
1972	Saltzman and Vernekar	4a	2b	$\mathbf{V}, T, \varepsilon, P, E, \overline{v'^2}$	2	3a	2	1a	1	—	5	Improved parameterizations over (Saltzman and Vernekar, 1971a), both hemispheres
1973	Hunt	0	1b	$\mathbf{V}, T, p_s$	3	2	1	1b	2a	*	0	No horizontal eddy fluxes, * moist vs. dry atmosphere
1973	Kurihara	6	2b	$\mathbf{V}, T, \varepsilon, P, E, \overline{v'^2}, \overline{T'^2}$	2	3a	2	2	1	—	4	(1970) model
1973	MacCracken	2	2b	$\mathbf{V}, T, \varepsilon, P, E, \sigma_t, \mathscr{C}$	3	3b	3	1b	1	—	1	ZAM2

Year	Author										Notes
1974	MacCracken and Luther	4b	2b	$\mathbf{V}, T, \varepsilon, P, E, \sigma_t, \mathscr{C}$	3	3b	3	1b	1, 2a	—	1	ZAM2 with Stone (1972), (1973) heat and mass transport parameterizations
1975	Egger	6	1b	$\mathbf{V}, T, \overline{v'^2}, \overline{T'^2}$	2	0	1	1b	1, 2a	C_D, r_s	4	Kurihara (1970) type model
1975	Lahiff	2	2b	$\mathbf{V}, T, \varepsilon$	2	3a	3	2	1, 2a	$S_0, T_{\text{mid-lat}}$	2	Tropics only, oceans included
1975	Saltzman and Vernekar	4a	2b	$\mathbf{V}, T, \varepsilon, P, E, \overline{v'^2}$	2	3a	2	1a	1, 2a	σ_t		(18,000 BP ice boundary conditions)
1975	Wiin-Nielsen and Fuenzalida	4b	2b	$\mathbf{V}, T$	2	—	3	2	1	—	6	Prescribed internal diabatic heating including latent heat release
1976	Petukhov	2	2b	$T_{M}, T_{As}, v, \sigma_t$	1	3b	3	3b	2a, b	—	3	Williams and Davies (1965) parameterization
1977	Webster and Lau	4b	2b	$T, \mathbf{V}, \sigma_t$	2	0	3	1b	1	—	6	Atmospheric–ocean continent model, (b) = $2 \frac{1}{2}$ -(ϕ, z)
1978	Ohring and Adler (see also Ohring and Gyoeri, 1977)	2	2a	$T, u, (\sigma_t, r_s)$	2	1	1	1b	1, 2a	$\xi(CO_2), S_0, \mathscr{C}$	6	Based on Sela and Wiin-Nielsen (1971) model, more detailed radiation parameterization, no poloidal motions calculated
1978	Taylor	4b	2a	$T, \mathbf{V}$	2	1	1	1b	1, 2	Ω	5	

[a] See Table III for definitions of classification indices (a)–(j).

composition and cloud structure. By means of such models one can account for the main features of the standard atmosphere which represents a significant component of the total spatial variability of climate (e.g., Manabe and Möller, 1961). Aside from listing these models in Table IV, we shall not go into any further details concerning this group in this review. A good survey discussion regarding applications for CO_2 sensitivity was given by Schneider (1975).

The vertically integrated, horizontally varying, thermodynamic models are designated under (b) by 1–(ϕ) which corresponds to averaging of the form $\langle \bar{\tilde{\psi}} \rangle$, and by 2–($\lambda$, ϕ) which corresponds to $\tilde{\psi}$. These models, which contain little resolution in the vertical beyond that arising from a differentiation between the atmospheric and subsurface domains, yield solutions for representative vertical mean temperatures that are often identified with the surface value. As noted above, because these models contain no explicit momentum dynamics, it is impossible to explicitly deduce from them the mean poloidal- and monsoonal-type circulations that are essential for the fullest description of climate (e.g., the climatic zonation forced by the presence of Hadley and Ferrel circulations). Nor is it possible to feed their important transport properties back into the calculation to improve the temperature estimate (though questionable attempts have been made to prescribe or parameterize these properties). However, as noted earlier, in spite of this limitation these models seem capable of yielding an acceptable first estimate of perhaps the most significant of all the climatic variables, temperature (Adem, 1962; Saltzman, 1967, 1968b) and hence of other highly temperature-dependent variables such as ice cover. Thus they provide a first-order approach to a theory of climate variability in time and space upon which a more rigorous statistical–dynamic theory can be built. For this reason in the next section we shall present the physical foundation for these models in some detail, and also shall review some of the main applications of these models.

Concerning the momentum group (IIB) we note again that the spatial resolution of these models generally falls into the (b) categories designated by 2 or $2\frac{1}{2}$–(ϕ, z) (i.e., models governing $\langle \psi \rangle$ without or with continent–ocean differentiation, respectively) and by 3–(λ, ϕ, z) (corresponding to models either governing the standing waves or the complete mean field $\bar{\psi}$). In Table IVB we list only the zonal average, (b) = 2 or $2\frac{1}{2}$–(ϕ, z), models, and hence we omit this index from the table. Comprehensive summaries of the standing-wave models for the $\bar{\psi}^*$ field have been given by Saltzman (1968a) and more recently by Manabe and Terpstra (1974).

In accordance with the definition at the beginning of this section we have also excluded from Table IVB all diagnostic models of the mean poloidal circulation in which the zonal mean temperature and/or internal atmospheric heat and momentum sources are prescribed from observations (e.g., Eliassen, 1951; Kuo, 1956; Davies and Oakes, 1962; Vernekar, 1967; Dickinson, 1971a,b; Derome and Wiin-Nielsen, 1972; Schneider and Lindzen, 1977) though we recognize that this type of study has been, and will continue to be, of immense diagnostic value in guiding the development of the more complete climate models.

Statistical–dynamical models from which the complete three-dimensional geographic distribution of the set of mean climatic statistics $\{\bar{\psi}\}$ (including mean variances and covariances as well as pure averages) can be deduced directly have yet to be advanced, though this is undoubtedly an ultimate objective of the development of climatic modeling. A plausible iterative scheme for achieving this synthesis, based on combining the standing-wave solutions $\bar{\psi}^*$ with the zonal average solutions $\langle\bar{\psi}\rangle$ (see Section 6.2) has been proposed by Saltzman (1968a), but one might also seek a more direct numerical means to this end by simply integrating the "primitive" mean equations (3.3), (3.4)–(3.6), and (3.12) at grid points with suitable transient eddy parameterizations of all the stresses (in much the same manner at GCMs are integrated, but using longer time steps).

We note, finally, concerning the classification scheme given in Table III, that for most of the categories, a higher index number indicates that a model is more general and has greater simulation potential than a model characterized by a lower index number. Generally speaking, it is implied that solutions corresponding to all index values lower than the one indicated for a given model listed in Table IV can also be deduced as special cases. For examples, under the thermodynamic group IIA, any model that contains a specific parameterization for horizontal heat transport (a) = 1–6 can be specialized to the no-transport case (a) = 0, and any model that yields the monthly mean climatic states, (g) = 3, automatically can yield the seasonal (g) = 2 and annual mean state (g) = 1.

5. The Vertically Integrated, Thermodynamic Models

5.1. Foundations

In this section we shall develop the general, vertically integrated, equations that are the basis of many of the so-called energy-balance or thermodynamic models included in Table IV, corresponding to the chain of models represented by the bold lines on Fig. 2. Since the terrestrial climatic system, and its energy balance in particular, is so strongly influenced by the presence and transformations of water in all three phases, we begin with a discussion of the vertically integrated mass balance of these phases, representing the hydrologic cycle.

5.1.1. The Water Mass Balance. The total water mass of the earth can be divided into an arbitrary number of components. For our purposes we shall distinguish between (1) atmospheric water vapor, (2) atmospheric water and ice particulates, i.e., clouds and hydrometeors, (3) liquid water in the form of (a) oceans, (b) land surface waters (rivers and lakes), (c) subsurface reservoirs (ground water and soil moisture), and (d) water contained in the biomass, and (4) solid phase surficial water in the form of snow, glaciers, ice sheets, ground ice, and sea, lake, and river ice. The continuity equations giving the rates of change of the ensemble monthly mean mass of each of these four components per fixed unit volume of the climatic system, can be written as specializations of (3.4) (cf. Tables I and II) in the following forms, respectively (see List of Basic Symbols):

(5.1) $\quad \partial \bar{\chi}_v / \partial t = -[\nabla \cdot (\bar{\chi}_v \bar{\mathbf{v}} + \overline{\chi'_v \mathbf{v}'} + \bar{\mathbf{f}}_v) + \partial/\partial z (\bar{\chi}_v \bar{w} + \overline{\chi'_v w'} + \bar{f}^{\uparrow}_v)] + \bar{\mathscr{E}}_n$

(5.2) $\quad \partial \bar{\chi}_n / \partial t = -[\nabla \cdot (\bar{\chi}_n \bar{\mathbf{v}} + \overline{\chi'_n \mathbf{v}'} + \bar{\mathbf{f}}_n) + \partial/\partial z (\bar{\chi}_n \bar{w} + \overline{\chi'_n w'} + \bar{f}^{\uparrow}_n)] - \bar{\mathscr{E}}_n$

(5.3) $\quad \partial \bar{\chi}_w / \partial t = -[\nabla \cdot (\bar{\chi}_w \bar{\mathbf{v}}_w + \overline{\chi'_w \mathbf{v}'_w} + \bar{\mathbf{f}}_w)$
$\qquad + \partial/\partial z (\bar{\chi}_w \bar{w}_w + \overline{\chi'_w w'_w} + \bar{f}^{\uparrow}_w)] + \bar{\mathscr{M}} - \bar{\mathscr{E}}$

(5.4) $\quad \partial \bar{\chi}_i / \partial t = -[\nabla \cdot (\bar{\chi}_i \bar{\mathbf{V}}_i + \overline{\chi'_i \mathbf{v}'_i} + \bar{\mathbf{f}}_i) + \partial/\partial z (\bar{\chi}_i \bar{w}_i + \overline{\chi'_i w'_i} + \bar{f}^{\uparrow}_i)] - \bar{\mathscr{M}}$

where

$$\mathbf{V}_j = \mathbf{v}_j + w_j \mathbf{k} \qquad (\mathbf{v} = u\mathbf{i} + v\mathbf{j})$$
$$\mathscr{F}_j = \mathbf{f}_j + f^{\uparrow}_j \mathbf{k} \qquad (\mathbf{f} = f_\lambda \mathbf{i} + f_\phi \mathbf{j})$$

and we have set $\mathbf{v}_A = \mathbf{v}$ and $w_A = w$. The symbols $\mathscr{E}$ and $\mathscr{M}$ represent the rates of evaporation and melting per unit volume, respectively.

With reference to Fig. 1, let us next integrate (5.1) and (5.2) vertically through the depth of the atmosphere [from the surface, where $z = z_s = (z_{sL} + h)$, $\bar{p} = \bar{p}_s$ and $w = \partial z_s / \partial t + \mathbf{v}_s \cdot \nabla z_s$, to the top ($z = \infty$, $p = 0$, $w = 0$)], and integrate (5.3) and (5.4) vertically from an arbitrary level $z = d$ ($= Z$ for the oceans), below which there is no liquid water or ice, to the surface where $z = \bar{z}_s$ ($= z_{sL}$ for the oceans).

We then have for any geographic location,

(5.5) $\qquad \partial \bar{m}_v / \partial t = -\nabla \cdot \bar{\mathbf{J}}_v + \bar{E} - \bar{C}$

(5.6) $\qquad \partial \bar{m}_n / \partial t = -\nabla \cdot \bar{\mathbf{J}}_n - \bar{P} + \bar{C}$

(5.7) $\qquad \partial \bar{m}_w / \partial t = -\nabla \cdot \bar{\mathbf{J}}_w + \bar{P}_w - \bar{E} - \bar{F}$

(5.8) $\qquad \partial \bar{m}_i / \partial t = -\nabla \cdot \bar{\mathbf{J}}_i + \bar{P}_i + \bar{F}$

where

(5.9) $\qquad \bar{m}_v = \int_{z_s}^{\infty} \bar{\chi}_v \, dz \approx \int_0^{\bar{p}_s} \bar{\xi}_v (dp/g) = (\bar{p}_s/g) \bar{\tilde{\xi}}_v$

is the mean mass of water vapor in the atmospheric column per unit horizontal area, i.e., the "precipitable water;"

$$(5.10) \quad \bar{m}_n = \int_{z_s}^{\infty} \bar{\chi}_n \, dz \approx \int_0^{\bar{p}_s} \bar{\xi}_n (dp/g) = (\bar{p}_s/g)\bar{\bar{\xi}}_n$$

$$= \bar{m}_{nw} + \bar{m}_{ni}$$

is the mean mass of water in the atmospheric column per unit horizontal area, in the form of liquid drops ($\bar{m}_{nw}$) and ice particles ($\bar{m}_{ni}$);

$$(5.11) \quad \bar{m}_w = \int_d^{z_s} \bar{\chi}_w \, dz$$

$$= \bar{m}_{wM} + \bar{m}_{wL}$$

is the mean mass of liquid water on the earth per unit horizontal area, in the form of ocean, $\bar{m}_{wM} = \int_Z^{z_{sL}} \bar{\chi}_M \, dz = (z_{sL} - Z)\bar{\bar{\rho}}_M$, and continental water, $\bar{m}_{wL} = \int_d^{z_s} \bar{\chi}_{wL} \, dz = (z_s - d)\bar{\bar{\chi}}_{wL}$;

$$(5.12) \quad \bar{m}_i = \int_d^{z_s} \bar{\chi}_i \, dz = (z_s - d)\bar{\bar{\chi}}_i$$

$$= \bar{m}_{iM} + \bar{m}_{iL}$$

is the mean mass of surface ice and snow, per unit horizontal surface area; and

$$\bar{\mathbf{J}}_v = \int_{z_s}^{\infty} \overline{(\chi_v \mathbf{v} + \mathbf{f}_v)} \, dz$$

$$= \int_0^{\bar{p}_s} \overline{(\xi_v \mathbf{v} + \rho^{-1}\mathbf{f}_v)} \frac{dp}{g} = \frac{\bar{p}_s}{g} \overline{\overline{(\xi_v \mathbf{v} + \rho^{-1}\mathbf{f}_v)}}$$

is the mean horizontal flux of water vapor due to motions on all scales;

$$\bar{\mathbf{J}}_n = \int_{z_s}^{\infty} \overline{(\chi_n \mathbf{v}_n + \mathbf{f}_n)} \, dz$$

$$= \int_d^{z_s} \overline{(\xi_n \mathbf{v}_n + \rho^{-1}\mathbf{f}_n)} \frac{dp}{g} = \frac{\bar{p}_s}{g} \overline{\overline{(\xi_n \mathbf{v}_n + \rho^{-1}\mathbf{f}_n)}}$$

is the mean horizontal flux of cloud elements;

$$\bar{\mathbf{J}}_w = \int_d^{z_s} \overline{\chi_w \mathbf{v}_w} \, dz = (z_s - d)\overline{\overline{\chi_w \mathbf{v}_w}}$$

is the mean horizontal flux of the liquid waters of the earth due to surface and subsurface flow;

$$\mathbf{J}_i = \int_d^{z_s} \overline{\chi_i \mathbf{v}_i} \, dz = (z_s - d)\overline{\chi_i \mathbf{v}_i}$$

is the horizontal flux of ice due to flow of glaciers, ice sheets, and sea ice, calving, blowing and drifting snow, and avalanches; and

$$E = \int_d^{z_s} \mathscr{E} \, dz = (f_v^\uparrow - \mathbf{f}_v \cdot \nabla z)_s \qquad C = -\int_{z_s}^{\infty} \mathscr{E}_n \, dz$$

$$P = -(f_n^\uparrow - \mathbf{f}_n \cdot \nabla z)_s \quad \text{and} \quad F = -\int_d^{z_s} \mathscr{M} \, dz$$

Note that $(m_v + m_n)$ represents the total water in the atmospheric column, and $(m_w + m_i)$ represents the total water in the subsurface column. Also, it will be of relevance to separate (5.6) into an equation governing *water* cloud and precipitation (m_{nw}) and an equation governing *ice* cloud and precipitation (m_{ni}), as follows

(5.13) $$\frac{\partial \bar{m}_{nw}}{\partial t} = -\nabla \cdot \bar{\mathbf{J}}_{nw} - \bar{P}_w + \bar{C} - \bar{F}_n$$

(5.14) $$\frac{\partial \bar{m}_{ni}}{\partial t} = -\nabla \cdot \bar{\mathbf{J}}_{ni} - \bar{P}_i + \bar{F}_n$$

where

$$F_n = -\int_{z_s}^{\infty} \mathscr{M}_{ni} \, dz$$

Next, if we average (5.5)–(5.8) zonally (i.e., around a latitude circle) these equations take the form

(5.15) $$\frac{\partial \langle \bar{m}_v \rangle}{\partial t} = -\frac{\partial (\langle \bar{J}_{v\phi} \rangle \cos \phi)}{a \cos \phi \, \partial \phi} + \langle \bar{E} \rangle - \langle \bar{C} \rangle$$

(5.16) $$\frac{\partial \langle \bar{m}_n \rangle}{\partial t} = -\frac{\partial (\langle \bar{J}_{n\phi} \rangle \cos \phi)}{a \cos \phi \, \partial \phi} - \langle \bar{P} \rangle + \langle \bar{C} \rangle$$

(5.17) $$\frac{\partial \langle \bar{m}_w \rangle}{\partial t} = -\frac{\partial (\langle \bar{J}_{w\phi} \rangle \cos \phi)}{a \cos \phi \, \partial \phi} + \langle \bar{P}_w \rangle - \langle \bar{E} \rangle - \langle \bar{F} \rangle$$

(5.18) $$\frac{\partial \langle \bar{m}_i \rangle}{\partial t} = -\frac{\partial (\langle \bar{J}_{i\phi} \rangle \cos \phi)}{a \cos \phi \, \partial \phi} + \langle \bar{P}_i \rangle + \langle \bar{F} \rangle$$

where $\bar{J}_{v\phi} \approx \bar{p}_s g^{-1} \overline{\varepsilon v}$, $\bar{J}_{n\phi} \approx \bar{p}_s g^{-1} \overline{\xi_n v_n}$ (assuming the meridional fluxes of water vapor and hydrometeors due to subsynoptic motions are negligible compared to the synoptic fluxes),

$$\bar{J}_{w\phi} = (z_s - d)\overline{\widetilde{\chi_w v_w}} \quad \text{and} \quad \bar{J}_{i\phi} = (z_s - d)\overline{(\widetilde{\chi_i v_i})}$$

Finally, if we average (5.15)–(5.18) globally over all latitudes, in accordance with (3.23), we obtain

(5.19) $\qquad \partial \widetilde{\bar{m}}_v/\partial t = \widetilde{\bar{E}} - \widetilde{\bar{C}}$

(5.20) $\qquad \partial \widetilde{\bar{m}}_n/\partial t = -\widetilde{\bar{P}} + \widetilde{\bar{C}}$

(5.21) $\qquad \partial \widetilde{\bar{m}}_w/\partial t = \widetilde{\bar{P}}_w - \widetilde{\bar{E}} - \widetilde{\bar{F}}$

(5.22) $\qquad \partial \widetilde{\bar{m}}_i/\partial t = \widetilde{\bar{P}}_i + \widetilde{\bar{F}}$

which, in turn, can be summed to yield the statement of global conservation of water in all forms,

(5.23) $\qquad (\partial/\partial t)(\widetilde{\bar{m}}_v + \widetilde{\bar{m}}_n + \widetilde{\bar{m}}_w + \widetilde{\bar{m}}_i) = 0$

Observational studies of the distribution and transport of water in all forms, covering both the atmospheric and surficial branches of the hydrologic cycle, are reviewed in Chorley (1969, see especially, articles by Barry, Nace, and Marcus) and Peixoto (1970), and details of ice distribution variation are given by Flint (1971, pp. 83–85) and Shumskiy et al. (1964).

5.1.2. The Energy Balance. Let us separate the climatic system into an *atmospheric* part, (denoted by subscript A), and a *subsurface* part that encompasses the ocean, including sea ice (denoted by the subscript M), and the nonoceanic subsurface components including all land masses, oceanic basement, and continental water in all forms (denoted by the subscript N). Starting with (3.13) we shall first develop the vertically integrated, total energy equation for each of these three separate components, and ultimately combine them into a single equation for the total energy contained in the complete climatic system. In each equation presented, the terms that are underlined are generally at least one order of magnitude less than the remaining terms.

5.1.2.1. The Atmosphere. To specialize (3.13) for the atmosphere, we note that (Table I)

$$\rho \equiv \rho_A \approx p/RT$$

and

$$e_A \approx c_V T + \xi_v L_v - \xi_{ni} L_f$$

Then, integrating (3.13) from the surface, $z = z_s$ (where $\bar{p} = \bar{p}_s$, $w_s = \partial z_s/\partial t + \mathbf{v} \cdot \nabla z_s$, $\mathbf{H} = \mathbf{H}_s$, $\mathbf{B} = \mathbf{B}_s$, and $m_s^\uparrow = \mathbf{m} \cdot \nabla z_s$) to the "top" of the atmosphere, $z = z_T \to \infty$ (where $p = p_T \to 0$, $w = 0$, $b^\uparrow = 0$, $\mathbf{h} = 0$, $H^{(1,2)\uparrow} = H_T^{(1,2)\uparrow}$, and $\mathbf{H}^{(3)} = \mathbf{H}_{lat} = \mathbf{M}_{conv} = \mathbf{M}_{lat} = 0$, thus excluding consideration of possible molecular exchanges with outer space) we obtain,

$$(5.24) \quad \frac{\partial}{\partial t}\int_{z_s}^{\infty} \bar{\rho}_A(c_P\bar{T} + \underline{\mathscr{K}})_A\, dz$$

$$= -\nabla\cdot\int_{z_s}^{\infty}[\bar{\rho}_A(c_P\overline{T\mathbf{v}} + \overline{\Phi\mathbf{v}} + \underline{\overline{\mathscr{K}\mathbf{v}}} + \mathbf{m}_{(c)} + \underline{\mathbf{b}} + \underline{\mathbf{h}}^{(1,2,3)})]\, dz$$

$$- z_s(\partial\bar{p}_s/\partial t) + (b^\uparrow - \mathbf{b}\cdot\nabla z)_s + \int_{z_s}^{\infty}\overline{\rho Q}\, dz$$

$$+ \bar{H}_T^{(1,2)\downarrow} + \bar{H}_s^{(1,2,3)\uparrow} + \bar{H}_A^{(4)}$$

where we have defined,

$$\mathbf{B} \equiv \mathbf{b} + b^\uparrow\mathbf{k}$$

$$\mathbf{H}^{(a,b,\ldots)} \equiv \mathbf{H}^{(a)} + \mathbf{H}^{(b)} + \cdots$$

$$H_s^{\uparrow(1,2,3)} \equiv [\mathscr{H}_s^{\uparrow(1,2,3)} - (\mathbf{h}^{(1,2,3)}\cdot\nabla z)_s]$$

$$H_T^{\uparrow(1,2)} \equiv \mathscr{H}_T^{\uparrow(1,2)}$$

and

$$H_A^{(4)} \equiv L_v C + L_f F_n$$

the last expression representing the net heating of the atmospheric column, per unit area, due to water phase transformations. In deriving (5.24) we have used the ideal gas relation $c_P = c_V + R$ and the water mass continuity equations (5.5) and (5.14).

5.1.2.2. The Ocean. We now specialize (3.13) for the ocean (including sea ice) by noting from Table I that the density of sea water is

$$\rho_M = \chi_w + \chi_s + \sum_j \chi_{jM}$$

the density of sea ice is χ_{iM}, and the internal energy per unit mass of ocean is

$$e_M \approx \xi_M c_w \tau_M + \xi_{iM}(c_i \tau_{iM} - L_f)$$

If we integrate (3.13) from the ocean floor, $z = Z$ (where $w_Z = \partial Z/\partial t + \mathbf{v}_Z\cdot\nabla Z$, $\mathbf{B} = \mathbf{B}_Z$, $\mathbf{H}^{(1,2)} = \mathbf{H}_{lat} = 0$, $\mathbf{H}^{(3)} = \mathbf{h}_Z^{(3)} + \mathscr{H}_Z^{(3)\uparrow}$, and $m_Z^\uparrow = \mathbf{m}_Z\cdot\nabla Z$) to the ocean surface, $z = z_s$ (where $w_s = \partial z_s/\partial t + \mathbf{v}_s\cdot\nabla z_s$, $\mathbf{B} = \mathbf{B}_s$, $\mathbf{H} = \mathbf{H}_s$, and $m_s^\uparrow = \mathbf{m}_s\cdot\nabla z_s$), assuming hydrostatic equilibrium and using (5.8), we obtain,

$$(5.25) \quad \frac{\partial}{\partial t} \int_Z^{z_s} \bar{\rho}_M c_w \bar{\tau}_M \, dz + \frac{\partial}{\partial t} \int_Z^{z_s} [c_i \bar{\chi}_{iM} \bar{\tau}_{iM} + (\bar{\rho}\bar{\Phi})_M + (\bar{\rho}\bar{\mathcal{K}})_M] \, dz$$

$$= -\nabla \cdot \int_Z^{z_s} [\bar{\rho}_M c_w \bar{\tau}_M \bar{\mathbf{v}} + c_i \bar{\chi}_{Mi} \bar{\tau}_{iM} \bar{\mathbf{v}}_i + \mathbf{m}_{(c)M}$$

$$+ \overline{\rho \Phi' \mathbf{v}'} + \overline{(\rho \mathcal{K})_M \mathbf{v}} + \underline{\mathbf{b}} + \underline{\mathbf{h}^{(1,2,3)}}] \, dz$$

$$+ (gz_s + \bar{p}_s/\rho_M) \int_Z^{z_s} \bar{\rho}_M \bar{\mathbf{v}} \, dz - [\bar{p}(\partial z/\partial t) - \mathbf{b} \cdot \nabla z + b^\uparrow]_Z^{z_s}$$

$$+ \int_Z^{z_s} \bar{\rho} Q_M \, dz + \bar{H}_s^{(1,2,3)\downarrow} - \bar{H}_s^{(4)\uparrow} + \underline{\bar{H}_{ZM}^{(3)\uparrow}}$$

where

$$H_s^{(4)\uparrow} = L_v E - L_f F$$

represents the vertical heat loss from the surface due to phase changes, $H_{ZM}^{(3)\uparrow}$ represents the vertical geothermal flux of heat into the ocean floor from the lithosphere ocean basement (generally negligible compared to the surface fluxes), and we have assumed $\overline{c_i \chi_{iM} \tau_{iM}} \approx c_i \bar{\chi}_{iM} \bar{\tau}_{iM}$.

5.1.2.3. The Biolithosphere. The total density of the nonoceanic subsurface media, including continent, ocean basement, and the water in all forms they contain, can be expressed as

$$\rho_N = \chi_\ell + \chi_{wL} + \chi_{iL}$$

where χ_ℓ is the combined mass concentration of solid earth (e.g., rock, soil, and sediment) and biomass, χ_{wL} is the mass concentration of continental water in all forms (rivers, lakes, ground water), and χ_{iL} is the mass concentration of continental ice in all forms. From Table I we see that the internal energy of the biolithosphere is given approximately by

$$e_N \approx \xi_\ell c_\ell T_\ell + \xi_{wL} c_w \tau_{wL} + \xi_{iL}(c_i \tau_{iL} - L_f)$$

With this form for e_N, if we integrate (3.13) from the lower limit of the climatic system $z = d$ (where $w_d = \partial d/\partial t + \mathbf{v}_d \cdot \nabla d$, $\mathbf{b} = \mathbf{b}_d$, $\mathbf{H}^{(1,2)} = \mathbf{M} = 0$, and $\mathbf{H}^{(3)} = \mathbf{H}_d^{(3)}$) to the top of the biolithosphere, $z = z_s$ (continent), Z (ocean) [where $w_{s,Z} = \partial(z_s, Z)/\partial t + \mathbf{v} \cdot \nabla(z_s, Z)$, $\mathbf{B} = \mathbf{B}_{s,Z}$, $\mathbf{H}^{(1,2)} = \mathbf{H}_{\text{lat}} =$

$\mathbf{M} = 0$, $\mathbf{H}^{(3)} = \mathbf{H}^{(3)}_{s,Z}]$, using (5.8), we obtain

(5.26)
$$\frac{\partial}{\partial t} \int_d^{z_s} [(c_\ell \bar{\chi}_\ell \bar{T}_\ell + c_w \bar{\chi}_{wL} \bar{\tau}_w + c_i \bar{\chi}_i \bar{\tau}_i)_N + (\overline{\rho\Phi})_N + (\overline{\rho\mathcal{K}})_N]\, dz$$

$$= -\nabla \cdot \int_d^{z_s, Z} [(c_\ell \bar{\chi}_\ell \bar{T}_\ell + c_w \bar{\chi}_w \bar{\tau}_w + c_i \bar{\chi}_i \bar{\tau}_i)_N \bar{\mathbf{V}}_N$$
$$+ \overline{(\rho\Phi\mathbf{v} + \rho\mathcal{K}\mathbf{v})_N} + \mathbf{m}_{(c)N} + \overline{(p\mathbf{v} + \mathbf{b} + \mathbf{h}^{(3)})_N}]\, dz$$
$$- \left[\bar{p}\frac{\partial z}{\partial t} + (b^\uparrow - \mathbf{b}\cdot\nabla z)\right]_d^{z_s, Z} + \int_d^{z_s, Z} \overline{(\rho Q)}_L\, dz$$
$$+ \bar{H}_s^{(1,2,3)\downarrow} - \bar{H}_s^{(4)\uparrow} + \bar{H}_d^{(3)\uparrow} - \bar{H}_{ZM}^{(3)\uparrow}$$

where $H_d^{(3)\uparrow}$ represents the upward geothermal flux at $z = d$, and $\mathbf{P}$ (which enters the expression for $\mathbf{B}$) may be taken to include the "elastic stress" tensor for solid earth and glacial ice, and we have assumed $\overline{\chi_j \tau_j} \approx \bar{\chi}_j \bar{\tau}_j$.

5.1.2.4. The Complete Subsurface Domain (Ocean and/or Land): The Surface "Heat Balance" Condition. If we add (5.25) and (5.26) we obtain the following vertically integrated energy equation for an arbitrary subsurface column consisting of biolithosphere and/or ocean:

(5.27)
$$\frac{\partial}{\partial t} \int_d^{z_s} c_w \bar{\chi}_w \bar{\tau}_w\, dz + \frac{\partial}{\partial t} \int_d^{z_s} [c_i \bar{\chi}_i \bar{\tau}_i + c_\ell \bar{\chi}_\ell \bar{T}_\ell + (\overline{\rho\Phi} + \overline{\rho\mathcal{K}})_{M,N}]\, dz$$

$$= -\nabla \cdot \int_d^{z_s} \{(c_w \bar{\chi}_w \bar{\tau}_w \bar{\mathbf{V}}_w + \overline{c_i \bar{\chi}_i \bar{\tau}_i \bar{\mathbf{V}}_i} + \overline{c_\ell \bar{\chi}_\ell \bar{T}_\ell \bar{\mathbf{V}}_\ell})_{M,N}$$
$$+ \mathbf{m}_{(c)M} + \mathbf{m}_{(c)N} + \overline{(\rho\Phi\bar{\mathbf{v}} + \rho\mathcal{K}\mathbf{v})_N}$$
$$+ \overline{(\rho\mathcal{K}\mathbf{v})_M} + [\rho(\bar{\Phi}_s + \bar{p}_s/\rho_M)\mathbf{v}]_M$$
$$+ \overline{(\rho\Phi'\mathbf{v}')}_{M,N} + \mathbf{b}_M + \mathbf{h}_M^{(1,2,3)}$$
$$+ \overline{(p\bar{\mathbf{v}} + \mathbf{b} + \mathbf{h}^{(1,2,3)})_N}\}\, dz$$
$$- \left[\bar{p}\frac{\partial z}{\partial t} + (b^\uparrow - \mathbf{b}\cdot\nabla z)\right]_d^{z_s} + \int_d^{z_s} \overline{\rho Q}\, dz$$
$$+ \bar{H}_s^{(1,2,3)\downarrow} - \bar{H}_s^{(4)\uparrow} + \bar{H}_d^{(3)\uparrow}$$

Let us now let d approach z_s to within an arbitrarily small distance δ_s, representing the so-called active layer (Budyko, 1974) within which nearly all the incoming solar radiation is absorbed and water phase transformations take place. Then all terms involving z_s and d as limits approach zero and we arrive at the surface heat balance condition,

(5.28)
$$\bar{H}_s^{(1)\downarrow} + \bar{H}_s^{(2)\downarrow} + \bar{H}_s^{(3)\downarrow} + \bar{H}_s^{(4)\downarrow} = \bar{H}_s^{(5)\downarrow}$$

where $H_s^{(5)\uparrow} \equiv H_{d \to s}^{(3)\uparrow}$ represents the upward subsurface heat flux near the earth's surface. This equation states that the sum of the heat fluxes toward the surface interface, due to all causes, is zero. It is seen, therefore, that this surface heat balance condition is essentially an alternate form of the vertically integrated energy equation for the subsurface, and models based on this equation fall within the class of vertically integrated models. Equation (5.28) applies also to the instantaneous or synoptic values of heat fluxes, as well as to the ensemble monthly mean values.

5.1.2.5. Energy Equation for a Column of the Complete Climatic System. If, finally, we add (5.27) and the atmospheric energy equation (5.24), we obtain the equation governing the total energy in a column extending through the depth of the complete climatic system:

$$(5.29) \quad \frac{\partial}{\partial t} \int_d^\infty (c_w \bar{\chi}_w \bar{\tau}_w + c_P \bar{\chi}_A \bar{T}) \, dz$$

$$+ \frac{\partial}{\partial t} \int_d^\infty \{(c_i \bar{\chi}_i \bar{\tau}_i + c_\ell \bar{\chi}_\ell \bar{T}_\ell) + [\bar{\rho}(\bar{\Phi} + \bar{\mathscr{K}})]_{M,N} + (\rho \bar{\mathscr{K}})_A\} \, dz$$

$$= -\nabla \cdot \int_d^\infty \{[\rho(c_P \overline{T \mathbf{v}} + \overline{\Phi \mathbf{v}} + \overline{\mathscr{K} \mathbf{v}}) + \mathbf{b} + \underline{\mathbf{h}^{(1,2,3)}}]_A$$

$$+ \mathbf{m}_{(c)A,M} + \underline{\mathbf{m}_{(c)N}} + (c_w \bar{\chi}_w \bar{\tau}_w \bar{\mathbf{v}}_w$$

$$+ c_i \bar{\chi}_i \bar{\tau}_i \bar{\mathbf{v}}_i + \underline{c_\ell \bar{\chi}_\ell \bar{T}_\ell \bar{\mathbf{v}}_\ell})$$

$$+ [\bar{\rho}(\bar{\Phi} + \bar{\mathscr{K}})\mathbf{v}]_N + (\overline{\rho \mathscr{K} \mathbf{v}})_M + [\bar{\rho}(\bar{\Phi}_s + p_s/\rho_M)\bar{\mathbf{v}}]_M$$

$$+ \underline{(\overline{\rho \Phi' \mathbf{v}'})_M} + \mathbf{b}_M + \underline{\mathbf{h}_M^{(1,2,3)}} + (\bar{\rho}\bar{\mathbf{v}} + \mathbf{b} + \underline{\mathbf{h}^{(3)}})_N\} \, dz$$

$$- \frac{\partial}{\partial t}(\bar{p}_s \bar{z}_s) + \left[p_d \frac{\partial d}{\partial t} + (\mathbf{b}_d^\uparrow - \mathbf{b}_d \cdot \nabla d) \right] + \int_d^\infty \overline{\rho Q} \, dz$$

$$+ \bar{H}_T^{(1)\downarrow} + \bar{H}_T^{(2)\downarrow} + (\bar{H}_A^{(4)} - \bar{H}_s^{(4)\uparrow}) + \underline{\bar{H}_d^{(3)\uparrow}}$$

where

$$(5.30) \quad (\bar{H}_A^{(4)} - \bar{H}_s^{(4)\uparrow}) = L_v(\bar{C} - \bar{E}) + L_f(\bar{F}_n + \bar{F})$$

$$= -L_v\left(\frac{\partial \bar{m}_v}{\partial t} + \nabla \cdot \bar{\mathbf{J}}_v\right)$$

$$+ L_f\left[\frac{\partial(\bar{m}_i + \bar{m}_{ni})}{\partial t} + \nabla \cdot (\bar{\mathbf{J}}_i + \bar{\mathbf{J}}_{ni})\right]$$

5.1.2.6. Simplifications. As noted at the beginning of this section, the terms underlined in (5.24)–(5.29) are generally at least one order of magnitude smaller than the remaining terms in each equation. If we neglect these terms, and use our definitions of the mass-weighted vertical averages (cf. Section 3), we can write these equations in the following approximate forms, for

(1) the *atmosphere*:

(5.31) $$\frac{\partial}{\partial t}(\bar{m}_A c_P \bar{\bar{T}}^{(A)}) \approx -\nabla \cdot [\bar{m}_A(c_P \overline{T\mathbf{v}}^{(A)} + \overline{\Phi\mathbf{v}}^{(A)})]$$
$$+ \bar{H}_T^{(1,2)\downarrow} + \bar{H}_s^{(1,2,3)\uparrow} + \bar{H}_A^{(4)}$$

where $H_A^{(4)} = L_v C + L_f F_n$

(2) the *oceans*:

(5.32) $$\frac{\partial}{\partial t}(\bar{m}_M c_w \bar{\bar{T}}^{(M)}) \approx -\nabla \cdot (\bar{m}_M c_w \overline{T\mathbf{v}}^{(M)}) + \bar{H}_s^{(5)\downarrow}$$

where from the heat balance condition (5.28),

$$\bar{H}_s^{(5)\downarrow} = \bar{H}_s^{(1,2,3,4)\downarrow}$$
$$= \bar{H}_s^{(1,2,3)\downarrow} - (L_v \bar{E} - L_f \bar{F})$$

(3) the *biolithosphere*:

(5.33) $$\frac{\partial}{\partial t}(\bar{m}_\ell c_\ell \bar{\bar{T}}^{(N)} + \bar{m}_{wL} c_w \bar{\bar{t}}_w^{(N)} + \bar{m}_{iL} c_i \bar{\bar{t}}_i^{(N)})_N \approx \bar{H}_s^{(5)\downarrow}$$

(4) the *complete subsurface domain* [(5.32) + (5.33)]:

(5.34a) $$\frac{\partial}{\partial t}(\bar{m}_w c_w \bar{\bar{t}}_w^{(M,N)} + \underline{\bar{m}_i c_i \bar{\bar{t}}_i^{(M,N)}} + \underline{\bar{m}_\ell c_\ell \bar{\bar{T}}^{(N)}})$$
$$\approx -\nabla \cdot (\bar{m}_w c_w \overline{t_w \mathbf{v}_w}^{(M,N)}) + \bar{H}_s^{(5)\downarrow}$$

which as noted above is equivalent to the surface heat balance condition

(5.34b) $$\bar{H}_s^{(1)\downarrow} + \bar{H}_s^{(2)\downarrow} + \bar{H}_s^{(3)\downarrow} + \bar{H}_s^{(4)\downarrow} = \bar{H}_s^{(5)\downarrow}$$

and (5) the total *climatic system* [(5.31) + (5.34)]:

(5.35a) $$\frac{\partial}{\partial t}(\bar{m}_A c_P \bar{\bar{T}}^{(A)} + \bar{m}_w c_w \bar{\bar{t}}_w^{(M,N)} + \underline{\bar{m}_i c_i \bar{\bar{t}}_i^{(M,N)}} + \underline{\bar{m}_\ell c_\ell \bar{\bar{T}}^{(N)}})$$
$$= -\nabla \cdot [\bar{m}_A(c_P \overline{T\mathbf{v}}^{(A)} + \overline{\Phi\mathbf{v}}^{(A)}) + \bar{m}_w c_w \overline{t_w \mathbf{v}_w}^{(M,N)}]$$
$$+ \bar{H}_T^{(1,2)\downarrow} + L_v(\bar{C} - \bar{E}) + L_f(\bar{F} + \bar{F}_n),$$

or, using (5.5), (5.8), and (5.14), and neglecting the rate of change of sensible heat of the ice and biolithosphere (underlined terms) relative to that of the atmosphere and oceans,

$$\text{(5.35b)} \quad \frac{\partial}{\partial t}[(\bar{m}_A c_P \bar{\bar{T}}^{(A)} + \bar{m}_w c_w \bar{\bar{\tau}}_w^{(M)}) + L_v \bar{m}_v - L_f(\bar{m}_i + \bar{m}_{ni})]$$

$$\approx -\nabla \cdot [\bar{m}_A(c_P \overline{T\mathbf{v}}^{(A)} + \overline{\Phi \mathbf{v}}^{(A)})$$

$$+ \bar{m}_w c_w \overline{\tau_w \mathbf{v}_w}^{(M)} + L_v \bar{\mathbf{J}}_v - L_f(\bar{\mathbf{J}}_i + \bar{\mathbf{J}}_{ni}) + \bar{H}_T^{(1,2)\downarrow}]$$

Because the heat fluxes H_T, H_s, and $\overline{\mathbf{v}T}$ are highly temperature dependent (Adem, 1962; Saltzman, 1964, 1967), these energy equations can be reduced through parameterization to approximate equations governing the mean or surface temperature of the climatic system. It is this property that has led to the prominent role being played in recent years by the energy balance models based on the above equations.

5.1.2.7. *Horizontal Space Averaging of the Simplified Energy Equation.* It is a simple matter now to apply the zonal-averaging operator, defined in Section 3, to make equations (5.31)–(5.35) applicable to the axially symmetric component of the total macroclimatic variability. Thus, for example, (5.31) and (5.34a or b) which govern the atmosphere and the subsurface (ocean and biolithosphere, including surface ice and continental waters in all forms) yield the following, respectively:

$$\text{(5.36)} \quad \frac{\partial}{\partial t}(c_P\langle \bar{m}_A \bar{\bar{T}}^{(A)}\rangle) \approx -\frac{\partial}{a \cos \phi \, \partial \phi}[\langle \bar{m}_A(c_P\overline{Tv}^{(A)} + \overline{\Phi v}^{(A)})\rangle \cos \phi]$$

$$+ \langle \bar{H}_T^{(1,2)\downarrow}\rangle + \langle \bar{H}_s^{(1,2,3)\uparrow}\rangle + \langle \bar{H}_A^{(4)}\rangle$$

$$\text{(5.37a)} \quad \frac{\partial}{\partial t}(c_w\langle \bar{m}_w \bar{\bar{\tau}}_w^{(M,N)}\rangle + c_i\langle \bar{m}_i \bar{\bar{\tau}}_i^{(M,N)}\rangle + c_\ell\langle \bar{m}_\ell \bar{\bar{T}}^{(N)}\rangle)$$

$$\approx -\frac{\partial}{a \cos \phi \, \partial \phi}(c_w\langle \bar{m}_w \overline{\tau_w v_w}^{(M,N)}\rangle \cos \phi) + \langle \bar{H}_s^{(5)\downarrow}\rangle$$

or,

$$\text{(5.37b)} \quad \langle \bar{H}_s^{(1)\downarrow}\rangle + \langle \bar{H}_s^{(2)\downarrow}\rangle + \langle \bar{H}_s^{(3)\downarrow}\rangle + \langle \bar{H}_s^{(4)\downarrow}\rangle = \langle \bar{H}_s^{(5)\downarrow}\rangle$$

and (5.35b), governing the complete climatic system, becomes

$$\text{(5.38)} \quad \frac{\partial}{\partial t}[(c_P\langle \bar{m}_A \bar{\bar{T}}^{(A)}\rangle + c_w\langle \bar{m}_w \bar{\bar{\tau}}_w^{(M)}\rangle) + (L_v\langle \bar{m}_v\rangle - L_f\langle \bar{m}_i + \bar{m}_{ni}\rangle)]$$

$$\approx -\frac{\partial}{a \cos \phi \, \partial \phi}[(c_P\langle \bar{m}_A \overline{Tv}^{(A)}\rangle + \langle \bar{m}_A \overline{\Phi v}^{(A)}\rangle$$

$$+ c_w\langle \bar{m}_w \overline{\tau_w v_w}^{(M)}\rangle + L_v\langle \bar{J}_{v\phi}\rangle - L_f\langle \bar{J}_{i\phi}\rangle) \cos \phi]$$

$$+ \langle \bar{H}_T^{(1)\downarrow}\rangle + \langle \bar{H}_T^{(2)\downarrow}\rangle$$

The zonal mean fluxes across latitude circles appearing in these equations, of the forms $\langle \bar{J}_{j\phi} \rangle = \langle \bar{m}_j \bar{\psi}_j \overline{v^{(j)}} \rangle$, can be resolved into components representing the flux by mean poloidal motions and by large-scale standing eddies and transient eddies—i.e., if we assume mass variations around latitude circles are neglectable,

(5.39) $\qquad \langle \bar{J}_{j\phi} \rangle = \langle \overline{\bar{m}_j \bar{\psi}_j \bar{v}^{(j)}} \rangle \approx \langle \bar{J}_{j\phi} \rangle_P + \langle \bar{J}_{j\phi} \rangle_E$

where

$$\langle \bar{J}_{j\phi} \rangle_P = \bar{m}_j \langle \bar{\psi}_j \rangle \langle \bar{v} \rangle$$
$$= \bar{m}_j \overline{\bar{\psi}_0 \bar{v}_0}^{(j)}$$

(transport by mean poloidal motions $\langle \bar{v} \rangle$) and

$$\langle \bar{J}_{j\phi} \rangle_E = \bar{m}_j (\langle \overline{\bar{\psi}_j^* \bar{v}^*} \rangle + \langle \overline{\psi'_j v'} \rangle^{(j)})$$
$$= \bar{m}_j [\overline{(\psi_{j1} v_1)_0 + (\psi'_j v')_0}]^{(j)}$$

(combined transport by standing eddy motions $\bar{v}^*$ and transient eddy motions v').

If we next integrate (5.36) and (5.37) over all latitudes, in accordance with (3.23) and (3.29), we obtain for the global atmospheric and subsurface domains, respectively,

(5.40) $\qquad (\partial/\partial t)(M_A c_P \hat{T}^{(A)}) \approx \sigma(\tilde{\bar{H}}_T^{(1,2)\downarrow} + \tilde{\bar{H}}_s^{(1,2,3)\uparrow} + \tilde{\bar{H}}_A^{(4)})$

and

(5.41) $\qquad (\partial/\partial t)[(M_M + M_N)(c_w \hat{\tau}_w^{(M,N)} + c_i \hat{\tau}_i^{(M,N)} + c_\ell \hat{T}_\ell^{(M,N)})] \approx \sigma \tilde{\bar{H}}_s^{(5)\downarrow}$

where M_A, M_M, and M_N are, respectively, the total masses of atmosphere, ocean, and biolithosphere contained in the climatic system (i.e., $M = M_A + M_M + M_N$).

Adding (5.40) and (5.41), we obtain the following statement of conservation of energy for the complete climatic system,

(5.42) $\qquad (\partial/\partial t)[M(c\hat{T})_{A,M,N} + L_v M_v - L_f(M_i + M_{ni})] \approx \sigma \tilde{\bar{H}}_T^{(1,2)\downarrow}$

since

$$M\hat{\psi} = M_A \hat{\psi}^{(A)} + M_M \hat{\psi}^{(M)} + M_N \hat{\psi}^{(N)}$$

and, from (5.30)

$$\tilde{\bar{H}}_A^{(4)} = \tilde{\bar{H}}_s^{(4)\uparrow} - L_v(\partial \tilde{\bar{m}}_v/\partial t) + L_f(\partial/\partial t)(\tilde{\bar{m}}_i + \tilde{\bar{m}}_{ni})$$

It follows from (5.42) that if the complete climatic system were in a steady state (i.e., in equilibrium) we would have

(5.43) $\qquad \tilde{\bar{H}}_T^{(1,2)\downarrow} = 0$

The existence of such an exact radiative equilibrium at any time seems highly unlikely (Saltzman, 1977), though for many considerations this assumption may be acceptable.

5.1.2.8. Observational Studies. Numerous studies have been made to determine from global observations the true nature of the terrestrial energy balance (i.e., the values of the energy reservoirs, their changes, and the energy fluxes and transformations represented by the terms in the foregoing integral equations). We shall not review these observations here, but we note that excellent surveys have been given by Lorenz (1967) and Newell *et al.* (1969) and, more recently, new results have been given by Kubota (1970, 1972) and Oort and Vonder Haar (1976).

5.2. Applications

The vertically integrated energy balance models provide the basis for obtaining first estimates of temperature of the planet, given the incoming radiation as a boundary condition. Thus, from the most severely averaged and simplified form of the energy equation for the complete climatic system

(5.43) $$\widetilde{\widetilde{H}}_T^{(1)\downarrow} - \widetilde{\widetilde{H}}_T^{(2)\uparrow} \approx 0,$$

together with the parameterizations,

(5.44) $$\widetilde{\widetilde{H}}_T^{(1)\downarrow} = \frac{(1 - \tilde{\alpha}_P)}{4} S_0$$

(5.45) $$\widetilde{\widetilde{H}}_T^{(2)\uparrow} = \sigma T_e^4$$

where $\tilde{\alpha}_P$ is the "planetary albedo," σ is the Stefan–Boltzmann constant, and T_e is an "effective" planetary radiative equilibrium temperature, we obtain

(5.46) $$T_e = \left[\frac{(1 - \tilde{\alpha}_P)S_0}{4\sigma}\right]^{1/4}$$

For the earth, where $S_0 = 1340$ Wm^{-2} and $\alpha_P \approx 0.34$, (5.46) gives $T_e \approx 246$ K (a value representative of the mid-troposphere), indicating that a large part of the long-wave radiation originates from the atmosphere and not the earth's surface which has a mean temperature of 288 K. This illustrates the critical role of the atmospheric greenhouse effect in maintaining a relatively warmer, more hospitable, surface temperature, which in its absence would be close to T_e (assuming the same value of α_P).

In order to use (5.43) to determine the surface temperature, the long-wave radiative flux must be parameterized in an empirical form dependent on $\tilde{T}_s$ which takes this greenhouse effect into account (Bryson, 1968; Budyko,

1969; Sellers, 1969). Models for deducing $\bar{T}_s$, and its sensitivity to changed values of the solar constant or atmospheric composition, are listed in Table IV under the categories (a) = 0, (b) = 0. Some of these include ice-albedo feedback (see Section 5.2.1 below) and/or cloud-albedo feedback.

A calculation that is of greater generality and interest than the above is one based on the zonal average models [(b) = 1 − (ϕ)] giving the meridional variation of the temperature. Such calculations, in which no horizontal heat transport is allowed [(a) = 0], have been made by Milankovitch (1930), Saltzman (1968b, cases A, B, and C) and Kurihara (1971), for examples. As would be expected, these all show that in the absence of horizontal heat transport an unreasonably large temperature gradient is established that is never realized because horizontal waves and mean poloidal motions develop in the atmosphere and oceans to transport heat poleward.

Atmospheric climate models based on highly simplified forms of (5.31), taking into account this poleward transport of heat by an assumed Austausch mechanism [(a) = 2], were originated by Defant (1921), and developed further by Ångström (1926), Hess and Frank (1953), Fritz (1960), Adem (1962), and Saltzman (1968b), among others. These showed that a reasonable simulation of the observed mean meridional distribution of a representative surface or atmospheric temperature could be obtained, accounting for a large part of the spatial and seasonal variance of this important climatic element.

5.2.1. The Budyko–Sellers Ice-Albedo Feedback Models. A significant innovation in the above zonal-average models was made by Budyko (1969) and Sellers (1969), who introduced simple representations of the dependence of ice coverage on the annual mean surface temperature $\bar{T}_s$, and of albedo on the ice coverage, thereby allowing the possibility for a destabilizing positive feedback under altered boundary conditions (e.g., a reduced value of the solar constant S_0). The basic equation for these models is (5.38), which, using (5.39), we can write in the form,

$$(5.38') \quad \frac{\partial}{\partial t}\langle U_1 \rangle + \frac{\partial}{\partial t}\langle U_2 \rangle \approx -\frac{\partial(\langle J_1 + J_2 \rangle \cos \phi)}{a \cos \phi \, \partial \phi} + \langle \bar{H}_T^{(1)\downarrow} \rangle - \langle \bar{H}_T^{(2)\uparrow} \rangle$$

where

$$U_1 = c_P \bar{\bar{m}}_A \bar{\bar{T}}^{(A)} + c_w \bar{\bar{m}}_w \bar{\bar{\tau}}_w^{(M)} \quad \text{(sensible heat)}$$

$$U_2 = L_v \bar{\bar{m}}_v - L_f(\bar{\bar{m}}_i + \bar{\bar{m}}_{ni}) \quad \text{(latent heat)}$$

$$\langle J_1 \rangle = c_P \bar{\bar{m}}_A \langle \overline{\overline{Tv}} \rangle^{(A)} + \bar{\bar{m}}_A \langle \overline{\overline{\Phi v}} \rangle^{(A)} + c_w \bar{\bar{m}}_w \langle \overline{\overline{\tau_w v_w}} \rangle^{(M)}$$

$$= \langle J_1 \rangle_P + \langle J_1 \rangle_E$$

(meridional flux of sensible heat and potential energy by poloidal and eddy motions) and

$$\langle J_2 \rangle = L_v \langle J_{v\phi} \rangle - L_f \langle J_{i\phi} \rangle$$
$$= \langle J_2 \rangle_P + \langle J_2 \rangle_E$$

(meridional flux of latent heat by poloidal and eddy motions) to which are added parameterization relations of the form

(5.47) $$H_T^{(1)\downarrow} = R_0(1 - \alpha)$$

where

$$\alpha = \alpha(\sigma_i(\bar{\bar{T}}_s)) = \alpha(\bar{\bar{T}}_s) \quad \text{(planetary albedo)}$$

and

(5.48) $$H_T^{(2)\uparrow} = f(T_s)$$

In Table V we list the particular parameterizations and simplifications of (5.38) made by Budyko and Sellers.

Note that the basic assumptions made in both models are that, (1) an equilibrium steady state exists with no changing storage of energy in the climatic system $[\partial(U_1 + U_2)/\partial t = 0]$, (2) all representations and parameterizations can be expressed in terms of the annual mean surface temperature $\bar{T}_s$, and, (3) the horizontal convergence of energy can be expressed in a simple Newtonian form (Budyko) or by a simple eddy diffusive mixing formula (Sellers).

Sellers also attempts to model crudely the transports due to mean poloidal motions, by the relation,

$$\overline{\langle \bar{v} \rangle \langle \bar{\psi} \rangle} = \begin{cases} a(\phi)\left(\dfrac{\partial \bar{\bar{T}}_s}{\partial \phi} + \left|\dfrac{\partial \bar{\bar{T}}_s}{\partial \phi}\right|\right)\langle \bar{\psi} \rangle_s, & (\phi > 5°\text{N}) \\[2ex] -a(\phi)\left(\dfrac{\partial \bar{\bar{T}}_s}{\partial \phi} - \left|\dfrac{\partial \bar{\bar{T}}_s}{\partial \phi}\right|\right)\langle \bar{\psi} \rangle_s, & (0 < \phi < 5°\text{N}) \end{cases}$$

where $a(\phi)$ is empirically determined. We shall have more to say about this attempt in Section 5.4.

In spite of the extreme crudeness of the parameterizations, the Budyko–Sellers models yield acceptable first-order estimates of the annual zonal mean surface temperature. The main results, however, concern the sensitivity (cf. Section 7.2) of the solutions for ice coverage, associated with the surface temperature, to altered incoming radiation (e.g., an altered solar constant S_0). It is shown that, because of the destabilizing effect of the positive ice-albedo feedback, relatively small changes in the solar constant

TABLE V. The Ice-Albedo Feedback Model Parameterization of Budyko (1969) and Sellers (1969)

$$\frac{\partial \langle \langle J_1 + J_2 \rangle \cos \phi \rangle}{a \cos \phi \, \partial \phi}$$

Author	$\frac{\partial \langle U_1 + U_2 \rangle}{\partial t}$	$\langle J_1 \rangle_E$	$\langle J_2 \rangle_E$	$\langle J_1 \rangle_P$	$\langle J_2 \rangle_P$	$\alpha(T_s)$	$\langle \hat{H}_T^{(21)} \rangle$
Sellers (1969)	0	$-k_T \dfrac{\partial T_s}{\partial \phi}$	$-k_\varepsilon \dfrac{\partial \varepsilon(T_s)}{\partial \phi}$	$-c_1 \left(\dfrac{\partial T_s}{\partial \phi} \pm \left\| \dfrac{\partial T_s}{\partial \phi} \right\| \right) T_s$	$-c_2 \left(\dfrac{\partial T}{\partial \phi} \pm \left\| \dfrac{\partial T_s}{\partial \phi} \right\| \right) \varepsilon(T_s)$	$b_1(\phi) - a_1 T_g,\ (T_g < 283.16)$ $b_1(\phi) - a_2,\ (T_g > 283.16)$ $T_g = T_s - a_3 \langle h \rangle$	$\sigma T_s^4 (1 - m \tanh a_4 T_s^6)$
				+: $\phi > 5°N$			
				−: $\phi < 5°N$			
Budyko (1969)	0			$\beta(T_s - \hat{T}_s)$		0.32 (ice), $T_s \leq 263$ K 0.62 (no ice), $T_s > 263$ K	$a_2 + b_2 T_s - (a_3 + b_3 T_s) \bar{n}$

can lead to relatively large changes in ice coverage. For a critical decrease of S_0 of roughly 2 % an equilibrium solution corresponding to total glaciation of the planet is obtained (i.e., a runaway glacier). Such possibilities are generally absent in models not containing this feedback. Although this effect is undoubtedly exaggerated in these simple models with their limited degrees of freedom and limited number of negative feedbacks that can act as buffers, the models do suggest the importance of including this particular feedback in assessing climatic thermal responses. For example, it seems clear that the response to earth-orbital radiation variations should be underestimated to some degree if this feedback is not incorporated, as in Shaw and Donn (1968) and Saltzman and Vernekar (1971b) (cf. Suarez and Held, 1976; Section 7.2.2). On the other hand, we must recognize that the growth of ice not only affects the surface albedo (r_s) leading to a positive feedback, but also affects other surface state parameters, i.e., the subsurface conductive capacity k, and the water available for evaporation w, both of which can lead to negative feedbacks. Sea ice, for example, acts as an effective insulator preventing the vertical flux of sensible and latent heat from the oceans possibly leading to a net warming of the oceans (see the Appendix). From another viewpoint, the Ewing and Donn (1956) scenario regarding ice-age oscillations rests largely on a negative feedback between ice coverage and water availability for supplying the hydrologic cycle.

Many further studies and extensions of the Budyko–Sellers ice-albedo feedback models, have been made recently. These include investigations of: (1) alternate parameterization of the ice-albedo feedback and long-wave radiation (e.g., Faegre, 1972; Gordon and Davies, 1974; Lian and Cess, 1977), (2) alternate considerations of the heat transport (e.g., Lindzen and Farrell, 1977), (3) time-dependent versions of the models in which $\partial(U_1 + U_2)/\partial t$ is approximated by $c\, \partial \bar{\bar{T}}_s/\partial t$ (e.g., Dwyer and Petersen, 1973; Schneider and Gal-Chen, 1973; Held and Suarez, 1974; North, 1975a,b; Frederiksen, 1976; Ghil, 1976; Su and Hsieh, 1976), revealing the sensitivity of the solutions to initial conditions (e.g., their transitivity–intransitivity properties, Section 7.1.1.2) and, (4) the analytical properties of these models, particularly their stability with regard to small displacements of the $(T_s - \alpha - \sigma_i)$ system from its equilibrium positions (e.g., with regard to the Budyko model, see Held and Suarez, 1974; Chýlek and Coakley, 1975; North, 1975a,b; Frederiksen, 1976; Su and Hsieh, 1976; with regard to Sellers' model, see Ghil, 1976; Drazin and Griffel, 1977). (A background discussion concerning the time dependent considerations involved in items 3 and 4 above is given in Section 7.1.)

It is shown in these studies that the degree of destabilization that can lead to a runaway glaciation for changes in external parameters (e.g., S_0) can be sensitive to the particular forms of parameterization used, but is relatively insensitive to *small* perturbations in initial conditions.

5.2.2. Multidomain Zonal Average Models. The Budyko–Sellers type models described above, based on (5.38), pertain to the complete, vertically integrated climatic system [(c) = 1a]. Models based on a two-domain resolution, in which the atmosphere and subsurface are treated separately but coupled by heat flux processes at their interface (i.e., at the surface), provide the possibility for independently calculating an atmospheric temperature T_a and a surface temperature T_s. Such models involve two energy equations, one for the atmosphere (5.36) and one for the subsurface (5.37a,b), and are designated by (c) = 2a or 2b in Tables III and IV. Examples of models based on such pairs of equations are found in the studies of Adem (1963, 1964a, 1965), Saltzman (1968b), Kubota (1972), Petukhov and Feygel'son (1973), Petukhov (1974), Paltridge (1975), Sergin and Sergin (1976), Suarez and Held (1976), and Taylor (1976).

These two-domain models require parameterization of the vertical heat fluxes not only at the top of the atmosphere (Budyko–Sellers parameterization of $H_T^{(1)\downarrow}$ and $H_T^{(2)\uparrow}$ given in Table V), but also at or near the earth's surface, i.e., $H_s^{(1)\uparrow}$, $H_s^{(2)\uparrow}$, $H_s^{(3)\uparrow}$, $H_s^{(4)\uparrow}$, and $H_s^{(5)\uparrow}$. A summary of various commonly used forms for these parameterizations is given by Budyko (1974) (see also, Table VI).

No purely thermodynamic models in which an attempt is made to differentiate the deep ocean processes from the active upper ocean layer processes have yet been constructed. Such a three-domain model [(c) = 3] will be of great importance in studying longer term climatic variations and will depend critically on adequate parameterization of the vertical heat fluxes between the abyssal and upper layers of the ocean.

5.2.3. Other Refinements and Applications. In addition to the ice-albedo feedback and multilayer formulations described above, other levels of complexity have been fruitfully introduced within the framework of the vertically integrated thermodynamic models. These may best be summarized by referring, in order, to the list of model properties given in Table III.

5.2.3.1. Representation of the Horizontal Eddy Flux of Heat. Of particular relevance in this connection are schemes for deducing rather than prescribing the large-scale eddy diffusion K (see item 4). Such formulations represent an important step forward in increasing the degrees of freedom of the model, allowing for a significant feedback in the climatic system. Most of the attempts along this line have been based on theoretical results from classical baroclinic wave theory (Charney, 1947; Eady, 1949; Kuo, 1952). From this theory heat transport properties representative of the initial growth phases of the major midlatitude storm systems can be deduced, albeit for idealized zonally and meridionally uniform conditions (Saltzman, 1968b; Green, 1970; Saltzman and Vernekar, 1971a; Stone, 1972; see also Schneider and

Dickinson, 1974, for an excellent review). These studies point to a primary dependence of K on the meridional temperature gradient ($K \sim \partial T/\partial \phi$), a result that is in general agreement with the observed seasonal variation of K in middle latitudes, but is in only partial agreement with the observed latitudinal and vertical variations.

In view of this questionable spatial agreement, Saltzman (1968b) and Saltzman and Vernekar (1971a) used a simplified formulation of the diffusive heat flux, in which K is internally deduced as a spatial constant satisfying an integral constraint on the total energy of the atmosphere. The necessity for further theoretical work in developing an improved parameterization of the eddy heat flux, taking into account the finite amplitude properties of the evolving baroclinic waves in a meridionally nonuniform zonal mean state, on a sphere, is indicated. Some recent progress is achieving a better representation of $\overline{v'T'}$, taking some of these factors into account, has been obtained by Stone (1974) but has yet to be incorporated in climate models.

5.2.3.2. Spatial Resolution. In several studies the consideration of zonally averaged conditions has been augmented by a rough differentiation between continent and ocean properties (Kubota, 1972; Sellers, 1973; Sergin and Sergin, 1976; Suarez and Held, 1976). Treatments that are more truly two-dimensional in the horizontal [(b) = 2 − (λ, ϕ)], yielding solutions for the geographic distribution of temperature $T(\lambda, \phi)$, have been considered by Adem (1964a, 1965, see also Shaw and Donn, 1968, 1971), Vernekar (1975), and Sellers (1976).

5.2.3.3. Climatic Domains. In Section 5.2.2 we have already discussed this important category of extensions of the vertically integrated thermodynamic models.

5.2.3.4. Dependent Variables and Feedbacks. Models which govern variables additional to the surface temperature T_s, represent significant enlargements of energy balance modeling capability. Such an introduction of new variables, and of new equations governing them, leads to the possibility for new feedbacks that can have profound effects on the sensitivity of the solution to changes in external parameters. Two examples that we have already discussed are (i) the Budyko–Sellers introduction of a variables surface albedo or ice coverage, $r_s(\sigma_i(T_s))$ (see Section 5.2.1), and (ii) the introduction of a variable eddy diffusivity, $K(T)$ [see Section 5.2.3.1 above].

As a further example, we note that the planetary albedo α_P is affected not only by the surface ice coverage but also by cloud coverage and type. Although it has proved difficult to introduce the cloud properties $\mathscr{C}$ as new variables that can be parameterized in terms of temperature alone, some preliminary attempts have been made to achieve such parameterizations in terms of other highly temperature-dependent hydrologic variables such as relative humidity r_v, and evaporation E (Paltridge, 1974; Petukhov, 1974; Weare and Snell, 1974; Sasamori, 1975; Sellers, 1976; Sergin and Sergin, 1976).

The precise nature of the total feedback between cloud and surface temperature has also been difficult to assess, not only because of the difficulty of relating cloud properties to temperature, but also because of the close tradeoff between opposing effects of short-wave radiative reflection and long-wave radiative emission (Schneider, 1972). In this connection, numerous studies have been made to test the sensitivity of thermodynamic models to prescribed cloud distributions, a subject more properly discussed in Section 5.2.3.10 below. In a recent review, Cess (1976, 1977) concludes that to a good first approximation the opposing long- and short-wave effects very nearly cancel, leaving the climatic system relatively insensitive to cloud amount (an approximation explicitly made by Budyko 1969, for example) but this has been disputed by Hoyt (1977). In another study Cess (1975) suggests that a consideration of cloud altitude introduces a strong positive feedback with surface temperature. In any event, the mere diagnostic deduction of so primary a climatic element as cloud amount, independent of any feedback consideration, is a major step forward to a complete climatic theory.

As indicated above, other quantities (e.g., r_v, E) have been introduced as new variables in connection with the problem of parameterizing the cloud coverage. These and additional hydrologic variables such as water vapor mixing ratio ε and precipitation P are also of great climatological significance in their own right, independent of any feedback considerations, but take on added importance when such feedbacks are also modelled. For example, the water vapor content ε is proportional to temperature, which in turn is proportional to water vapor content due to its long-wave emissivity properties (the greenhouse effect). Thus a positive feedback prevails that in an extreme limiting case can lead to the runaway effect conjectured to have occurred on Venus (Ingersoll, 1969). Conversely, the phase change components of the hydrologic cycle, precipitation and evaporation, tend to act as negative feedbacks, warming the cooler atmospheric domain of the climatic system by condensational release of latent heat and cooling the warmer surface domain (generally in proportion to its relative warmth). Vertically integrated thermodynamic models by means of which components of the hydrologic cycle other than clouds have been deduced have been presented by Sellers (1973, 1976), Paltridge (1974), Petukhov (1974), and Sergin and Sergin (1976).

Another variable inferred diagnostically in some vertically integrated thermodynamic models is the geostrophic zonal wind field and its associated zonally averaged pressure field. This side calculation is based on the thermal wind equation applied to the deduced meridional temperature gradient, but requires an approximation regarding the surface values of the zonal wind or pressure field (Adem, 1962).

5.2.3.5. Vertical Resolution of Atmospheric Variables. By definition, vertically integrated models contain no vertical resolution within each domain over which the averaging takes place (e.g., the atmosphere, or subsurface ocean). However, it is possible to use the surface value of temperature and the mean mid-tropospheric atmospheric temperature, determined from vertically integrated equations, to estimate an atmospheric lapse rate (Saltzman, 1968b). In some studies, attempts have been made to parameterize the vertical profiles of variables such as temperature, water vapor, and cloud in terms of the surface temperature, particularly for use in explicit radiation calculations (Weare and Snell, 1974), but these models would no longer be categorized as purely vertically integrated.

5.2.3.6. Hydrologic Cycle. Although it is possible to deduce some of the hydrologic variables in an ad hoc manner, as described in Section 5.2.3.4 above, a fuller inclusion of the complete closed cycle, within a purely thermodynamic model, is not possible. This is because the cycle depends critically on the water transport properties of the mean poloidal motions which can only be deduced through a momentum equation. This question is discussed in Section 5.3.

5.2.3.7. The Averaging Period. Vertically integrated thermodynamic models have been solved for annual, seasonal, and monthly mean conditions. Clearly, the solutions for monthly mean conditions are the most desirable, allowing for the treatment of more extreme radiative conditions while still permitting annual mean solutions to be obtained by post-averaging in time. The Budyko–Sellers type models discussed in Section 5.2.1 have all been applied to annual mean conditions, however, leading to inadequacies in modeling the highly seasonal process of ice formation and decay (Kukla, 1975; Suarez and Held, 1976). Many of the multidomain studies discussed in Section 5.2.2 deal with seasonal or monthly means. Studies aimed at elucidating the seasonal-cyclic nature of the climatic response are more appropriately discussed under the next category.

5.2.3.8. Time Dependence of Solution. Most vertically integrated thermodynamic models have been designed to yield equilibrium solutions corresponding to the monthly, seasonal, or annual means, deduced as a pure boundary value problem [(h) = 1a, e.g., the Budyko–Sellers type models] or as the asymptotic steady state of a mixed boundary and critical value problem [(h) = 1b, e.g., Dwyer and Petersen (1973) and Schneider and Gal-Chen (1973)]. Some of the difficulties in such a steady state deduction of the seasonal or monthly mean have been discussed by Taylor (1976), who showed how the phase lags introduced by subsurface thermal inertia can lead to significant inaccuracies. The important problem of deducing the seasonal cycle of monthly mean values [(h) = 2] has been treated by only a few investigators, all with considerable simplifications (Hess and Frank, 1953; Fritz, 1960; Adem, 1963, 1964a, 1965, 1970; Wiin-Nielsen, 1970; Kurihara, 1971; Kubota, 1972; Taylor, 1976). The deduction of the longer term evolutionary behavior of the climatic system [(h) = 3], for which the vertically integrated thermodynamic models might be highly appropriate, has as yet hardly been approached, there being only a few preliminary attempts (Petukhov, 1974; Bryson and Dittberner, 1976; Sergin and Sergin, 1976; Suarez and Held, 1976). Methodologically, the study of Sergin and Sergin (1976) seems most novel, being based on concepts from systems analysis and nonlinear control theory and utilizing electrical analogue methods for solution (in this latter regard, see also Kraus and Lorenz, 1966).

5.2.3.9. Application of the Model. Energy balance models have generally been applied with the aim of (1) accounting for the presently observed climatic distribution [i.e., simulation, (i) = 1], (2) gaining understanding about the operation of special mechanisms of a thermodynamic nature under idealized conditions, and (3) testing the sensitivity of the idealized model climatic system to changes in parameters [(i) = 2], with possible inferences for future climatic change sometimes made on this basis (Budyko, 1977). Included in (3) above are some attempts to calculate the atmospheric climate that would be in equilibrium with known past boundary conditions such as radiation input associated with previous earth-orbital positions (Shaw and Donn, 1968) or previous land–sea distributions (Donn and Shaw, 1977).

No attempt has been made to use these simple models to make time dependent predictions of the future course of climatic change, though, as remarked previously, attempts have been made to account for past evolutionary variations [Bryson and Dittberner (1976) with regard to secular variations forced by postulated atmospheric composition changes over the past 80 years, and Suarez and Held (1976) with regard to earth-orbital changes over the past 150,000 years].

5.2.3.10. Prescribed Parameters Varied. The sensitivity studies alluded to above have dealt mainly with the response to variations of the solar constant, but the effects of variations of many other parameters have also been studied. These include the earth-orbital parameters (Shaw and Donn, 1968; Budyko, 1969; Sellers, 1970a; Suarez and Held, 1976; see also Saltzman and Vernekar, 1971b, which is not a purely thermodynamic model); surface ice coverage (Shaw and Donn, 1971; see also Saltzman and Vernekar, 1975); surface conductive capacity (Saltzman, 1967, 1968b; Taylor, 1976); land–sea distributions (Saltzman, 1967; Sellers and Meadows,

1975); and large-scale eddy diffusivity (Adem, 1962, 1963; Stone, 1973; Held and Suarez, 1974; North, 1975a,b; Gal-Chen and Schneider, 1976).

The main testing ground for studying the effects of variations in atmospheric constituents, such as water vapor (r_v or ε), clouds, aerosols, CO_2, and hydrofluorocarbons, has been the vertical column models [(a) = 0, (b) = 0, 1 − (z), or 2 − (ϕ, z)] wherein detailed multilevel radiation flux calculations are made [as indicated in Section 5.2.3.4, numerous sensitivity studies have been made of the climatic response to prescribed variations in cloud properties (see Möller, 1963; Ohring and Mariano, 1964; Manabe and Wetherald, 1967; Kubota, 1972; Schneider, 1972; Cess, 1974, 1975; Temkin et al., 1975; Ramamathan, 1976)]. Some attempts have been made, however, to use simpler radiation parameterizations within the framework of the vertically integrated models to test the climatic sensitivity to atmospheric composition changes, in some cases allowing for the possibility of horizontal heat transport (Sellers, 1969, 1973, 1974; Bryson and Dittberner, 1976; Temkin and Snell, 1976; Lee and Snell, 1977).

As we would expect, the results of these studies of prescribed variations of parameters show that the sensitivities are crucially affected by the degree to which relevant positive and negative feedbacks are included in the model. For example, as mentioned previously (Section 5.2.1 and 5.2.3.7 above), the direct response to earth-orbital changes is small in the absence of an ice-albedo feedback, but when this feedback is included along with a representation of seasonal fluctuations, the response can be enhanced significantly. On the other hand, a fuller inclusion of the negative feedbacks that prevail in the climatic system [such as the absorption and release of heat by the subsurface media portrayed in multidomain models (cf. Section 5.2.2), requiring a more detailed representation of heat fluxes at the surface], leads to more muted sensitivities [cf. Gordon and Davies (1974) study of the effects of the Smagorinsky (1963) representation of the long-wave radiative flux at both the surface and atmosphere rather than the simpler forms used by Budyko and Sellers].

All of the prescribed parameters listed above have been shown to be capable of having significant effects on the climate. It is noteworthy, however, that the climatic solutions appear to be relatively insensitive to the form of representation of the eddy heat flux [see Table III, (a) = 1 − 4] though its complete omission [(a) = 0] will of course lead to a vastly different climate (Saltzman, 1968b). Other specific sensitivity results can be found in the references given; some of the effects of cloud variations were described above. We note, finally, that further sensitivity studies of the effects of parameter changes have been made using the more advanced momentum models discussed in Section 6.

5.3. Critique

There can be little doubt that the simple vertically integrated, energy balance models discussed in this section constitute an important evolutionary stage in the development of climate models. At the least, they can give an acceptable first estimate of the temperature field within the climatic system, and they give some preliminary understanding of the role that special mechanisms of a primarily thermodynamical nature can play in the real climatic system under certain idealized conditions. This is bound to be of value in formulating and interpreting the results of more realistic statistical-dynamical models and the explicit–dynamical models. At the most, a vertically integrated thermodynamic model might provide a near-optimal description and theory of the gross globally integrated planetary mean conditions and their very long term variations, since the internal momentum dynamics are probably of less significance in this globally integrated case.

If we are interested in the geographic and vertical variation of the climatic state and in the prediction of natural and man-forced changes in climate at specific localities, however, we must seek more realistic and detailed models than these. Even after including all the levels of complexity and enriching the feedback possibilities as described in Section 5.2, serious inadequacies remain in all purely thermodynamic models, particularly the vertically integrated forms we have been discussing in this section.

We discuss next some major terrestrial climatic features that cannot be adequately treated by the thermodynamic models—i.e., the meridional zonation and associated mean poloidal circulations, the nonzonal monsoonal circulation, and the general hydrologic cycle which depends so strongly on these climatic features.

5.3.1. Climatic Zonation and Mean Poloidal Motions.

One of the outstanding observational facts concerning the earth's climate is its banded zonal structure: the low pressure rainbelt associated with converging easterly trade winds (the ITCZ) in the tropics; high pressures, with excess evaporation over precipitation resulting in deserts, in the subtropics; a low pressure, prevailing westerly, storm belt in the middle latitudes; and a predominately easterly zone in the polar regions. Associated with these low level wind and pressure features is a meridionally nonuniform zonation of the basic north–south temperature field characterized by a weak gradient in the tropical and polar regions, and a strong gradient in subtropical through middle latitudes in geotrophic equilibrium with the mean zonal westerly "jet stream" of the upper troposphere. The existence of this climatic zonation is intimately connected to the existence of mean poloidal motions in the atmosphere (i.e., the mean Hadley and Ferrel circulations) which owe their origin to global dynamical processes governed by the equations of motion and involving meridional eddy transports of momentum as well as of heat (Eliassen, 1951; Kuo, 1956).

Although some attempts have been made to parameterize the transport of heat and water vapor by these mean poloidal motions within the context of vertically integrated thermodynamic models (Sellers, 1969; see Section 5.2.1; see also Temkin and Snell, 1976), these attempts are almost purely empirical, containing little of the essential physics by means of which one can deduce these motions and the changes in them that are bound to arise from altered external parameters.

5.3.2. Mean Nonzonal Circulations. Another outstanding feature of the earth's climate is its variability around a latitude circle (i.e., its departure from pure zonality) arising from the nonhomogeneous nature of the earth's surface—the topographically irregular continent and ocean distribution. The changes in all the climatic elements (temperature, wind, rainfall) within and between continents and oceans, the seasonal shifts of atmospheric mass observed on monthly mean pressure charts, and the associated reversals in relative temperature between continents and oceans, all represent a significant part of the total spatial and temporal variance of climate. These features are also closely connected with mean vertical circulations, in zonal as well as meridional planes (the monsoons), that require a consideration of the equations of motion for their deduction.

5.3.3. The Hydrologic Cycle. Without an adequate representation of the mean poloidal and the mean nonzonal circulations referred to above, it is impossible to deduce properly the hydrologic cycle which constitutes such an important part of the terrestrial climate and which plays such an important role in determining the heat sources and sinks in the climatic system. The trade winds, representing the surface branch of the tropical Hadley circulation, are by far the major mode by which water vapor is transported over nearly half of the globe, giving rise to the ITCZ rainband that is the main feature of the near-equatorial climate. Likewise, the seasonal wind and mass shifts over the globe, of which the Southeast Asian monsoon is the most dramatic example, is a major determinant of the global flow of water vapor and of the field of condensation. Thus, there can be no pretensions that a purely thermodynamic model can provide an adequate deductive theory for the global distribution of rainfall, evaporation, or the difference between the two which is the determinant of whether arid or moist climatic conditions will prevail.

We are thus led to the conclusion that only a climate model based on the complete system of equations including the equations of motion [i.e.,

(3.3)–(3.6), (3.12), (3.14), and (3.15)] will be adequate if we are interested in a prognostic theory governing the geographical distribution and seasonal variability of all the important climatic elements.

6. Momentum Models

In order to represent the mean poloidal and monsoonal circulations discussed in Section 5.3, it is clear that we must permit some vertical resolution of the atmosphere $[(e) \neq 1]$. Thus, all the models described in Section 5 are inadequate for deducing these circulations not only because they treat only the thermodynamics, but also because they are based on vertical averaging that eliminates any atmospheric resolution. In fact, only by abandoning the scheme A system of models described in Sections 3 and 4, which depend on an initial vertical averaging, in favor of scheme B, can the momentum dynamics be accommodated.

As we have said, in scheme B an initial separation is made between the zonally symmetric mean state $\psi_0 \equiv \langle \bar{\psi} \rangle$ (encompassing the mean poloidal and toroidal motions) and a zonally asymmetric mean state $\psi_1 \equiv \bar{\psi}^*$ (encompassing the mean nonpoloidal motions induced by the continent–ocean structure). An exposition of this resolution and of a systematic procedure for combining these two basic components into a comprehensive theory of the climatic-mean state, $\bar{\psi}$, was given by Saltzman (1968a). We next present the basic atmospheric equations for each of these component problems.

6.1. Equations for the Axially Symmetric and Asymmetric Mean States

The climatic-mean equations for the atmosphere are given in Section 3 (Table II). If we apply the zonal averaging operator defined by (3.18) to these equations we obtain the following system of equations governing the zonally symmetric mean variables $(\psi_0)_A$:

$$(6.1) \quad \frac{\partial u_0}{\partial t} = \left(f - \frac{\partial u_0 \cos \phi}{a \cos \phi \, \partial \phi} \right) v_0 - \omega_0 \frac{\partial u_0}{\partial p}$$
$$+ F_{\lambda 0} - \left\{ \nabla_3 \cdot [(\mathbf{V}'u')_0 + (\mathbf{V}_1 u_1)_0] - \frac{\tan \phi}{a} [(u'v')_0 + (u_1 v_1)_0] \right\}$$

$$(6.2) \quad \frac{\partial v_0}{\partial t} = -\left(v_0 \frac{\partial v_0}{a \, \partial \phi} + \omega_0 \frac{\partial v_0}{\partial p} \right) - \left(f + \frac{\tan \phi}{a} u_0 \right) u_0 - \frac{\partial \Phi_0}{a \, \partial \phi} + F_{\phi 0}$$
$$- \left\{ \nabla_3 \cdot [(\mathbf{V}'v')_0 + (\mathbf{V}_1 v_1)_0] - \frac{\tan \phi}{a} [(u'^2)_0 + (u_1^2)_0] \right\}$$

$$(6.3) \quad \partial \Phi_0 / \partial p + (R/p) T_0 = 0$$

$$(6.4) \quad \frac{\partial T_0}{\partial t} = -v_0 \frac{\partial T_0}{a\, \partial \phi} + \left(\frac{RT_0}{c_p p} - \frac{\partial T_0}{\partial p} \right) \omega_0 + \frac{q_0}{c_p}$$

$$- \left\{ \nabla_3 \cdot [(\mathbf{V}'T')_0 + (\mathbf{V}_1 T_1)_0] - \frac{R}{c_p p} [(\omega'T')_0 + (\omega_1 T_1)_0] \right\}$$

$$(6.5) \quad \frac{\partial \omega_0}{\partial p} + \frac{\partial v_0 \cos \phi}{a \cos \phi\, \partial \phi} = 0$$

$$(6.6) \quad \frac{\partial \varepsilon_0}{\partial t} = -\left(v_0 \frac{\partial \varepsilon_0}{a\, \partial \phi} + \omega_0 \frac{\partial \varepsilon_0}{\partial p} \right) - \frac{1}{\rho} \mathscr{C}_0$$

$$- \nabla \cdot \left[(\varepsilon' \mathbf{v}')_0 + (\varepsilon_1 \mathbf{v}_1)_0 + \frac{\mathbf{f}_{v0}}{\rho} \right]$$

$$+ g \frac{\partial}{\partial p} [\bar{\rho}(\varepsilon' w')_0 + (\varepsilon_1 w_1)_0 + f^{\uparrow}_{v0}]$$

$$(6.7) \quad \frac{\partial \xi_{j0}}{\partial t} \approx -\left(v_0 \frac{\partial \xi_{j0}}{a\, \partial \phi} + \omega_0 \frac{\partial \xi_{j0}}{\partial p} \right) + \frac{\mathscr{S}_{j0}}{\bar{\rho}}$$

$$- \nabla_3 \cdot \left[(\xi'_j \mathbf{V}') + (\xi_{j1} \mathbf{V}_1)_0 + \frac{\mathscr{F}_{j0}}{\bar{\rho}} \right]$$

If we subtract this system from the original nonzonally averaged system given in Table II, using the definition $\psi_1 \equiv \bar{\psi} - \psi_0$, we obtain the following set of equations governing the asymmetric mean atmospheric variables $(\psi_1)_A$:

$$(6.8) \quad \frac{\partial u_1}{\partial t} = -u_0 \frac{\partial u_1}{a \cos \phi\, \partial \lambda} + \left(f - \frac{\partial u_0 \cos \phi}{a \cos \phi\, \partial \phi} \right) v_1 - \omega_1 \frac{\partial u_0}{\partial p}$$

$$- v_0 \frac{\partial u_1 \cos \phi}{a \cos \phi\, \partial \phi} - \omega_0 \frac{\partial u_1}{\partial p} - \frac{\partial \Phi_1}{a \cos \phi\, \partial \lambda} + F_{\lambda 1}$$

$$- \left\{ \nabla_3 \cdot [(\mathbf{V}'u')_1 + (\mathbf{V}_1 u_1)_1] - \frac{\tan \phi}{a} [(u'v')_1 + (u_1 v_1)_1] \right\}$$

$$(6.9) \quad \frac{\partial v_1}{\partial t} = -u_0 \frac{\partial v_1}{a \cos \phi\, \partial \lambda} - \left(f + \frac{2 \tan \phi}{a} u_0 \right) u_1 - v_0 \frac{\partial v_1}{a\, \partial \phi}$$

$$- v_1 \frac{\partial v_0}{a\, \partial \phi} - \omega_0 \frac{\partial v_1}{\partial p} - \omega_1 \frac{\partial v_0}{\partial p} - \frac{\partial \Phi_1}{a\, \partial \phi} + F_{\phi 1}$$

$$- \left\{ \nabla_3 \cdot [(\mathbf{V}'v')_1 + (\mathbf{V}_1 v_1)_1] - \frac{\tan \phi}{a} [(u'^2)_1 + (u_1^2)_1] \right\}$$

(6.10) $$\partial \Phi_1/\partial p + (R/p)T_1 = 0$$

(6.11) $$\frac{\partial T_1}{\partial t} = -u_0 \frac{\partial T_1}{a \cos\phi\, \partial\lambda} - v_1 \frac{\partial T_0}{a\, \partial\phi} - v_0 \frac{\partial T_1}{a\, \partial\phi}$$
$$+ \omega_1 \left(\frac{RT_0}{c_p p} - \frac{\partial T_0}{\partial p}\right) + \omega_0 \left(\frac{RT_1}{c_p p} - \frac{\partial T_1}{\partial p}\right)$$
$$+ \frac{q_1}{c_p} - \left\{\nabla_3 \cdot [(\mathbf{V}'T')_1 + (\mathbf{V}_1 T_1)_1] - \frac{R}{c_p p}[(\omega' T')_1 + (\omega_1 T_1)_1]\right\}$$

(6.12) $$(\partial\omega_1/\partial p) + \nabla \cdot \mathbf{v}_1 = 0$$

(6.13) $$\frac{\partial \varepsilon_1}{\partial t} = -\left(u_0 \frac{\partial \varepsilon_1}{a \cos\phi\, \partial\lambda} + v_1 \frac{\partial \varepsilon_0}{a\, \partial\phi} + \omega_1 \frac{\partial \varepsilon_0}{\partial p}\right.$$
$$\left. + v_0 \frac{\partial \varepsilon_1}{a\, \partial\phi} + \omega_0 \frac{\partial \varepsilon_1}{\partial p}\right) - \frac{1}{\bar{\rho}} \mathscr{C}_1$$
$$- \nabla \cdot \left[(\varepsilon' \mathbf{V}')_1 + (\varepsilon_1 \mathbf{v}_1)_1 + \frac{\mathbf{f}_{v1}}{\bar{\rho}}\right]$$
$$+ g \frac{\partial}{\partial p}[\bar{\rho}(\varepsilon' w')_1 + (\varepsilon_1 w_1)_1 + f_{v1}^\uparrow]$$

(6.14) $$\frac{\partial \xi_{j1}}{\partial t} = -\left(u_0 \frac{\partial \xi_{j1}}{a \cos\phi\, \partial\lambda} + v_1 \frac{\partial \xi_{j1}}{a\, \partial\phi} + \omega_1 \frac{\partial \xi_{j0}}{\partial p}\right.$$
$$\left. + v_0 \frac{\partial \xi_{j1}}{a\, \partial\phi} + \omega_0 \frac{\partial \xi_{j1}}{\partial p}\right) + \frac{1}{\bar{\rho}} \mathscr{S}_{j1}$$
$$- \nabla_3 \cdot \left[(\xi_j' \mathbf{V}')_1 + (\xi_{j1} \mathbf{V}_1)_1 + \frac{\mathscr{F}_{j1}}{\bar{\rho}}\right]$$

where (see Sections 2.1 and 2.3),

$$\mathscr{F}_j = \mathscr{E}_j + \mathscr{D}_j^{(s)} + \mathscr{W}_j^{(s)} \equiv \mathbf{f}_j + f_j^\uparrow \mathbf{k}$$

Note that in addition to eddy stress terms representing the transport of heat, we now have eddy stress terms in the equations of motion representing the flux of momentum. We shall next discuss, separately, models based on each of the above component systems for ψ_0 and ψ_1, with emphasis on some of the recent work of the writer and his colleagues.

6.2. Symmetric (Zonal) Models

In Table IVB we have attempted to present a complete list of the studies made thus far in which zonally symmetric momentum models and their solutions are discussed, excluding those in which the momentum and thermal eddy stresses are prescribed from observations (cf. Section 4). The crucial new requirement for these models is, therefore, a parameterization scheme for the meridional eddy flux of momentum. The various approximate forms used by different investigators are itemized in Table III under index (k). Of these, only (k) = 4–8 have sound physical bases. The diffusion approximation (k) = 1, and approximation (k) = 2, cannot possibly represent the countergradient flux characteristic of large-scale atmospheric behavior (Starr, 1968). To the extent that the motions are quasi-geostrophic and the thermal wind equation applies, approximation (k) = 3 is nearly the same as (k) = 2 and has the same deficiency with regard to representing the countergradient flux. However, because the latitude phase difference between $(u'v')_0$ and u_0 is relatively small, approximations (k) = 2 and 3 can give fairly reasonable-looking distributions of $(u'v')_0$. The lack of a physical basis for these parameterizations should caution us, though, regarding their more general applicability with altered external forcing, model parameters, or boundary conditions.

One momentum model containing most of the important elements of a complete theory of the zonal mean state, including a hydrologic cycle, in a fairly simple though self-consistent way is that developed over the past 10 years by Saltzman and Vernekar (1971a,b, 1972, 1975). This model is based on the momentum transport parameterization cited as (k) = 4 (Saltzman and Vernekar, 1968). We shall now review this model briefly, demonstrate its reducibility to the simpler vertically integrated thermodynamic models discussed in Section 5, describe some of the results obtained (including some new results to be published shortly), and finally indicate some further improvements that we plan to make.

6.2.1. The Saltzman–Vernekar (SV) Model.

To form this model the following approximations are applied to the fundamental zonally averaged mean atmospheric equations (6.1)–(6.6) [cf. Table IVB]:

1. The climatic average is taken for a winter and summer half-year [(g) = 2],
2. For this averaging period the atmosphere (but not necessarily the subsurface) is assumed to be in a quasi-steady (equilibrium) state, $\partial \psi_{0A}/\partial t = 0$,
3. Transports due to stationary waves are neglected [$(\mathbf{V}_1 \psi_1)_0 = 0$],
4. Mean zonal motions are assumed to be quasi-geostrophic so that the second equation of motion (6.2) takes the form,

$$fu_0 + (\partial \Phi_0 / a\, \partial \phi) = 0$$

When this is combined with the hydrostatic equation (6.3), together with the definition of potential temperature,

$$\theta_0 = (p_s/p)^\kappa T_0 \qquad (\kappa = R/c_P)$$

we have the thermal wind equation,

$$\frac{\partial u_0}{\partial p} = \frac{R}{fp}\left(\frac{p}{p_s}\right)^\kappa \frac{\partial \theta_0}{a\,\partial \phi}$$

5. We assume $\omega = 0$ at the top and bottom of the atmosphere ($p = 0, p_s$).

6. Near the surface we assume that the synoptic eddy stresses are negligible compared to the surface friction which we write in the form

$$F_{\lambda s 0} = -g(\partial \tau_{\lambda 0}/\partial p)_s \approx -(g/\Delta p)\tau_{\lambda s 0}$$

where

$$\tau_{\lambda s 0} = \rho_s C_F u_{0s}^3 \qquad (p = p_s)$$

$$\Delta p \approx 100 \text{ mb}$$

(pressure depth of the planetary boundary layer). Thus for the planetary boundary layer we may write (6.1) in the approximate form,

$$\left(f - \frac{\partial u_{0s} \cos \phi}{a \cos \phi\, \partial \phi}\right) v_{0s} \approx \frac{g \rho_s C_F u_{0s}^3}{\Delta p}$$

7. Near the equator ($\phi < 15°$), where the Coriolis parameter $f \to 0$, we use some quadratic smoothing to determine u_0 and v_{0s}.

8. The climatic variables are assumed to have the following simple vertical distributions, where the values at the midtropospheric level (at $p = p_a \approx 500$ mb $\approx p_s/2$) are taken to be the vertically averaged values so that $\psi_0(p_a) = \overline{\psi}_0$:

(6.15) $$u_0(p) = u_{0s} + (p - p_s)\overline{\left(\frac{\partial u_0}{\partial p}\right)}$$

(6.16) $$v_0(p) = v_{0s}\left(\frac{2p - p_s}{p_s}\right)$$

(6.17) $$\theta_0(p) = \overline{\theta}_0 + (p - p_a)\left(\frac{\partial \theta_0}{\partial p}\right)_A$$

where

(6.18) $$\left(\frac{\partial \theta_0}{\partial p}\right)_A = \frac{2(\theta_{0s} - \overline{\theta}_0)}{p_s}$$

$$\varepsilon_0(p) = \varepsilon_{0s}\left(\frac{p}{p_s}\right)^\gamma$$

where
$$\varepsilon_{0s} = \tilde{r}_{vs}[\varepsilon_{0s}^{(sat)}(\theta_{0s})]$$

These vertical distributions are portrayed schematically in Fig. 3. We also set

(6.19) $$q_0(p) = \mu(p)\overline{q_0} \equiv \mu(p)(g/p_s)H_{0A}$$

where $\mu(p)$ represents the vertical profile of $q_0[\overline{\mu(p)} = 1)]$.

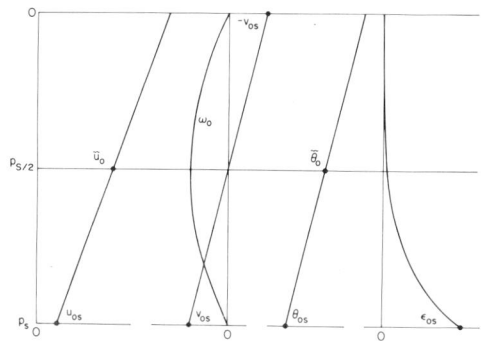

FIG. 3. Assumed vertical variations of zonal mean climatic variables in the SV model.

With these approximations, Eqs. (6.1)–(6.6), expressing conservation of momentum, mass, and energy for the atmosphere become

Momentum:

(6.20) $$-\frac{\partial}{a \cos^2 \phi \, \partial \phi} \{[\overline{\tilde{u}_0 \tilde{v}_0} + \overline{(u'v')_0}] \cos^2 \phi\} - (g/p_s)\rho_s C_F u_{0s}^3 = 0$$

(6.21) $$\frac{\partial u_0}{\partial p} = \frac{R}{fp_a}\left(\frac{p_a}{p_s}\right)^\kappa \frac{\partial \theta_{0a}}{a \, \partial \phi}$$

(6.22) $$\left[f - \frac{\partial(u_{0s} \cos \phi)}{a \cos \phi \, \partial \phi}\right]v_{0s} - \frac{g\rho_s C_F u_{0s}^3}{\Delta p} = 0$$

Mass:

(6.23) $$\omega_0 = \frac{\partial v_{0s} \cos \phi}{a \cos \phi \, \partial \phi}\left[\frac{p(p - p_s)}{p_s}\right], \quad \overline{v_0} = 0$$

(6.24) $$-\frac{p_s}{g \, a \cos \phi} \frac{\partial}{\partial \phi} \{[\overline{\tilde{v}_0 \tilde{\varepsilon}_0} + \overline{(v'\varepsilon')_0}] \cos \phi\} + E_0 - P_0 = 0$$

where

(6.25) $$E_0 = -H_{s0}^{(4)}/L_v$$

(6.26) $$P_0 = H_{A0}^{(4)}/L_v$$

TABLE VI. Heating Parameterizations Used in the SV Model

	Surface heat flux	Parameterization	Reference
$H_s^{(1)\downarrow}$	Solar radiation	$(1-\chi)(1-r_s)(1-r_a)R_0$	Smagorinsky (1963)
		$r_a = (a+bn)n$	Berliand (1960)
$H_s^{(2)\uparrow}$	Terrestrial radiation	$\sigma(v^\downarrow T_a^4 - T_s^4)$	Smagorinsky (1963)
$H_s^{(3)\downarrow}$	Convection	$-[c_1(T_s - T_a) - c_2]$	Saltzman (1967)
$H_s^{(4\mathrm{E})\downarrow}$	Evaporation	$w[c_3 H_s^{(3)\downarrow} - c_4]$	Saltzman (1967)
$H_s^{(5)\downarrow}$	Subsurface flux	$k(T_s - T_D)$	Saltzman (1967)
		$k_M = c_5\left[1 + \dfrac{(v'^2)_0}{(\tilde{v}'^2)_0}\right] - c_6(T_s - T_D),\ k_L = \text{const.},\ k_{Mi} = \text{const.}$	Saltzman and Vernekar (1971a, 1972)

Atmospheric diabatic heating

$H_A^{(1)}$	Solar radiation	$\chi(1-r_a)R_0$	Smagorinsky (1963)
$H_A^{(2)}$	Terrestrial radiation	$\sigma[\Gamma T_s^4 - (v^\downarrow + v^\uparrow)T_a^4]$	Smagorinsky (1963)
$H_A^{(3)}$	Convection	$H_s^{(3)\uparrow}$	Saltzman (1967)

Energy:

(6.27) $$-\frac{p_s c_P}{g}\left(\frac{p_a}{p_s}\right)^\kappa \frac{\partial}{a \cos \phi \, \partial \phi}\{[\widetilde{v_0 \theta_0} + \widetilde{(v'\theta')}_0] \cos \phi\}$$
$$+ H_A^{(1)} + H_A^{(2)} + H_A^{(3)} + H_A^{(4)} = 0$$

To these atmospheric equations we add the statement of conservation of energy for the complete vertically integrated subsurface medium (ocean, ice, land) in the form of the surface heat balance condition (5.28),

(6.28) $$H_{s0}^{(1)\downarrow} + H_{s0}^{(2)\downarrow} + H_{s0}^{(3)\downarrow} + H_{s0}^{(4)\downarrow} = H_{s0}^{(5)\downarrow}$$

Equations (6.15)–(6.27) will form a closed system for the important climatic variables $\theta_0, u_0, v_0, \omega_0, \varepsilon_0$ (and their vertical variations), as well as for E_0 and P_0, providing we can express (parameterize) the heat fluxes $H_s^{(1)\downarrow} \cdots H_s^{(5)\downarrow}$, $H_A^{(1)} \cdots H_A^{(3)}$, and the transient-eddy stress terms of the form $(\widetilde{a'b'})_0$, in terms of these variables. The particular forms used are given in Tables VI and VII. Note that the parameterization for the surface zonal frictional stress was already incorporated (see 6. above).

To complete the specifications, values must be assigned to the parameters appearing in these parameterizations [i.e., the constants $(c_1 - c_{10},$

TABLE VII. Parameterizations of Atmospheric Transient Eddy Stresses Associated with Weather Variations

Stress		Parameterization
Heat transport	$\widetilde{(v'\theta')}_0$	$-K\dfrac{\partial \widetilde{\theta}_0}{a \, \partial \phi}, \quad K = \left[\left(\dfrac{\partial \widetilde{\theta}_0}{a \, \partial \phi}\right)^2 \left(\dfrac{\partial \widetilde{\theta}_0}{\partial p}\right)^{-1}\right] \left[\dfrac{g 4^\kappa}{2c_P} \sum_{n=1}^{4} \widetilde{\mu}_u^{(T)} \widetilde{H}_{0A}^{(n)} - \widetilde{(\omega_0 \theta_0)}\right]$
		$\widetilde{\psi}^{(T)} = p_a^{-1} \displaystyle\int_0^{p_a} \psi \, dp$
	$\widetilde{(\omega'\theta')}_0$	$K\left(\dfrac{\partial \widetilde{\theta}_0}{\partial p}\right)^{-1}\left(\dfrac{\partial \widetilde{\theta}_0}{a \, \partial \phi}\right)^2$
Meridional wind variance	$\widetilde{(v'^2)}_0$	$A\dfrac{\partial \widetilde{\theta}_0}{a \, \partial \phi}, \quad A = K\left[\left(\dfrac{\partial \widetilde{\theta}_0}{a \, \partial \phi}\right)^2\right]\left[c_8\left(\dfrac{\partial \widetilde{\theta}_0}{a \, \partial \phi}\right)\left(-\rho \theta_0 \dfrac{\partial \widetilde{\theta}_0}{\partial p}\right)\right]^{-1}$
Momentum transport	$\widetilde{(u'v')}_0$	$T\widetilde{(v^2)}_0 \cos \phi \dfrac{d\mu}{d\phi}, \quad \mu = \dfrac{\widetilde{u}_0}{a \cos \phi} - a \cos \phi \left[\beta - \dfrac{\partial}{a^2 \, \partial \phi}\left(\dfrac{\partial u_0 \cos \phi}{\cos \phi \, \partial \phi}\right)\right] n^{*-2}$
		$T = \left[c_{10}\widetilde{u_0^2} - \left(f + \dfrac{\tan \phi}{a}\widetilde{u}_0\right)\widetilde{u_0 v_0}\right]\left[\widetilde{(v'^2)}_0 \cos^2 \phi \dfrac{d\mu}{d\phi}\dfrac{\partial(\widetilde{u}_0/\cos \phi)}{a \, \partial \phi}\right]^{-1}$
		n^* = characteristic wave number for baroclinic instability
Water vapor transport	$\widetilde{(v'\varepsilon')}_0$	$-B\widetilde{(v'^2)}_0 \dfrac{\partial \varepsilon_0}{a \, \partial \phi}, \quad B = \text{const.}$

$\bar{\mu}_1^{(T)} \cdots \bar{\mu}_4^{(T)}$, r_{v0}, B, n^*), the radiation parameters [$v^\downarrow$, $v^\uparrow$, Γ, χ, and $r_A(n_0)$], and the surface state parameters (k, r_s, w)]. Since the earth's surface consists of (1) ocean, (2) sea ice, (3) land, (4) snow, and (5) glacier, the zonal mean values of the surface state parameters can be estimated from

$$(k, r_s, w) = \sum_{N=1}^{5} j_N(\phi)(k_N, r_{sN}, w_N)$$

where $N = 1$–5 denotes the five surface physiogeographic types itemized above and j_N represents the fraction of latitude ϕ occupied by each type ($\sum_{N=1}^{5} j_N = 1$). The basic external forcing is imposed on the model climatic system by the specification of the radiative flux at the top of the atmosphere $R_0(\phi)$, and the temperature at the base of the seasonal thermocline $T_{MD}(\phi)$.

With the parameterizations given in Tables VI and VII the atmospheric energy equation (6.27) can be reduced to the form,

(6.29) $$K \frac{d}{d\xi}\left[(1 - \xi^2)\frac{d\bar{\theta}_0}{d\xi}\right] - \lambda(\xi)\bar{\theta}_0 = F_0(\xi)$$

where $\xi = \sin \phi$, the Austausch coefficient K is a function of $\bar{\theta}_0$ and the other parameters included in the integral expression given in Table VII, $\lambda(\xi)$ is a function of the radiation parameters and surface state parameters, and the forcing function F_0 is also a function of these radiation and surface state parameters as well as of the incoming radiation $R_0(\phi)$, and the subsurface temperature $T_D(\phi)$ and the transports by poloidal motions, $(\widetilde{v_0 \theta_0})$ and $(\widetilde{v_0 \varepsilon_0})$ and by transient eddies $(\widetilde{v'\varepsilon'})_0$. Given K, $\lambda(\xi)$ and $F_0(\xi)$, this equation can be solved numerically by the Gauss elimination method applied to tri-diagonal matrices (Richtmyer, 1957). The solution so obtained for $\bar{\theta}_0$ allows the calculation of $\theta_{0s}(\equiv T_{0s})$ from the surface heat balance condition (6.28), and also of ε_0, $\partial u_0/\partial p$ and the transient eddy statistics $(\overline{v'^2})_0$, $(\overline{v'\theta'})_0$, $(\overline{\omega'\theta'})_0$, and $(\overline{v'\varepsilon'})_0$. In turn, these permit the calculation of the zonal mean motion u_0, v_0, and ω_0 by means of the momentum equations. These values of the mean poloidal motions then permit an improved estimate of F_0 which can be used to obtain a better solution for θ_0. When the process is repeated iteratively a convergent solution for all the variables can be obtained. A schematic portrayal of the iterative scheme is shown in Fig. 4. As it turns out, about 95 cycles through this system, taking roughly 6 minutes of high-speed computer time, are required to achieve convergence.

Before discussing the results obtained, we note that by setting $v_0 = \omega_0 = 0$ and eliminating the momentum subsystem, it is possible to reduce the above system to a purely thermodynamic model. Such a model, (with the radiation and surface state parameters taken to be uniform with latitude for separate land, ocean, and swamplike conditions and with a simplified hydrologic

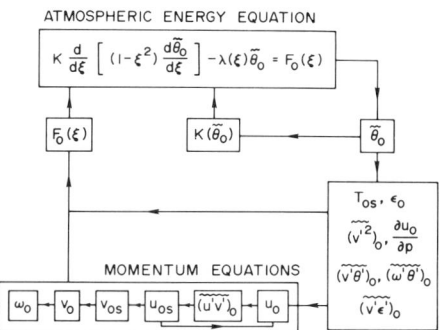

FIG. 4. Schematic representation of the iterative cycle of calculations by which the Saltzman and Vernekar (1971a) model is solved for climatic equilibrium. In the atmospheric energy equation, $\lambda(\xi) = \lambda[(r_s, w, k), (v^\downarrow, v^\uparrow, \Gamma, \chi, r_a)]$ and $F_0(\xi) = F[R_0, T_D, (r_s, w, k), (v^\downarrow, v^\uparrow, \Gamma, \chi, r_a), \overline{(v_0'\theta_0')}, \overline{(v_0'\varepsilon_0')}, \overline{(v'\varepsilon')}_0]$.

cycle in which local precipitation is given by $H_A^{(4)} = \tilde{H}_s^{(4)\uparrow}$) was described by Saltzman (1968b). For this case an analytical solution of the atmospheric energy equation in terms of Legendre functions of zero order P_j can be obtained, as shown in Fig. 5 which schematically portrays the complete iterative solution process. Further simplifications and special cases are discussed in the paper cited, including the results for pure radiative equilibrium ($H_{s,A}^{(3)} = H_{s,A}^{(4)} = H_{s,A}^{(5)} = K = 0$), and "local" equilibrium ($K = 0$, $H_s^{(4)\uparrow} = H_A^{(4)}$). A globally averaged solution similar to the one described by Paltridge (1974), for example, is easily obtained in the form

$$\tilde{\theta}_{0A} = \tilde{F}_0/c^2$$

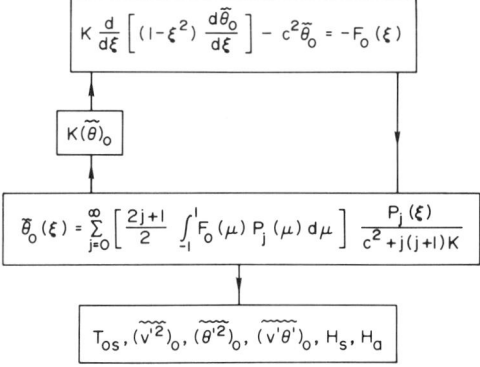

FIG. 5. Schematic representation of the iterative cycle of calculations by which a simplified, purely thermodynamic, version of the SV model (Saltzman, 1968b) is solved for a climatic equilibrium.

from which $\hat{T}_{0s}$, $\hat{H}_s$, and $\hat{H}_A$ can also be calculated in a straightforward manner from the surface heat balance equations.

6.2.1.1. General Results. We return now to the full momentum model schematized in Fig. 4. The results of such calculations for present global boundary conditions show that a good first approximation to the observed climatic zonation can be deduced, including the main differences between the seasons and the hemispheres (Saltzman and Vernekar, 1971a, 1972). Additionally, it is found that in the absence of an ice-albedo feedback, which was not yet included in the model, only minor and probably insignificant changes in global climatic solution occur when the earth-orbital parameters (e, ε, Π) are varied to the extreme values that prevailed at 10,000 and 25,000 years BP (Saltzman and Vernekar, 1971b). Finally, when the model is applied to the surface ice conditions that probably prevailed during the glacial maximum at 17,000 years BP (Saltzman and Vernekar, 1975), we obtain as a solution a new climatic equilibrium state that qualitatively agrees with many inferences made about past climatic zonation from other lines of evidence as well as from other dynamical modeling experiments (e.g., Williams *et al.*, 1974; Gates, 1976).

One aspect of this glacial solution may be of significance for future studies of the ice ages. When the heat fluxes at the surface of the earth are integrated over the globe and over both the summer and winter half-year periods (an annual average), we find that during the glacial maximum there is a net global flux *downward* into the oceans, and during the present period of relatively low glaciation there is a net *upward* flux from the oceans. This result is at least partly a consequence of the fact that sea ice tends to insulate the high latitude ocean waters from heat loss, and its latitudinal extent can be decisive in determining whether a net positive or negative balance will prevail for the globe. Since the model is based on the assumption that the atmospheric climatic adjustment time is small enough that it achieves equilibrium with the much slower varying lower boundary sea surface temperature and ice coverage ($\partial \psi_{0A}/\partial t = 0$), these net surface fluxes must be accompanied by equal and similarly directed net fluxes of radiation at the top of the atmosphere. The implications of such fluxes for long term climatic evolution are discussed by Saltzman (1977). Briefly, if the above results are valid, it is implied by the energy equation for the complete climatic system (5.42) that average ocean temperatures should reach a maximum sometime after maximum glaciation (which may be instrumental in ultimately destroying the glaciation) and reach a minimum after minimum glaciation (which may be the precursor of a new glaciation). This conforms with the ice-age scenario proposed by Newell (1974), suggesting that sea ice and ocean temperature form an oscillatory system. In the Appendix we present some preliminary results concerning an attempt to model such an oscillation.

6.2.1.2. Recent Extensions and Planned Improvements. (a) Ice-albedo feedback: A deficiency of the model discussed thus far is the lack of a formulation for deducing ice coverage and the attendant changes in surface albedo (r_s) and in the other surface state parameters (i.e., conductive capacity k and water availability w), that can lead to significant feedbacks in the climatic system. This is not serious for attempts simply to deduce equilibrium climates corresponding to prescribed surface boundary conditions including snow and ice coverage (as done in Saltzman and Vernekar, 1971a,b, 1972, 1975), but, as implied by the Budyko–Sellers results, can be serious if one wishes to estimate the overall climate response to altered fundamental parameters such as the solar constant. As noted in Section 5.2.1, however, ice coverage leads not only to a positive feedback involving short-wave radiation, but also can lead to other, probably negative, feedbacks through modification of the two other surface state variables, k and w. Moreover, when parameterizations of the radiative heat transfers are used that are more detailed than Budyko's, for example (Gordon and Davies, 1974), a reduced ice-albedo feedback sensitivity is found; and it can be expected that when vertical heat exchange between the atmosphere and the ocean, and horizontal heat exchange due to poloidal motions, are permitted, there will be a significant further buffering of the complete system mitigating positive feedbacks. The SV model, which contains many of these added details and mechanisms, can be used to test these possibilities, providing an ice-extent formulation is included in the model. With this aim in mind, and in view of the intrinsic importance of deducing ice cover as a climatic variable, we have recently added a Budyko–Sellers-type ice representation to the SV model. Specifically, we have postulated that where the temperature is equal to or lower than 268 K in the seasonal average, ice cover exists for that season. Preliminary results indicate that, contrary to the results of Budyko and Sellers, not before the solar constant is reduced by over 20% will permanent glaciation of the entire planet result.

(b) Parameterization of heat fluxes: Clearly, there is much room for improvement of all of the heating parameterizations listed in Table IV. Some of the more straightforward possibilities for improvement that we plan to incorporate are the following:

(i) Short-wave radiation, $H_{A,s}^{(1)\downarrow}$: As noted by Maykut and Untersteiner (1969) and developed further by Schneider and Dickinson (1976), it is important (especially in polar regions) to account for atmospheric absorption of reflected radiation from the surface and for multiply reflected radiation from clouds and the surface. This can be included in a relatively straightforward manner. Furthermore, it will be of importance for sensitivity studies to express the atmospheric short-wave absorptivity χ as a function of the concentration of atmospheric trace constituents such as water vapor, cloud, aerosols, and ozone. Very simple relations of this type for some of these constituents have already been proposed and used by Sellers (1973, 1974) and Bryson and Dittberner (1976), for example. The atmospheric albedo r_a has already been expressed as a function of cloud amount n. A reliable parameterization of cloud amount that can be incorporated in this model will be important not only in this context but also because cloud amount is a primary climatic variable that should be deduced along with the other variables [cf., Section 5.2.3.4].

(ii) Long-wave radiation, $H_{A,s}^{(2)\downarrow}$: In this case also, it will be of importance for studies of climatic sensitivity to atmospheric composition changes, to express the long-wave absorptivity Γ, and the atmospheric emissivity factors $(v^\downarrow, v^\uparrow)$ in terms of atmospheric composition $[\xi(CO_2, \mathscr{A}), \mathscr{C}]$. Again, simple prescriptions, such as those of Sellers (1973, 1974) and Bryson and Dittberner (1976), for $\xi(CO_2)$, may provide acceptable first estimates. In addition, a factor representing the effective surface emissivity v_s can easily be included, though the indications are that this would not be of as much significance as the other surface state parameters, k, r_s, and w.

(iii) Vertical sensible heat flux, $H_s^{(3)\uparrow}$: An improved parameterization of this heat flux including the rectifier effects associated with diurnal and synoptic frequency oscillations near the earth's surface was recently proposed by Saltzman and Ashe (1976). It is relatively easy to use this new parameterization in place of the simple Newtonian form used previously, listed in Table VI.

(iv) Vertical latent heat flux, $H_s^{(4)\uparrow}$: The parameterization listed in Table VI can be shown to be a highly simplified form of Penman's (1948) formula in which the coefficients c_3 and c_4 are taken as constants appropriate for latitude 45° where a mean temperature of roughly 283 K prevails. More generally, c_3 is a function of surface temperature $[c_3 = c_3(T_s)]$ and c_4 is a function of surface temperature, relative humidity, and mean wind speed $[c_4 = c_4(T_s, r_v, |V_s|)]$. The inclusion of these functional dependences would constitute a major improvement in modeling evaporation and the associated latent heat flux.

As a related matter, it would also improve our capability to study climatic sensitivity to changes in ice coverage and continent–ocean distribution to express the mean surface relative humidity $\tilde{r}_{vs}$, appearing in (6.18), as a variable, if not in space, at least in time. It would appear that this quantity might bear some relation to the ratio of water to land and ice for the globe $(\tilde{r}_{vs} \approx \sigma_M/\sigma)$.

(v) Atmospheric latent heat release, $H_A^{(4)}$: One mode of atmospheric heating that is neglected in this model (as well as in most others) is the latent heat of fusion associated with snow formation $(L_f F_{ni})$. To incorporate this additional heating some criterion must be formulated, presumably in terms of temperature, by means of which a distinction between mean precipitation in the form of rain and snow can be made.

(c) Nonequilibrium formulation: Up to now the SV model has been designed only to yield steady-state equilibrium solutions. A natural extension is to seek to include time dependence so that the evolving seasonal and longer term variations can be deduced. Taylor (1976) emphasized the importance of considering the seasonal time dependence and phase lags between radiation and temperature, even if one is only interested in winter and summer steady-state values. We shall have more to say about modeling the time-dependent evolutionary process in Section 7.

The suggested improvements itemized under (a)–(c) above would all serve to increase the simulation and sensitivity capability of the SV model. It is hoped that the results of experiments embodying such improvements will be reported in the near future.

6.2.2. Other Zonally Averaged Models. Noteworthy improvements over some of the features of the SV model described above, as well as over other major momentum model groups [the Williams and Davies (1965)-type models (see also Egger, 1975), the Wiin–Nielsen (1970, 1972; Wiin-Nielsen and Fuenzalida, 1975)-type models (see also Sela and Wiin-Nielsen, 1971), and the MacCracken (1973; MacCracken and Luther, 1974)-type model (see Table IVB)] have been made quite recently by Ohring and Adler (1978) [see also Ohring and Gyoeri (1977)], Webster and Lau (1977) and Taylor (1978).

The Ohring and Adler (1978) model combines the two-layer potential vorticity formulation of the momentum dynamics introduced by Sela and Wiin-Nielsen (1971), the heating formulations for $H_{s,A}^{(3)}$ and $H_{s,A}^{(4)}$ of Saltzman (1968b), the short-wave $(H_{s,A}^{(1)})$ flux formulation of Lacis and Hansen (1974), and the long-wave $(H_{s,A}^{(2)})$ flux formulation of Sasamori (1968, 1970). These radiation parameterizations represent the most detailed treatments of these fluxes yet included in a momentum model, allowing for their dependence on cloud amount and structure as well as on atmospheric constituents. The result is an enhanced capability for making global sensitivity studies on cloud and compositional changes of the atmosphere.

The work of Webster and Lau (1977), in which not only the axially symmetric effects but also the asymmetric effects are considered, represents one of the first serious attempts to treat the dynamics of the ocean domain (including sea ice formation) within the framework of a statistical–dynamical model (see also Pike, 1971).

A significant improvement over the quasi-geotrophic SV formulation for modeling the mean poloidal and zonal (toroidal) motions has recently been achieved by Taylor (1978), who has constructed an essentially three-level (surface, 750 mb, 250 mb) primitive equation dynamical model employing the SV diabatic and eddy transfer parameterizations given in Tables VI and VII. By means of this model the tropical climate, in particular, and the general distribution and seasonal evolution of the climatic zonation described in Section 5.3.1 can be represented with improved fidelity. It should also be possible to consider planetary climates characterized by substantially different fundamental external constants, such as rates of heating and rotation, that can in principle lead to highly nongeostrophic conditions.

6.3. Asymmetric (Nonzonal) Models

As an amplification of the comments made in Section 5.3.2, it is clear from any map of surface climatic fields (temperature) that the departures from zonal symmetry associated with the continent-ocean distribution are of the same order as the meridional variations of the zonally averaged climatic fields. Moreover, when one examines the present pattern of glaciation, and the patterns of glaciation that probably prevailed during the last great ice age at 17,000 BP (CLIMAP, 1976), we see a marked asymmetry around latitude circles. Over shorter, secular, time scales, we note that for regional changes in climate over the last 10 to 100 years, observational evidence points to the importance of shifts in the standing-wave patterns that are probably associated with changes in mean asymmetric heat source and sink distributions (changes in snow cover and sea-surface temperatures, Namias, 1974).

For all these reasons it follows that a theory for the axially asymmetric climatic state is at least as important as the previously discussed theory of the axially symmetric climatic state. Although reasonably good deductions of asymmetric surface temperatures are possible with purely thermodynamic models (Saltzman, 1967; Vernekar, 1975), a fuller theory by means of which mean winds, the hydrologic cycle, and the upper air climate can also be deduced, requires a consideration of the equations of motion also (momentum models). This is because all these features are intimately related to circulations largely in vertical latitudinal planes (the monsoons) as well as to the associated standing horizontal wave motions.

As remarked in Section 4, reviews of linear statistical–dynamical models of the mean asymmetric flow, in which the symmetric mean flow is prescribed, have been presented by Saltzman (1968b) and Manabe and Terpstra (1974). We shall not repeat these summaries here. Suffice it to note that all of the previous studies reviewed, and most of the subsequent studies on this topic (Derome and Wiin-Nielsen, 1972; Webster, 1972, 1973a,b; Egger, 1976a,b,c, 1977; Bates, 1977), have been based on (1) prescribed diabatic heat sources [or prescribed surface temperatures from which heat sources were obtained from a Newtonian-type parameterization (Döös,

1962; Sankar-Rao and Saltzman, 1969)], and (2) prescribed asymmetric transient eddy sources of heat and momentum (Saltzman, 1963; Ashe, 1977), the probable importance of which was noted by Saltzman (1962) and commented on quite recently by Green (1977). As such, these models are not full-fledged climate models in the sense of the working definition discussed at the beginning of Section 4. Moreover, in contrast to most of the climate studies discussed thus far, these asymmetric studies have been concerned almost exclusively with deducing the wind field rather than the temperature field (the emphasis has been on the dynamics rather than the thermodynamics).

In spite of their shortcomings, the above studies have been of great value as diagnostic guides to a more valid climate theory. For example, it is clear from these studies that the effects of mountains and heating in generating the asymmetric climate patterns are of comparable magnitude. One implication is that the topographic blocking effect of ice sheets (such as the Laurentide of 17,000 BP) on the prevailing atmospheric global climate may have been as large as the thermal effects associated with the increased ice coverage. Also, it is clear that quasi-resonant horizontal modes are possible that can become excited or damped depending on the distribution of continent and ocean. This will be of interest in considering the terrestrial climates that prevailed millions of years ago when land–sea distributions were much different.

First attempts, within the framework of a statistical–dynamical model, to deduce the asymmetric mean surface temperature field along with the free atmospheric wave structure, forced by an internally determined (not a prescribed) heating field, have been made very recently by Ashe (1977) and Vernekar and Chang (1978). In treating the heating as a parameterized function of the other climatic variables, these studies open up for consideration a wide variety of possible feedbacks that have hitherto been ignored. Of special interest will be the feedbacks between atmospheric temperature or snow cover mentioned above, though there are many other interesting possibilities.

6.4. *The Complete Time Average State*

In the last analysis, we are interested not in the separate axially symmetric (zonally averaged) climatic component or the axially asymmetric component, but in the sum of these two representing the complete climate, (the temporal means and variances) prevailing at all geographic locations. A theory for this complete time average state can be achieved by coupling models for the two separate components discussed in Sections 6.2 and 6.3, in the manner suggested by Saltzman (1968a) and illustrated schematically in Fig. 6. In this figure, A is the set of free-atmospheric climatic statistics, B is

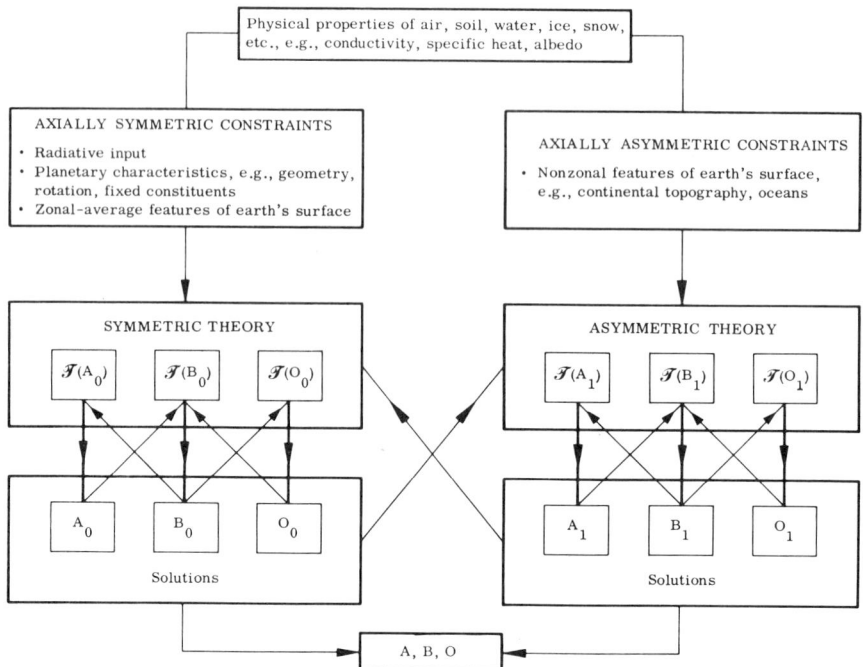

Fig. 6. Schematic structure for an overall climate theory based on coupled symmetric and asymmetric models. Bold arrows indicate solutions and crossed light arrows indicate feedbacks required between the components. (From Saltzman, 1968a.) (Copyright by the American Meteorological Society. Used with permission.)

the set of surface and boundary-layer climatic statistics near the atmosphere–ocean–land interface, O the set of free-oceanic climatic statistics, and $\mathscr{T}$ denotes the appropriate sets of governing equations and boundary conditions. The bold arrows indicate solutions, and the required feedbacks between the components are indicated by the crossed light arrows. As yet no attempts have been made to combine the two problems into a comprehensive statistical–dynamical theory of the global climate in the manner indicated.

As mentioned in Section 4, it may in fact turn out to be more convenient to bypass this scheme entirely in favor of solving the complete time average equations directly, by numerical process, for a global grid network. Steps in this direction have already been taken by Sellers (1976), using a thermodynamic model, and Webster and Lau (1977), using a more complete momentum model. However, it should be recognized that the information gained from the separate problems will be invaluable in formulating and interpreting such a 'primitive climatic equation' model for the globe.

7. Modeling the Evolution of Climate

7.1. General Concepts

We noted in Section 4 that climate models can be classified according to the manner in which they treat time dependence. In particular, we distinguished between models from which we can deduce (1) only steady-state or equilibrium solutions, as a pure boundary-value problem [(h) = 1a] or as an asymptotic mixed initial and boundary-value problem [(h) = 1b], (2) the annual cycle [(h) = 2], or (3) the longer term variations of climate as a fully time-dependent problem [(h) = 3].

7.1.1. Equilibrium Solutions. Most of the studies listed in Table IV, including those discussed in the previous section, are concerned with the steady-state or equilibrium solutions, corresponding to prescribed boundary conditions and some prescribed internal parameters or variables. The equilibria obtained as solutions, which constitute the monthly or annual means climate defined in Section 3, are really functions of time, varying slowly over longer time periods than the averaging period, and hence should more appropriately be called " quasi-steady state " or " quasi-equilibrium " solutions. They are generally based on the assumption that time derivatives of the climatic means can be neglected entirely, or become negligible asymptotically, compared to the other terms in the climatological mean equations given in Section 3.2 [an approximation that can lead to systematic errors due to seasonal phase lags between the atmosphere and subsurface when applied to monthly means (e.g., Taylor, 1976), but is more valid for longer term variations of annual means].

A major goal of these equilibrium studies, and also of the studies aimed at deducing the seasonal cycle of the monthly means, is to account for the presently observed global climate, i.e., simulation [(i) = 1], though many equilibrium models have been developed for the more limited goal of exploring specific mechanisms in an idealized form (radiative processes) [(i) = 2]. These latter have been classified as "mechanistic" models rather than simulation models by Schneider and Dickinson (1974). We must recognize that a reasonable degree of agreement of the solutions with observations (simulation) is a necessary condition for accepting the model as a theory from which the natural variability of climate and the consequences of altered external or internal forcing (natural or man-made) can be deduced or predicted. Although there is much to be proud of, from an intellectual point of view, in being able to deduce the present climate as a quasi-equilibrium condition from the laws of physics and a minimum number of constraints, it is these latter prognostic aims involving nonequilibrium conditions that are the ultimate practical objectives of all our climate modeling efforts [(i) = 3].

7.1.2. Sensitivity Analyses. A first step toward achieving these prognostic aims is to gain understanding of the possibilities for climatic variability through sensitivity analysis of the quasi-equilibrium and seasonal cycle models [(i) = 2a]. By sensitivity analysis we mean, in general, the systematic determination of the changes in the model climate (i.e., equilibrium solutions, seasonal cycle, or the longer term evolution) that result from changes in individual parameters on which such solutions depend, thus revealing the relative importance of these parameters in influencing the climatic system (or model thereof). We may classify these parameters as follows: (i) parameterization constants (or phenomenological coefficients) that are determined semi-empirically (cf. c_1–c_{10} in Table VI); (ii) external forcing parameters (S_0, (e, ε, Π), $h(\lambda, \phi)$, L/M); (iii) internal parameters, or climate variables that are prescribed in a particular model [$\xi(CO_2, O_3$, etc.), $\mathscr{C}$, σ_i, k, r_s, w, K, T_D]; and, possibly, (iv) initial values, or starting values for iterative equilibration.

We may further distinguish between a *partial* sensitivity obtained by fixing the values of parameters other than the one being studied, and a *total* sensitivity obtained by allowing for possible simultaneous changes in the other parameters (Paltridge, 1974; Coakley, 1977; Saltzman and Pollack, 1977).

The sensitivity analysis of the parameterization constants used in a model provides an important test of the validity of deductions from the model when applied to changed physical parameters such as are included under (ii) and (iii) above. Since these constants are generally determined semi-empirically and are therefore known only to some finite accuracy (say, 10 %), changes in them of this magnitude should not lead to significant variations of the solutions (i.e., the model output should not be very sensitive to the fixed model parameters). Such an analysis can be of great value in improving a model by revealing which constant parameters influence solutions the most and hence must be determined with the greatest accuracy.

An important quantity that arises in connection with the sensitivity analysis of the external and internal parameters (ii) and (iii), is the response time for the different components of the climatic system (sometimes also called the adjustment time, equilibration time, or the time constant). This can be defined as the typical time it takes a component domain to equilibrate to a new set of climatic statistics after a change on external forcing or in the climatic values in other domains (the atmospheric response to ocean temperature or ice coverage changes). Observational and theoretical estimates of these for the various domains are given in Table VIII, showing for example, that the atmosphere and surface boundary-layer domains equilibrate quite rapidly compared to the deep ocean and glacial domains. Thus, these latter domains are potentially the deterministic carriers of long-term

TABLE VIII. Estimated Response Time for Various Components of the Climatic System

Climatic domain	Response time
Atmosphere	Weeks–month
Lithospheric surface layer	Month
Sea ice	Month
Inland waters (surface hydrologic cycle) and vegetation	Months–years
Ocean-upper mixed layer	Months–years
Mountain glaciers	100 years
Ocean-abyssal	1000 years
Ice sheets	10,000 years

climatic change, with the statistics of the atmosphere and boundary layer adjusting "quasi-statically" with this longer term change. Some interesting mathematical ramifications of the wide disparity in time scales between the atmospheric and surface components of the climatic system have been discussed recently by Hasselmann (1977), and related discussions of the stochastic behavior of the atmospheric weather in relation to the more slowly varying surface and subsurface climate-bearing components have been discussed in a series of articles by Hasselmann (1976), Frankignoul (1977), Frankignoul and Hasselmann (1977), and Lemke (1977).

7.1.2.1. Uniqueness and Stability of Equilibrium. An important further question regarding uniqueness of the solution arises when considering the sensitivity of model solutions (steady-state solutions, in particular) to initial conditions (item iv above): that is, whether, depending on the initial conditions, there can be more than one equilibrium state satisfying the same model equations and boundary conditions.

This uniqueness question is related to a second important question for a model containing internal feedbacks [particularly of the positive variety, e.g., $r_s(\sigma_i)$], namely, that of the stability of the equilibrium solution. At issue is whether a small or finite perturbation of the steady state will damp (stable case), amplify (unstable case) possibly resulting in some markedly different new state, or lead to an oscillatory response perhaps identifyable with observed semi-oscillatory variations of climate. To investigate these possible dynamic feedback effects one must, of course, consider the full time-dependent equations that were reduced to steady-state equations by setting $\partial/\partial t = 0$.

7.1.1.2. Climatic Determinism. The above uniqueness and stability considerations have much relevance for more general questions of climatic determinism. In the terminology of Lorenz (1964, 1968, 1970), if the set of equilibrium climatic statistics is both stable and unique for all initial conditions when the climatic system (or model thereof) is allowed to evolve over an infinite time period, the system is said to be *transitive* (or *ergodic*). If no such unique equilibrium exists the system is said to be *intransitive*. As a third possibility, if the system is transitive when allowed to evolve over an infinite time period, but in the course of this evolution demonstrates finite periods during which distinctly different climatic regimes prevail (presumably due to internal feedbacks involving different components of the complete climatic system), the system is said to be *almost-intransitive*. The glacial and interglacial periods that prevailed in the past on the earth could be manifestations of such an almost-intransitive system.

7.1.3. Evolution of Climate. We can refer to studies aimed at exploring differences in equilibrium climates obtained from different models, and with different values of the parameters (i)–(iv) for a particular model (sensitivity studies), as exercises in *comparative equilibria*. In such studies the goal is to deduce the climate without concern for the path leading to the equilibria or from one equilibrium to another, or concern for the temporal evolution of the quasi-equilibrium over longer time periods.

However, the terrestrial climate is known to vary significantly and continuously on many time scales, ranging from a year to the age of the earth. Thus, as noted at the end of Section 7.1.1, our interest lies more with the time-dependent path of climatic evolution, especially the future course of climatic change (prediction), than with the equilibria we have been discussing for the most part. This involves an interest not only in the sensitivity of equilibria to changes in parameters (i)–(iv) but also sensitivity of the path of evolution to changes in these parameters. The study of the dependence of such evolution paths on model structure and parameter assignments might be called exercises in *comparative evolution*. We shall be concerned largely with this topic in the remainder of this section.

7.1.3.1. Forced and Free Time-dependent Behavior. There are two basic modes of climatic variability: (1) *forced* variability due to changing external factors that affect the climatic state but are themselves unaffected by the climatic state (astronomical factors such as the solar output and earth-orbital parameters; solid-earth tectonic factors such as continental drift, mountain building, and volcanic activity; human factors such as some forms of pollution and altered land use), and (2) *free* variations due to internal instabilities and feedbacks, usually involving nonlinear interaction among different components of the climatic system, that can occur even if there are no forcing changes.

Although it is natural to look first for possible changes in external forcing as a cause of climatic change, it is entirely possible that the swings in terrestrial climate are more of the free type, reflecting the behavior of an inherently unstable system that never achieves a true equilibrium between all its components. In actuality, the observed climatic variability probably results from a complex interaction of forced and free effects (i.e., we are dealing with a forced, nonlinear, system with dissipative damping and many possible sources of instability).

A simple illustration of the difference between an externally forced variation and one that results from internal instabilities can readily be found at the higher frequency end of the meteorological spectrum containing phenomena with which we are all familiar. In particular, the annual seasonal variation of the climate is a clear example of an externally forced variation imposed by the astronomical variation in the geometry of the earth relative to the sun. As is the case in most of the proposed theories of climatic change due to external causes, a simple linear cause and effect relationship is involved, with the response being of exactly the same frequency as the forcing. However, we know that there are other very significant changes of the environmental variables which bear no direct relation to this fundamental annual forcing or, for that matter, to the diurnal forcing. These are the irregular fluctuations, having frequencies of several days to a week, that are associated with storm movements, that is, with the passage of the highs and lows of the weather map. In a single day, with the passage of a strong weather front, the temperature can change at a station as much as the *mean* temperature changes between summer and winter. This kind of change is the consequence of inherent instabilities in the atmosphere which lead to responses of far different frequency than the forcing. What results is a form of turbulence on a global scale. In much the same way, it could be that what we are looking at in the very long term climatic record over millenia is in the nature of "climatic turbulence." This climatic record may indeed be viewed as a response to the external forcing of the system, such as solar radiative input, but it may bear little similarity in phase, frequency, or amplitude to this forcing.

In the next two subsections we shall discuss, separately, statistical–dynamical modeling studies of forced and free climatic change.

7.2. Studies of Forced Climatic Change

The most regular and best known climatic forcing variations to which the earth is subjected result from the earth's changing orbital relation to the sun: these are primarily (1) the familiar annual or seasonal variation, and (2) the longer term so-called Milankovitch variations associated with changes in the eccentricity e, obliquity ε, and longitude of perihelion Π (i.e., precession) of the earth. We shall begin by discussing studies aimed at deducing the response to these two astronomical forcings, and then we shall discuss very briefly studies aimed at deducing possible responses to two major terrestrial forcings of climatic change: (1) variations in atmospheric composition (e.g., CO_2, aerosols) due to volcanic and human activity, and (2) very long term variations of land–sea and topographic distributions due to continental drift, polar wandering, and mountain-building processes.

There are many other possible astronomical and terrestrial forcing mechanisms, the past or future variations of which we know too little to study in a time-dependent framework. These include such possibilities as changing solar output (e.g., Öpik, 1965), earth collision with interplanetary matter leading to meteor showers (Talbot *et al.*, 1976), and changing geothermal and volcanic heat fluxes. Also, a progressive decrease in the rate of the earth's rotation has been inferred from tidal theory (MacDonald, 1964) and from paleontological studies (Wells, 1963), but no modeling studies of the response to this phenomenon have been performed thus far.

We note also that not always does climatic forcing need to take the form of a continuous periodic or irregularly varying process. It is possible that events of finite duration, perhaps of a catastrophic nature such as volcanic eruptions may occur that can trigger either longer term responses (Bray, 1976) or simply a marked climatic aberration over a short time span (the Krakatoa volcanic explosion).

Similarly, the free internal behavior of the climatic system may include events in one domain that are of a catastrophic nature (e.g., ice sheet surges, see Wilson, 1964), and it can be of value to view these events as a quasi-external forcing of all the other components of the climatic system. More generally, it is always possible to relax our definition of forced climatic change to include the changes imposed by one climatic domain on those of another. Thus, for example, we may consider the atmospheric climate to be forced by prescribed changes in sea-surface temperature or snow coverage, or consider changes in glaciers to be forced by atmospheric conditions (Nye, 1963) and thereby represent a sort of thermometer for the global atmospheric mean temperature. In this discussion we shall generally adhere to the stricter view discussed in Section 7.1.3.1 that forcing of climate can only come from purely external sources.

7.2.1. The Seasonal Cycle. The annual variation of surface and tropospheric temperature has been studied, and simulated to a fairly good first approximation, by means of time-dependent thermodynamic models. The first attempt appears to have been made by Hess and Frank (1953), who included only radiative diabatic processes, and a diffusive meridional heat transport applied to near-surface temperatures (much as was done later by Sellers, 1969). Not included were latent heat processes or subsurface heat fluxes (by consideration of which the nature of the earth's surface enters the problem). A reasonable amplitude of the seasonal cycle at all latitudes was obtained, but the observed phase lag between solar heating and the temperature response was absent.

A further discussion of the annual cycle of zonal mean temperature, based on the diffusion equation with a prescribed heating function approximating the observed annual and meridional variation, was made by Fritz (1960). His main goal was to examine the sensitivity of the solution to changes in the magnitude of the heating (the solar constant) taking into account possible simultaneous changes in the Austausch coefficient.

A more recent, purely thermodynamic, study of the seasonal variation of the zonal mean temperature was made by Kurihara (1971). Here horizontal heat transport was neglected entirely, but more detailed heating parameterizations were used than previously and, in particular, the role of a nonlinear rather than linear representation of long-wave radiation in causing the observed heating temperature phase lag was explored. Subsurface heat fluxes and storage were omitted, however, the importance of which was studied and emphasized later by Taylor (1976). An even more complete recent thermodynamic study than Kurihara's, taking into account horizontal heat transport and subsurface fluxes and showing the sensitivities of the annual temperature cycle to the type of subsurface medium (land vs. ocean), surface albedo, and cloud coverage, was made by Kubota (1972). Although there is no discussion of the results, a similarly complete deduction of the seasonal cycle is implicit in the work of Suarez and Held (1976) (see Section 7.2.2).

We note also that some attempts to use a time-dependent, multidomain, thermodynamic model to predict the future month-to-month variations of the mean temperature starting with known initial conditions have been made by Adem (1964a, 1965) for the complete climatic system, and (Adem, 1970a,b, 1971, 1975) for the ocean surface only.

The seasonal climatic cycle is, of course, of great interest and importance in its own right, but in addition, it has had great significance for all longer term climatic variations. Of particular concern is the seasonal cycle of snow and ice coverage, with progressive longer term changes being critically dependent on small differences between winter growth and summer decay (cf. Section 7.2.2). In this connection, a noteworthy study of the seasonal cycle of sea ice in the Arctic was made by Maykut and Untersteiner (1971) based on a purely thermodynamic model.

Up to now we have discussed only temperature and ice mass variations governed by thermodynamic models. To study the seasonal variations of the complete climate, including zonal and meridional winds and the full hydrologic cycle, requires a consideration of a momentum model. Such applications were introduced by Wiin-Nielsen (1970) who used a two-level quasi-geostrophic, zonal mean, atmospheric model, with a simple Newtonian form of heating as forcing and with no eddy momentum fluxes or hydrologic considerations. Further improvements in this model were made with respect to inclusion of momentum fluxes (Sela and Wiin-Nielsen, 1971), and more rigorous parameterization of diabatic heating (Wiin-Nielsen and Fuenzalida, 1975), leading to reasonably good simulations of the observed annual variations of atmospheric temperature, circulation, and energetics.

Another significant contribution to modeling the seasonal variation of the zonal mean general circulation was made by Kurihara (1973) using his two-level statistical–dynamical, primitive equation momentum model (Kurihara, 1970). Interesting results for a separate land- and ocean-covered earth were obtained.

Finally, we note two momentum model studies specifically concerned with annual variations in the tropics. Pike (1972) introduced an annually varying solar heat input into his interacting atmosphere–ocean model of the tropics (Pike, 1971) yielding solutions for the seasonal migration of the ITCZ which exercises such a large control on the overall tropical climate. A similar study of tropical seasonal variations, using more constrained modeling approximations was made more recently by Lahiff (1975).

7.2.2. Long-Term Earth Orbital Variations. It was suggested qualitatively by Adhémar (1842) and Croll (1875), and later determined quantitatively by Milankovitch (1920, 1930, 1941) that systematic variations of the earth orbital elements (e, ε, Π) having periods of the order of 10,000–100,000 years would lead to varying meridional distributions of insolation that could force climatic changes on the earth and possibly account for the late Cenozoic ice ages. More comprehensive, computer based, calculations of the past and future changes of these elements, and of the implied insolation variations, have been made recently by Vernekar (1972) and in an improved form by Berger (1976a, 1977a,b). Following Milankovitch, all these calculations have been reduced to the form of caloric half-year winter and summer averages, obtained by integrating the insolation over equal time lengths ($\sim$ 182.5 days) centered on the solstices. This use of caloric half-years avoids problems of comparison that would arise if one considered the astronomical half-years (the times between the equinoxes) which are generally of varying unequal length over geologic time. More recently, Berger (1976b, 1977a,c) has tabulated the midmonthly insolation as a function of latitude for the past 1 million years, noting the possible importance of treating these, rather than the more highly smoothed caloric half-year values. This importance is due to the strong probability that seasonally tuned feedbacks and correlations with other variables such as ice cover may produce amplified responses (Bowling, 1971; Kukla, 1975; Suarez and Held, 1976).

The first attempt to deduce the thermal response to the known earth-orbital forcing was made by Milankovitch (1941) using the crudest of possible climate models; i.e., one postulating simple *in situ* radiative equilibrium only, with no possibility for redistribution of heat by any other means. Naturally, a large sensitivity to the forcing was found which appeared to Milankovitch (and many other adherents of his theory) to correlate nicely with the observed Pleistocene temperature and ice variations that were inferred at the time.

This treatment and all those that have followed since have essentially considered the long-term time-dependent climate to be obtainable as a sequence of equilibrium responses to a slowly evolving forcing function [(h) = 3a]. In fact, because of the great amount of computation involved, the first attempts to use more sophisticated thermodynamic models than Milankovitch's aimed only at determining "snapshot" equilibrium solutions for selected dates corresponding to extreme anomalies from present conditions, and hence could be called studies in comparative equilibria (cf. Section 7.1.3). For example, Budyko (1969) applied his annual mean model to 22,000 BP insolation conditions (when high latitude Northern Hemisphere summer radiation was relatively low), and Sellers (1970a) used his (Sellers, 1969) model to deduce the annual mean response to changes in obliquity ε only, for two times (251,000 BP and 230,000 BP) other than the present, keeping all other model parameters constant. Although these Budyko and Sellers models contain an ice-albedo feedback, a very small surface temperature response of less than 1 K for a given latitude was obtained, corresponding to ice boundary movement of about one degree of latitude. However, the use of an annual mean model was questionable, as was the consideration in Seller's case of only the obliquity variations when the full radiation variations due to the combined e, ε, Π effects were equally available.

These deficiencies were absent in Saltzman and Vernekar's (1971b) application of their more elaborate momentum model (Saltzman and Vernekar, 1971a) for half-year summer and winter zonal means to radiative conditions that prevailed at 10,000 and 25,000 BP (corresponding to a maximum and minimum, respectively, at 60°N). However, no ice-albedo feedback was included, and again the surface temperature and other climatic variables showed only relatively small maximum changes in the equilibria. Using Adem's (1965) model, which also excluded the ice-albedo feedback, Shaw and Donn (1968) had previously obtained a similar result as part of a more continuously time-dependent sequence of equilibrium calculations for every 5000 years going back a total of 200,000 years. These calculations were for the geographic distribution of surface temperature and therefore could give the more extreme results for land. The maximum anomalies were of the order of 3 K in certain specific locations and times, but averaged less than 2 K.

More recently, Berger (1977a) reapplied the Sellers (1969) model to the continuous record of earth-orbital insolation variations over a period of 500,000 years centered on the present, and obtained small annual mean responses similar to those obtained by Sellers (1970). He pointed out, however, the desirability of considering the continuous seasonal oscillation of the radiation, feedbacks involving the oceans, and land–sea contrasts. These, in fact, were being taken into account by Suarez and Held (1976), who have made the most extensive calculations to date.

In the Suarez and Held model the atmosphere is represented thermodynamically by two layers, with a thermally active subsurface whereon sea ice and snow may fluctuate mainly as a function of the surface temperature. The ice-albedo feedback thus introduced plays a major role in determining the response to the insolation, which is obtained for every 5000 years of the past 100,000 years in the form of an asymptotically derived equilibrium seasonal cycle of land and sea ice extent. When compared with an index of summer sea surface temperatures derived from a North Atlantic sediment core, the results show an impressive amount of agreement, though there are obvious differences in phase (that may, however, be explainable by auto-oscillations involving the deep ocean-cryosphere system; see Section 7.3.2 and the Appendix). On the basis of this study there has been a renewed acceptance of the possibility that the Milankovitch mechanism plays a significant role in forcing the ice-age oscillations of the Quaternary. It will be of great interest to see whether these results are confirmed with more complete models (models containing momentum dynamics, cloud variation and feedback, deep ocean feedbacks, and better heating and flux parameterizations).

The past ice ages were in fact characterized by huge accumulations of ice, in the form of ice sheets possessing a complex dynamics and inertia of their own that is essentially neglected in all the above thermodynamic models. Two studies have recently been made (Weertman, 1976, Birchfield, 1977) to examine the response of such ice sheets to the Milankovitch insolation changes, assuming these insolation changes are parametrically related to the latitudinal distribution of snow accumulation. It was concluded that the insolation changes are large enough to have bearing on the growth and decay of the ice sheets, and perhaps of more significance, point to the importance of inertial and other nonlinear effects that can introduce phase shifts and strong periodic behavior at frequencies not containing the maximum forcing (e.g., the 100,000 year eccentricity-induced cycle, cf. Hays *et al.*, 1976; Wigley, 1976).

7.2.3. Nonastronomical Forcing. As we have noted there are two major *terrestrial* sources of forcing of climate: (1) fluxes of matter and heat into the climatic system due to tectonic processes (e.g., volcanism) and human activities, and (2) changes in continent–ocean distribution and relief due to a complex of geophysical processes.

Very few studies have been made of the time-dependent response to these forcings, though, in the case of atmospheric composition changes, there have been numerous comparative equilibrium sensitivity studies to determine the magnitude of the responses to be expected (Schneider, 1975, with regard to CO_2 changes). Only two attempts appear to have been made up to now, by Schneider and Mass (1975) and Bryson and Dittberner (1976, 1977), to deduce the continuous secular change in mean surface temperature due to the changing levels of CO_2 and particulates originating from volcanic and anthropogenic causes. Schneider and Mass also discuss the effects of astronomical forcing due to solar constant changes implied by sunspot variations. The results of their studies, which are based on a highly simplified thermodynamic model, seem to show reasonably good agreement with observed temperature trends over the past three centuries (Schneider and Mass, 1975) and the past 80 years (Bryson and Dittbener, 1976). This indicates that these external forcing mechanisms are at least competitive with other possible internal mechanisms in accounting for secular climatic changes, though questions remain regarding the approximations made [as discussed by the authors themselves and also by Woronko (1977) with regard to the Bryson–Dittbener paper].

A noteworthy study concerning the response to land–sea distributions over geologic time has recently been made by Donn and Shaw (1977), who applied Adem's (1965) model to obtain equilibrium solutions for a sequence of five epochs ranging from the early Triassic (200 million years ago) to the present. The results show that not until roughly 10–15 million years ago, when land masses had drifted poleward in the Northern Hemisphere to roughly their present positions, did polar surface temperatures drop below freezing and permit the formation of ice. Thus no extraterrestrial (i.e., astronomical) forcing would need to be invoked to account for the most recent onset of ice ages in the early Cenozoic. We may add, speculatively, that after this point of ice formation, the Milankovitch forcing could begin to play a role due to its sensitivity to the presence of ice-albedo feedback, and moreover, possible free oscillations of the sea-ice/deep-ocean system (such as described in the Appendix) could become significant, leading to the marked glacial oscillations of the late Cenozoic.

7.3. Studies of Free Climatic Change

The climatic system contains possibilities for many different types of positive and negative internal feedbacks that can lead to instabilities or oscillations acting independently of, or strongly modifying, any external forcings such as were described above. To a limited degree these were already included in some of the models described in Section 7.2 (models containing ice-albedo feedback).

7.3.1. Linear Analyses. Some insights regarding the time-dependent behavior resulting from these feedback possibilities can be gained from stability analyses of equilibrium climatic states subjected to small perturbations. The characteristic values (eigenvalues) obtained give information concerning (1) the response time of components of the climatic system (Section 7.1.2) when damped modes prevail, (2) the natural oscillatory periodicities that can be exhibited by the system, and (3) the existence of instabilities that, in the real climatic system, must be checked by negative feedbacks not included in the idealized system treated or by nonlinearities that become important as the small perturbation grows.

Only a few such analyses appear to have been made thus far. One group, mentioned previously in Section 5.2.1, is concerned with the ice-albedo feedback system embodied in the simple Budyko–Sellers thermodynamic models. A second group, by Petukhov and Feygel'son (1973), and Petukhov (1974, 1976) is concerned with a special system involving air–sea interactions (i.e., heat and moisture fluxes) with cloud structure as a feedback regulator. It was found in these studies that, depending on the values of the basic physical parameters determining the equilibrium state, the air–sea system admits amplifying modes, damped modes with time constants of roughly three months, and oscillatory modes of periods also of roughly three months. The type of mechanisms modeled in these papers is therefore incapable of accounting for secular or longer term oscillations of climate. In the Appendix we treat another highly idealized system also involving heat and moisture fluxes between the ocean and air above, but with the much longer term process of annual mean sea-ice formation as the feedback regulator rather than cloud formation. In this case we find that the system can admit free oscillations of periods approaching those of Quaternary glaciations.

7.3.2. Nonlinear Analyses. The real climatic system is replete with possible nonlinearities that can greatly modify its idealized linear properties. The effects of these nonlinearities take several forms. When small (i.e., linear) climatic disturbances begin to grow to large finite amplitudes, new effects become important that may act to further amplify the growth of the disturbance through positive feedback (self-amplification), or damp the growth through negative feedback (possibly leading to quasi-periodic or auto-oscillatory behavior). Moreover, whereas an initial small disturbance, whose behavior is essentially linear, may have little effect on the course of climatic evolution, a large finite disturbance governed by nonlinear equations (e.g., an ice surge, or major volcanic eruptions) may cause a radically different evolution of climate. More generally, as noted in Section 7.1.3.1, the existence of dominant nonlinearity implies that the realized frequencies, phases, and amplitudes of climatic variability will be different from that implied by any external forcing. The general mathematical treatment of all these possibilities will be a subject of increasing climatological interest, bringing to the fore such fields of applied mathematics and engineering as dynamical systems analysis, feedback and control theory, and nonlinear vibrations (Minorsky, 1962).

At this time, theoretical analysis of the effects of nonlinearity on the time dependent behavior of the climatic system, within the context of statistical–dynamical models, has barely begun. As one example, the studies of Petukhov mentioned above included some extension to the nonlinear case. However, the most ambitious and interesting attempt to treat the nonlinearly varying climatic system is that of Sergin and Sergin (1976).

In the Sergins' work a full system of statistical–dynamical equations is developed for the main components of the climatic system (atmosphere, ocean, sea ice, ice sheets), including a highly parameterized representation of the global hydrologic cycle, and the complete time-dependent nonlinear system is solved by an electronic analog computer for a variety of parameter values. The results show that when the temperatures are high only aperiodic solutions are possible, corresponding to the existence of only a single inertia link or carrier of long-term evolution, i.e., the oceans. When temperatures can drop below a critical near-freezing value, however, sea ice and continental glaciers can form which represent a second inertia link. In this case, oscillatory solutions are obtained having properties that depend strongly on the assumed time constants for ocean temperature and ice mass changes. These oscillations may correspond to the ice ages that prevailed in the late Cenozoic. In a general way, this result is compatible with the scenario described by Newell (1974), and the results of Saltzman and Vernekar (1975) and Saltzman (1977) mentioned in Section 6.2.1.1, all of which point to the possibly important role of sea-ice formation and deep-ocean temperature variations as a feedback mechanism for ice-age oscillations. As noted, in the Appendix we present a preliminary development regarding this mechanism.

7.4. Climatic Prediction

The practical objective and ultimate end-product of all climate modeling efforts is the *prediction* of the future climate, including the capability for determining the consequences of purposeful changes in parameters which is the prerequisite for the *management* of climate. Our most immediate concerns are the shorter scale variations ranging from months to tens of years, but changes over hundreds of years and longer are also of immense interest from both practical and intellectual viewpoints. A good discussion of the need for, and benefits to be derived from, a general predictive model of climate is given by Schneider and Mesirow (1976).

It has been determined from sensitivity studies (i.e., studies of comparative equilibria and evolution) that most of the forcing and feedback mechanisms discussed thus far (atmospheric composition changes, earth-orbital and solar constant changes, atmosphere–ocean–cryosphere interactions) are capable of contributing significantly to the observed variability of climate over different, but generally overlapping, time scales. Thus, the prediction of future climate will require the development of models general enough to govern, with adequate fidelity, all such mechanisms operating simultaneously over the time scales for which the prediction is sought. This is a formidable task—one that we are as yet far from achieving. Even if such models were developed, there would still be definite limits to long-term predictability unless one could also predict the variations in solar output, or the occurrence of volcanoes for example.

The time-dependent models considered thus far have usually been concerned with one, or at most a few, of the relevant mechanisms. Some predictions based on such limited models have been ventured, but these have always had to be qualified with the proviso "all other factors being constant" (Bryson, 1968). For example, Budyko (1977) has recently suggested that temperatures are likely to rise over the next quarter century because his model shows that the warming effect of the projected CO_2 increase will be larger than the cooling effect of the projected increase in anthropogenic aerosols. On a longer time scale and from another set of considerations, however, Berger (1977a) has suggested that there should be a cooling trend due only to earth-orbital variations with an expected minimum at about 10,000 years from now. From still another point of view, if we are now in the phase of the sea–ice–ocean temperature oscillation described in the Appendix wherein the oceans are experiencing a net heat *loss* (with a simultaneous net radiative loss from the top of the atmosphere), the earth may be in a stage precursive of colder temperatures and more extensive glaciation (Saltzman, 1977), all other factors being constant.

Clearly, until more refined and general models than any we have discussed in this review are developed, there will remain much room for speculation about the future course of climate based on ad hoc mechanistic models. However, the rate of activity and progress in the development of climate models has been rapid (if not breathtaking) in recent years, and, as we have seen, this has been especially true of the statistical–dynamical models that are so relevant for considerations of longer term climatic variability.

Appendix. A Simple Sea Ice–Ocean Temperature Oscillator Model

A.1. Introduction

We shall now describe a highly idealized, mechanistic, purely thermodynamic model designed to illustrate one of many possible modes of oscillatory behavior that may exist in the terrestrial climatic system. This involves feedback between sea ice extent and mean ocean temperature as discussed in a qualitative way by Newell (1974) and Saltzman (1977).

Briefly stated, sea ice tends to "insulate" the ocean surface over which it forms, preventing the large heat loss that would otherwise occur (Bunker, 1976). Thus, when ice extent is large there will tend to be a net globally averaged surface flux into the ocean ($\tilde{H}_s^{(5)\downarrow} > 0$) warming the oceans as a whole, with the reverse net negative balance tending to prevail when ice extent is small. To the degree that the sea ice extent itself is dependent on the overall ocean temperature, we thus have the makings of a negative feedback loop capable of supporting oscillatory behavior.

To model this system in the simplest possible manner let us consider an all-ocean earth with zero obliquity (i.e., no seasonal variations), a cross section of which is shown in Fig. 7. Here $\xi = \sin \phi$, η is the value of ξ at the ice edge, and D is the mean ocean depth. Our first concern will be with the development of an equation governing the rate of change of the sea ice limit η.

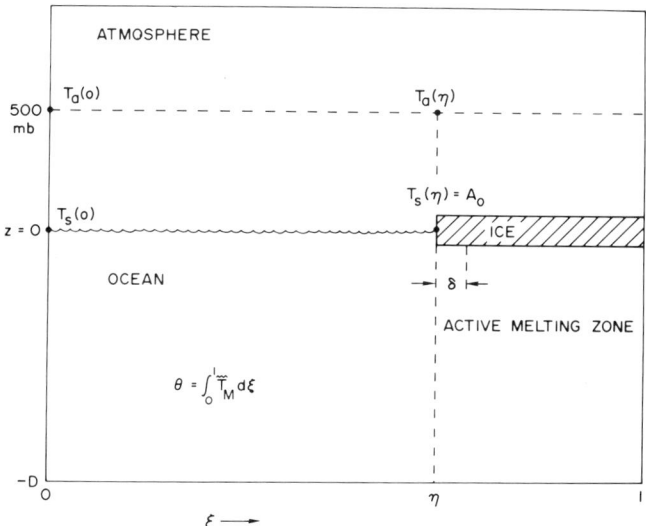

Fig. 7. Idealized atmosphere–ocean–sea-ice system.

A.2. The Ice Limit Equation

If we assume all sea ice changes are due to melting (or freezing) of sea water, neglecting accumulations of sea ice due to snowfall, we can write the continuity equation (5.22) for ice for our system in the form,

$$\text{(A.1)} \qquad \frac{d}{dt}\int_0^1 m_i \, d\xi = -\int_0^1 M \, d\xi$$

(In this Appendix we shall omit the overbar denoting a time average).

Let us assume, further, that (i) sea ice forms to a uniform thickness I as portrayed in Fig. 7; (ii) at the ice edge $\xi = \eta$ the surface temperature is always the same value $T_s(\eta) = A_0$, and (iii) melting (or freezing) of ice is a maximum at the ice edge η and falls off to zero at a distance δ poleward (or equatorward) of η. This distance δ, which defines the active melting/freezing zone, should be related to the distance measured from the ice edge at which the surface temperature *determined from the heat balance condition (5.28) appropriate to the prevailing ocean and atmosphere conditions near* $\xi = \eta$ equals A_0. That is, although $T_s(\eta) = A_0$ at any instant, this need not be an *equilibrium* condition and the value of $T_s = A_0$ as computed from (5.28) may lie poleward of η by a positive distance δ (indicating melting over this increment) or equatorward by a negative distance δ (indicating freezing).

Thus, noting from (i) that

$$m_i = \begin{cases} \rho_i I & \xi \geq \eta \\ 0 & \xi < \eta \end{cases}$$

we can write the left-hand side of (A.1) in the form,

$$\frac{d}{dt}\int_0^1 m_i \, d\xi = -\rho_i I \frac{d\eta}{dt}$$

and, assuming a linear decrease of M from the ice edge, we can write the righthand integral in the approximate form

$$\text{(A.2)} \qquad \int_0^1 M \, d\xi \approx \frac{M(\eta)\delta}{2}$$

(Note that negative values of M indicate freezing). Equation (A.1) therefore becomes

$$\text{(A.3)} \qquad \frac{d\eta}{dt} = \frac{\delta}{2\rho_i I} M(\eta)$$

As we have noted the rate of melting (or freezing) at the ice edge $M(\eta)$ is to be determined from the surface heat balance condition (5.28). We next

STATISTICAL-DYNAMICAL MODELS OF THE TERRESTRIAL CLIMATE   283

describe the approximations we shall make in specifying this condition at any point within the domain $\xi = 0$ to 1.

A.3. The Surface Heat Balance

The surface heat balance condition can be written in the alternate forms (5.28) or (5.32), which apply to every point (cf. Section 5.1.2.4),

(5.28) $\quad H_s^{(1)\downarrow} + H_s^{(2)\downarrow} + H_s^{(3)\downarrow} + H_s^{(4E)\downarrow} + H_s^{(4M)\downarrow} = H_s^{(5)\downarrow}$

(5.32) $\quad = \dfrac{\partial}{\partial t}(m_M c_w \tau) + \nabla \cdot (m_M c_w \overline{\tau \mathbf{v}})$

where $m_M = \rho_w D$, $\tau = T - 273$, $H_s^{(4E)\downarrow} = -L_v E$, and $H_s^{(4M)\downarrow} = -L_f M$.

To evaluate these heat fluxes over the ocean $(0 < \xi < \eta)$ and at the ice edge $(\xi = \eta)$, let us adopt the parameterizations given in Table VI, with the following additional modelling approximations.

1. Because the ocean surface albedo r_s is a prescribable function of latitude, as is the incoming radiation at the top of the atmosphere R_0, we shall combine these into a new function representing the fully transmitted radiation absorbed at the surface,

$$\mathcal{R}(\xi) = [1 - r_{sM}(\xi)]R_0(\xi) \qquad (0 < \xi < \eta)$$

and assume that it can be represented by a linear function of ξ, i.e.,

$$\mathcal{R}(\xi) = \mathcal{R}(0)(1 - a_1 \xi) \qquad (0 < \xi < \eta)$$

where a_1 is a constant. Thus, assuming further that the atmospheric short wave transmissivity, $\mu = (1 - \chi)(1 - r_a)$, can be assigned a uniform constant value, we have for $(0 < \xi \lesssim \eta)$,

(A.4) $\quad H_s^{(1)\downarrow}(\xi) = \mu \mathcal{R}(0)(1 - a_1 \xi)$

At the ice edge we set

$$r_s(\eta) = \dfrac{r_{sM}(\eta) + r_{si}(\eta)}{2}$$

and assume that $r_{sM} = 0.10$ and $r_{si} = 0.70$ so that

$$\mathcal{R}(\eta) = [1 - r_s(\eta)]R_0(\eta) = 0.6 R_0(\eta)$$

and

(A.5) $\quad H_s^{(1)\downarrow}(\eta) = \mu \mathcal{R}(0)(1 - a_1 \eta)\dfrac{0.6}{[1 - r_{sM}(\eta)]}$

$\qquad\qquad\quad = 0.67 \mu \mathcal{R}(0)(1 - a_1 \eta)$

2. In evaluating the long-wave radiative flux we assume the linear expansion,

$$T^4 \approx 4A^3T - 3A^4$$

where $A = 273$ K. Thus, for $(0 < \xi < 1)$,

(A.6) $\qquad H_s^{(2)\downarrow}(\xi) = \sigma\{4A^3[v^\downarrow T_a(\xi) - T_s(\xi)] - 3A^4(v^\downarrow - 1)\}$

in which we identify T_a with the 500 mb atmospheric temperature.

3. The expressions for $H_s^{(3)\downarrow}$ and $H_s^{(4E)\downarrow}(= -L_v E)$ in Table VI can be combined to yield for $(0 < \xi < 1)$,

(A.7) $\qquad [H_s^{(3)\downarrow} + H_s^{(4E)\downarrow}] = -b_1[T_s(\xi) - T_a(\xi)] + b_2$

where $b_1 = c_1(1 + wc_3)$ and $b_2 = [c_2(1 + wc_3) - wc_4]$. For the ocean $w = 1.0$, and for ice we take $w = 0.2$; so that at the ice edge ($\xi = \eta$) we can take the mean, $w_\eta = 0.6$.

4. The upward flux of heat from the ocean at the ice edge will be related linearly to the mean ocean temperature ϑ, i.e.,

(A.8) $\qquad H_s^{(5)\uparrow}(\eta) = b_3(\vartheta - A_0)$

where

(A.9) $\qquad \vartheta = \frac{1}{D}\int_0^1 \int_0^D T\, dz\, d\xi$

This is equivalent to assuming $T_D(\eta) = A_0 + \alpha(\vartheta - A_0)$ (where α is a fractional constant) and setting $k = a$ constant in the expression for $H_s^{(5)\downarrow}$ given in Table VI, which would imply that $b_3 = k\alpha$. Equation (A.8) represents a crude attempt to express the influence of the overall deep ocean temperature on the surface heat balance at the ice edge.

5. To complete the specifications we now postulate functional forms for $T_s(\xi)$ and $T_a(\xi)$. The ocean surface temperature is assumed to decrease linearly with ξ from the equator [where $T_s = T_s(0)$] to the ice edge where we have assumed $T_s = T_s(\eta) = A_0$. Thus,

(A.10) $\qquad T_s(\xi) = T_s(0)[1 - (\xi/\eta)] + A_0\,\xi/\eta$

The atmospheric temperature, a representative value of which is taken to apply at 500 mb, is also assumed to decrease linearly with ξ from the equator to the pole such that a constant vertical temperature difference (i.e., static stability) is maintained at all times over the ice edge $\Gamma(\eta) = [A_0 - T_a(\eta)]$, and over the equator $\Gamma(0) = [T_s(0) - T_a(0)]$. Thus, we have

(A.11) $\qquad T_a(\xi) = T_a(0) - a_2(\xi/\eta)$

where $T_a(0) = [T_s(0) - \Gamma(0)]$ and $a_2 = [T_s(0) - A_0 + \Gamma(\eta) - \Gamma(0)]$. It follows that the atmospheric temperature will vary as a function of sea surface temperature and ice coverage, behaving as a fast response part of the system (cf. Table VIII) that equilibrates quasi-statically with its underlying surface conditions.

We are now in a position to use (5.28) together with (A.5, A.6, A.7, and A.8) to evaluate $M(\eta)$ appearing on the right-hand side of (A.3). We obtain

(A.12) $$M(\eta) = \Lambda_1 \vartheta - \Lambda_2 \eta + \Lambda_3$$

where

$$\Lambda_1 = b_3/L_f$$

$$\Lambda_2 = (0.67\mu\mathcal{R}(0)a_1)/L_f$$

$$\Lambda_3 = \frac{1}{L_f}[0.67\mu\mathcal{R}(0) + \sigma\{4A^3[v^\downarrow T_a(\eta) - A_0] - 3A^4(v^\downarrow - 1)\}$$
$$- b_1(\eta)[A_0 - T_a(\eta)] + b_2(\eta) - b_3 A_0]$$

and can thus write the ice extent equation (A.3) in the final form

(A.13) $$d\eta/dt = X_1 \vartheta - X_2 \eta + X_3$$

where $X_1 = \Lambda_1 \delta/2\rho_i I$, $X_2 = \Lambda_2 \delta/2\rho_i I$, and $X_3 = \Lambda_3 \delta/2\rho_i I$ with δ and I remaining to be specified.

A.4. The Mean Ocean Temperature Equation

To close the system for η, we now integrate the ocean heat balance equation (5.32) from $\xi = 0$ to 1, to obtain an equation for the rate of change of the mean ocean temperature [cf. (A.9)],

$$\frac{d\vartheta}{dt} = \frac{1}{c_w \rho_w D}\left\{\int_0^\eta [H_s^{(1)\downarrow} + H_s^{(2)\downarrow} + (H_s^{(3)\downarrow} + H_s^{(4E)\downarrow})]\,d\xi\right.$$
$$\left. - \int_\eta^1 H_{si}^{(5)\uparrow}\,d\xi - L_f \int_0^1 M\,d\xi,\right.$$

where $H_{si}^{(5)\uparrow}$ is the heat flux through the ocean–sea ice interface.

On substituting from (A.2, A.4, A.6–A.8, A.10, and A.11), we can write this equation in the form

(A.14) $$\frac{d\vartheta}{dt} = Y_1 \eta - Y_2 \eta^2 - \frac{L_f \delta}{2\rho_w c_w D} M(\eta) - \frac{1}{\rho_w c_w D}\int_\eta^1 H_{si}^{(5)\uparrow}\,d\xi$$

where

$$Y_1 = (\rho_w c_w D)^{-1} \Big\{ \mu \mathcal{R}(0) - \frac{b_1}{2}[\Gamma(0) + \Gamma(\eta)] + b_2$$
$$+ 4A^3\sigma \Big[v^{\downarrow} T_a(0) - \Big(\frac{T_s(0) + A_0 + a_2 v^{\downarrow}}{2}\Big)\Big] - 3A^4\sigma(v^{\downarrow} - 1) \Big\}$$

and

$$Y_2 = (2\rho_w c_w D)^{-1}[\mu a_1 \mathcal{R}(0)]$$

In (A.14) we have assumed as a first approximation that we can take the equatorial surface temperature $T_s(0)$ to be a constant (CLIMAP, 1976), though it would be relatively easy to use the heat balance condition to evaluate this quantity in a slightly more elaborate model. Moreover, we shall neglect the last two terms in (A.14), which can be shown to be small compared to the first two. Again, it would pose no difficulty to retain these terms through the use of (A.12) to evaluate $M(\eta)$ and the assignment of some uniform constant value or simple latitude dependence, for $H_{si}^{(5)\uparrow}$.

Thus, our simplified equation for the rate of change of the mean ocean temperature is

(A.15) $$d\vartheta/dt = Y_1 \eta - Y_2 \eta^2$$

A.5. Equilibrium Conditions and Constants

We now examine the properties of the system (A.13) and (A.15),

$$d\eta/dt = X_1 \vartheta - X_2 \eta + X_3$$
$$d\vartheta/dt = Y_1 \eta - Y_2 \eta^2$$

First we set down the condition for equilibrium,

(A.16) $$X_1 \vartheta_0 - X_2 \eta_0 + X_3 = 0$$
(A.17) $$Y_1 \eta_0 - Y_2 \eta_0^2 = 0$$

where η_0 and ϑ_0 denote the steady-state values of the ice limit and mean ocean temperature, respectively. From (A.17) we find the nontrivial, physically significant solution

(A.18) $$\eta_0 = Y_1/Y_2$$

which implies, from (A.16) that

(A.19) $$\vartheta_0 = (X_2 Y_1 - X_3 Y_2)/X_1 Y_2$$

STATISTICAL-DYNAMICAL MODELS OF THE TERRESTRIAL CLIMATE   287

To evaluate these equilibrium values we must specify all the constants of the problems on which X_1, X_2, X_3, Y_1, and Y_2 depend. These include physical constants, prescribed empirical values of climatic variable appropriate for the annual mean terrestrial conditions, and the parameterization constants used by Saltzman and Vernekar (1971a). The values adopted are listed as follows.

Physical constants	
L_f	3.34×10^5 J kg^{-1}
σ	5.67×10^{-8} W m^{-2} K^{-4}
ρ_i	10^3 kg m^{-3}
ρ_w	10^3 kg m^{-3}
c_w	4.186×10^3 J kg^{-1} K^{-1}
A	273 K

Empirical values and parameterization constants			
χ	0.30	c_1	4.0 W m^{-2} K^{-1}
r_a	0.30	c_2	95.0 W m^{-2}
μ	0.50	c_3	1.27
$\mathcal{R}(0)$	480 W m^{-2}	c_4	38.9 W m^{-2}
a_1	0.65	b_1	9.1 W m^{-2} K^{-1}
v^i	1.30	$b_1(\eta)$	7.0 W m^{-2} K^{-1}
D	4×10^3 m	b_2	176.8 W m^{-2}
$T_a(0)$	267 K	$b_2(\eta)$	167.2 W m^{-2}
$T_a(\eta)$	245 K	l	2.5 m
$T_s(0)$	300 K	a_2	22 K
$T_s(\eta) \equiv A_0$	273 K	w	1.0
$\Gamma(0)$	33 K	w_η	0.6
$\Gamma(\eta)$	28 K		

Two constants appearing in the ice extent equation (A.13), about which we have relatively little empirical knowledge, remain to be determined, b_3 and δ.

An estimate of b_3 can be obtained from (A.19) providing we specify ϑ_0. Setting $\vartheta_0 = 278$ K (which is slightly warmer than the present mean ocean temperature of 276.8 K, see Dietrich, 1963, p. 185), we find

$$b_3 = 7.0 \text{ W m}^{-2} \text{ K}^{-1}$$

a value which seems consistent with the values of k (see Table VI) adopted for the oceans in other studies.

The width of the active melting/freezing zone δ can be estimated only roughly from its definition [see Section A.2, (iii)] using (A.12). We shall postulate that this value is in the vicinity of 10^{-3}, corresponding to about

0.1° latitude or about 10 km, but may in fact lie in the range between 10^{-2} and 10^{-4}.

In any event, the value of η_0 is independent of both b_3 and δ, and can be determined from (A.18) using the other constants. The value obtained is

$$\eta_0 = 0.885$$

which corresponds to $\phi = 62.5°$. When the sea ice extends to this latitude, the ocean as a whole is neither gaining nor losing heat, there being an exact balance between a gain in low latitudes and loss in high latitudes up to the ice edge. If the ice extended further than η_0 toward the equator, the balance would shift in favor of a net gain of heat by the ocean, resulting in a positive value of $\partial \vartheta / \partial t$, the reverse being true if the ice edge were further toward the pole.

The values of the coefficients appearing in (A.13) and (A.15) are

$$X_1 = 42\delta \times 10^{-10} \text{ K}^{-1} \text{ sec}^{-1} \qquad X_2 = 627\delta \times 10^{-10} \text{ sec}^{-1}$$
$$X_3 = -111\delta \times 10^{-8} \text{ sec}^{-1} \qquad Y_1 = 414 \times 10^{-11} \text{ K sec}^{-1}$$
$$Y_2 = 468 \times 10^{-11} \text{ K sec}^{-1}$$

in which δ is treated as a parameter that, most likely, lies in the range 10^{-2} to 10^{-4}.

A.6. Linear Analysis

Let us now decompose η and ϑ into their equilibrium values and departures therefrom (denoted by primes), i.e.,

$$\eta = \eta_0 + \eta' \qquad \vartheta = \vartheta_0 + \vartheta'$$

Introducing these expansions into (A.13) and (A.15), and assuming the departures are small so that we can neglect their products, we obtain the linear perturbation equations,

(A.20) $$d\eta'/dt = X_1 \vartheta' - X_2 \eta'$$

(A.21) $$d\vartheta'/dt = -(2Y_2 \eta_0 - Y_1)\eta'$$

These can be combined by eliminating ϑ' to yield the following second-order ordinary differential equation for the transients η':

(A.22) $$\frac{d^2 \eta'}{dt^2} + 2h \frac{d\eta'}{dt} + \lambda^2 \eta' = 0$$

where

$$h = X_2/2 \qquad \lambda^2 = X_1(2Y_2 \eta_0 - Y_1)$$

The properties of this equation are well known (Kaplan, 1958). Briefly, if we assume a solution of the form

(A.23) $$\eta' = Ne^{\omega t}$$

where $\omega = \omega_r + i\omega_i$, and N is the amplitude of the initial perturbation (which can arise from external forcing mechanisms and/or internal instabilities that prevail in the complete climatic system) it follows that,

$$\omega = -h \pm (h^2 - \lambda^2)^{1/2}$$

Since in our case $h > 0$, and $\lambda^2 > h^2$, the solution represents damped oscillatory behavior of period

$$\mathscr{P} = \frac{2\pi}{\omega_i} = \frac{2\pi}{(\lambda^2 - h^2)^{1/2}},$$

and time constant for damping,

$$\tau = -1/\omega_r = 1/h$$

In our case this damping is due to negative feedback between the amount of solar radiation at the ice edge and the latitude of the ice edge. We can also see from (A.21) that the angular phase difference by which η' lags θ' is given by

$$\varepsilon = \arctan(\omega_i/\omega_r)$$

and that, if the initial amplitude (before damping) of the temperature perturbation is Θ then the initial amplitude of the ice edge variation is

$$N = \frac{\Theta(\omega_r^2 + \omega_i^2)^{1/2}}{(2Y_2\eta_0 - Y_1)}$$

For the particular constants of our problem we have,

$$\mathscr{P} = \frac{10^2}{(4.4\delta - 248.0\delta^2)^{1/2}} \text{ (years)}$$

$$\approx \begin{cases} 720 \text{ yr} & (\delta = 10^{-2}) \\ 1550 \text{ yr} & (\delta = 10^{-3}) \\ 4800 \text{ yr} & (\delta = 10^{-4}) \end{cases}$$

and

$$\tau = 1.01\delta^{-1} \text{ (years)}$$

$$= \begin{cases} 100 \text{ yr} & (\delta = 10^{-2}) \\ 1000 \text{ yr} & (\delta = 10^{-3}) \\ 10{,}000 \text{ yr} & (\delta = 10^{-4}) \end{cases}$$

For $\delta = 10^{-3}$, we find that $\varepsilon = 77°$ and that 1 K amplitude of ϑ implies a 0.03 amplitude of n which corresponds to roughly 4 degrees of latitude near η_0.

A.7. Concluding Remarks

The above results for a highly idealized model of sea ice–ocean temperature interaction, are, at the least, suggestive of the possibility that the terrestrial climatic system can exhibit internal "free" oscillations having periods ranging up to several thousand years (i.e., periods approaching those of minor glacial interglacial episodes such as the little ice age and climatic optimum). The oscillations deduced are characterized by an increase of deep ocean temperature during the time of maximum ice extent and a decrease during minimum ice extent. It will be recognized that these, as well as other aspects of the model treated, represent some of the main elements of the ice-age scenario that was described by Newell (1974) for longer term variations and discussed from an energetical viewpoint by Saltzman (1977).

Clearly, there is much room for improvement and extension of this preliminary study (1) by testing the model for its sensitivity to the prescribed constants and parameterizations, (2) examining the nonlinear properties of the model, and (3) by including at least a few more of the many omitted physical processes that are also operating in the real climatic system—e.g., external forcing, other internal feedbacks (such as the destabilizing ice albedo feedback which is only partially taken into account through the prescribed form of T_a), land effects (including continental glaciation), and seasonal rectifier effects. One possibility that we are exploring is the inclusion of the mechanism illustrated here into a time-dependent version of the much more complete Saltzman and Vernekar (1971a) momentum model described in Section 6.2.1.

ACKNOWLEDGMENTS

Support for preparing this review was provided by the National Science Foundation under grant ATM 77-02497 of the Climate Dynamics Program, Climate Dynamics Research Section, Division of Atmospheric Sciences. I am grateful to Miss Betsy Dabakis for her careful typing of the manuscript.

LIST OF SYMBOLS

- a radius of the earth; subscript denoting 500 mb level of the atmosphere
- a_1, a_2 positive constants
- A subscript denoting atmospheric property; 273 K
- A_0 surface temperature at the sea ice edge, $T_s(\eta)$
- $\mathscr{A}$ set of variables describing aerosol properties in atmosphere (e.g., concentration, size distribution, chemical composition)

b_1, b_2 positive constants
B parameterization constant for meridional flux of water vapor
$\mathcal{B}$ set of variables describing biological properties of the surface (e.g., biomass, type, density, and dimensions of vegetation)
$\mathbf{B}$ $(\mathbf{D} + \mathbf{J} \cdot \mathbf{V}) = \mathbf{b} + b^{\uparrow}\mathbf{k}$
c specific heat; subscript denoting property of "carrier" fluid
c_1, c_2 positive constants
C $\int_{z_s}^{\infty} \mathscr{C} \, dz = C_w + C_i$, rate of condensation ($C_w$) and sublimation ($C_i$) of water vapor in an atmospheric column per unit area
C_F a surface friction coefficient
$\mathscr{C}$ $\mathscr{C}_w + \mathscr{C}_i$, rate of condensation of sublimation of water vapor into cloud drops ($\mathscr{C}_w$) and ice cloud particles ($\mathscr{C}_i$) per unit volume; set of variables describing cloud structure and distribution
d subscript denoting property of "dry air"; height of lower limit of climatic variability above reference level (cf. Fig. 1); rate of frictional dissipation of kinetic energy per unit mass
D depth of ocean
$\mathscr{D}_j$ $(\chi_j \mathscr{V}_j)$, flux of constituent j due to molecular diffusion
e internal energy per unit mass
e eccentricity of the earth's orbit; vapor pressure
E $(f_v^{\uparrow} - \mathbf{f}_v \cdot \nabla z)_s = \int_d^z \mathscr{E} \, dz$, rate of evaporation from the ocean and topographically irregular land surface, per unit horizontal area; subscript denoting contribution due to *eddy* motions
$\mathbf{E}$ external force
$\mathscr{E}$ rate of evaporation of liquid water per unit volume
$\mathscr{E}_j$ flux of constituent j due to subsynoptic eddies
f $2\Omega \sin \phi$, Coriolis parameter
$f_j^{\uparrow}$ vertical flux of constituent j due to all subsynoptic motions
$\mathbf{f}_j$ horizontal flux of constituent j due to all subsynoptic motions
F $-M$, rate of freezing of surface water per unit horizontal area
F magnitude of frictional force
$\mathbf{F}$ $\mathbf{F}_{(m)} + \mathbf{F}_{(E)}$, frictional force per unit mass due to molecular and eddy viscosity, respectively $(= F_\lambda \mathbf{i} + F_\phi \mathbf{j})$
$\mathscr{F}_j$ $\mathbf{f}_j + f^{\uparrow}\mathbf{k}$, flux of constituent denoted by j due to all subsynoptic motions (molecular diffusion, eddies, and particulate sedimentation), i.e., $(= \mathscr{D}_j + \mathscr{E}_j + \mathscr{W}_j)$
g acceleration of gravity
G tide-generating external gravitational force per unit mass
h height of the surface above sea level; enthalpy per unit mass
$\mathbf{h}$ horizontal heat flux
$H_s^{\uparrow}$ $\mathscr{H}_s^{\uparrow} - \mathbf{h}_s \cdot \nabla z_s$
$\mathscr{H}^{\uparrow}$ vertical heat flux
$\mathbf{H}$ $\mathbf{h} + \mathscr{H}^{\uparrow}\mathbf{k}$, heat flux
i subscript denoting ice
$\mathbf{i}$ unit vector, eastward
I thickness of ice
j generalized subscript denoting component or constituent of climatic system (e.g., i, v, w, ..., 1, 2, 3, ...)
$\mathbf{j}$ unit vector, northward
$\mathbf{J}_j$ $J_{j\lambda}\mathbf{i} + J_{j\phi}\mathbf{j}$, vertically integrated horizontal flux of constituent or property denoted by j due to all modes of motion (synoptic, eddy, diffusive)

J $\overline{\rho V'V'}$, Jeffreys stress tensor
k $V^2/2$, kinetic energy of synoptic motions per unit mass; molecular and/or eddy conduction coefficient for subsurface medium (cf. Table VI)
k_E $\overline{V'^2}/2$, large-scale transient eddy kinetic energy per unit mass
k_m $\overline{V}^2/2$, climatic mean kinetic energy per unit mass
k_T thermal conductivity
k   unit vector, vertically upward
K   eddy diffusivity for horizontal flux of heat by large-scale eddies
K   kelvin degrees
$\mathscr{K}$ $(k + \kappa)$, total synoptic plus subsynoptic energy per unit mass
l subscript denoting solid-earth property (rock, soil, sand, sediment, biomass)
L   subscript denoting property of the continents (i.e., biolithosphere and continental water in all forms)
L_v latent heat of vaporization at 273 K
L_f latent heat of fusion at 273 K
L_s $L_v + L_f$, latent heat of sublimation at 273 K
$\mathscr{L}$ set of variables describing lithospheric surface
m_j mass of constituent denoted by j per unit horizontal area
M   subscript denoting ocean (marine) property; rate of melting of surface ice per unit horizontal area; total mass
$\mathscr{M}$ rate of melting of ice per unit volume
$\mathbf{M}_j$ $\overline{\chi'_j \mathbf{V}'}$, mass flux of constituent j due to transient synoptic eddies $(= \mathbf{m}_j + m_j^\prime \mathbf{k})$
n   subscript denoting cloud property; fractional cloud coverage
n*   characteristic wave number for planetary flow
N   subscript denoting nonoceanic subsurface property (including continents, ocean basement, and the water in all forms they contain)
p pressure
P   annual period (1 year)
P $P_w + P_i = -(f_n^\uparrow - \mathbf{f}_n \cdot \nabla z)_s$, rate of precipitation reaching surface per unit area, in the form of water (P_w) and ice (P_i); subscript denoting contribution of mean poloidal motions
$\mathscr{P}$ $\mathscr{P}_r + \mathscr{P}_i$, rate of addition of water to a unit volume due to rainfall $(\mathscr{P}_r)$ and snowfall $(\mathscr{P}_i)$; period of oscillation
P   Navier–Stokes tensor
q rate of heat addition per unit mass due to all causes
Q   rate of heat addition per unit mass due only to internal heat sources, such as radioactivity or chemical reactions
r radial distance from center of planet
r_a atmospheric albedo
r_s surface albedo
r_v relative humidity
R   gas constant for air
R_v gas constant for water vapor
R_0 solar radiative flux normal to the earth's surface at the top of the atmosphere
$\mathscr{R}$ $(1 - r_s)R_0$, solar radiation absorbed at the earth's surface if fully transmitted through the atmosphere
R $-\overline{\rho \mathbf{V}^* \mathbf{V}^*}$, Reynolds stress tensor
S   subscript denoting value at earth's surface; subscript denoting salinity; χ_S/χ_M, salinity
S_0 solar constant
$\mathscr{S}_j$ rate of production of constituent j per unit volume (source function)

S	seasonal stress tensor
t	time
T	temperature (K)
T	climatic time averaging period (e.g., 1 month); subscript denoting top of the atmosphere
T_e	equivalent black body (i.e., "effective", or "planetary radiative equilibrium") temperature for the earth
T_{virt}	virtual temperature, $T(1 - .378e/p)^{-1}$
u	$r \cos \phi \, d\lambda/dt$, eastward speed
U_1, U_2	total sensible and latent heat, respectively, in the atmosphere–ocean system (Section 5.2.1)
v	$r \, d\phi/dt$, northward speed; subscript denoting water vapor
V	volume; subscript denoting constant volume
$\mathbf{v}$	$u\mathbf{i} + v\mathbf{j}$, horizontal velocity vector
$\mathbf{V}$	$\mathbf{v} + w\mathbf{k}$, three-dimensional velocity vector
$\mathscr{V}_j$	diffusion velocity of constituent j relative to the carrier current
w	dz/dt, upward speed; subscript denoting liquid water; surface water availability factor
W_j	fall, or sedimentation, speed of particulate j
$\mathscr{W}_j$	$-\chi_j W_j$, flux of particulate j due to sedimentation
x	distance eastward $(dx = a \cos \phi \, d\lambda)$
X_1, X_2	constant coefficients
y	distance northward $(dy = a \, d\phi)$
Y_1, Y_2, Y_3	constant coefficients
z	distance vertically upward; subscript denoting vertical component
z_s	height of surface above reference level
z_{sL}	height of sea level above reference level
z_r	roughness of surface
Z	height of the biolithosphere, excluding ice and water bodies, above the reference level; subscript denoting quantity measured at this level
α	ρ^{-1}, specific volume; planetary albedo
α_p	$\tilde{\alpha}$, mean planetary albedo for earth as a whole
β	$\partial f/a \, \partial \phi$, gradient of the Coriolis parameter, coefficient for Newtonian-type heating due to convergence of the large scale sensible heat flux
Γ	long-wave atmospheric absorptivity
δ	active melting-freezing zone at the sea ice edge in units of $\xi = \sin \phi$
δ_s	active layer depth of the earth's surface
Δ	$(Z - d)$, arbitrary fixed depth below the top of the biolithosphere (Z), that defines the base of the climatic system
ε	$\xi_v = \chi_v/\rho$, water vapor mixing ratio; obliquity of the earth's axis to the orbital plane; phase angle
$\varepsilon^{(\text{sat})}$	saturation mixing ratio
η	sine of the latitude of the ice edge
θ	potential temperature of the atmosphere
ϑ	mean temperature of the entire ocean
Θ	amplitude of ϑ
κ	$k/\rho c$, thermal diffusivity; R/c_P, Poisson ratio; subsynoptic kinetic energy per unit mass
λ	longitude; wavelength; subscript denoting eastward component of vector

μ transmissivity of atmosphere to short wave radiation $[(1-\chi)(1-r_a)]$, according to the parameterization in Table VI
μ_T coefficient of thermal expansion
μ_P coefficient of isothermal compressibility
μ_s coefficient of expansion due to salinity
μ_j coefficient of expansion due to trace substance j
ν emissivity; kinematic viscosity
ξ $\sin \phi$
ξ_j χ_j/ρ, mass fraction (or mixing ratio) of constituent denoted by j
Π longitude of perihelion measured from fixed point, associated with precession
ρ $\sum_j \chi_j$, total density
ρ_j density of homogeneous substance denoted by j
ρ_{i0} density of pure ice at 273 K and 1 atm (0.917×10^3 kg m^{-3})
ρ_{w0} density of pure water at 273 K and 1 atm (1.000 kg m^{-3})
ρ_{M0} density of sea water at 273 K, 1 atm, and 35‰ salinity
σ $4\pi a^2$, surface area of earth; Stefan–Boltzmann constant
σ_j area of the globe covered by constituent j
τ $(T - 273)$, temperature in °C; time constant
Υ $(k + \kappa + \Phi + e)$, total energy per unit mass
ϕ latitude; subscript denoting northward component of vector
Φ gz, geopotential energy per unit mass
χ opacity of the atmosphere to short-wave radiation
χ_j mass concentration ("density") of constituent denoted by j for a fixed volume
ψ arbitrary climatic variable
ω dp/dt, individual pressure change; complex angular frequency, $\omega_r + i\omega_i$
Ω angular speed of the earth's rotation (7.292×10^{-5} sec^{-1})
$\boldsymbol{\Omega}$ angular velocity vector for the earth

Mathematical operators and modes of averaging

$\mathbf{A}$ $A_\lambda \mathbf{i} + A_\phi \mathbf{j} + A_z \mathbf{k}$, arbitrary vector
$\nabla \cdot \mathbf{A}$ $(a \cos \phi)^{-1}(\partial A_\lambda/\partial \lambda + \partial A_\phi \cos \phi/\partial \phi)$
$\nabla \psi$ $\mathbf{i}\, \partial\psi/a \cos \phi\, \partial\lambda + \mathbf{j}\, \partial\psi/a\, \partial\phi$
$\nabla_3 \psi$ $\nabla \psi + \mathbf{k}\, \partial\psi/\partial z$
$\bar{\psi}$ $T^{-1} \int_{-T/2}^{T/2} \psi\, dt$, time average over climatic sampling period T (one month)
$\bar{\bar{\psi}}$ time average over one year
$\bar{\psi}^{(s)}$ time average over synoptic sampling period (15 minutes)
$\langle \psi \rangle$ $(2\pi)^{-1} \int_0^{2\pi} \psi\, d\lambda$, zonal average around a latitude circle
$\{\psi\}$ $(\xi_2 - \xi_1)^{-1} \int_{\xi_1}^{\xi_2} \psi\, d\xi$, area weighted meridional average between latitudes ϕ_1 and ϕ_2
$\tilde{\psi}$ $\{\langle \psi \rangle\} = (4\pi)^{-1} \int_{-\pi/2}^{\pi/2} \int_0^{2\pi} \psi \cos \phi\, d\lambda\, d\phi$, area average over globe
$\underline{\psi}$ mass-weighted vertical average through a domain of the climatic system $[p_s^{-1} \int_0^{p_s} \psi\, dp$ for the atmosphere, $(z_2 - z_1)^{-1} \int_{z_1}^{z_2} \psi\, dz$ for a homogeneous subsurface medium of depth $(z_2 - z_1)]$
$\hat{\psi}$ mass weighted average over the complete climatic system
ψ^* $\psi - \bar{\psi}^{(s)}$, departure from synoptic average
ψ' $\psi - \bar{\psi}$, departure from climatic average (monthly)
ψ'' $\psi - \bar{\bar{\psi}}$, departure from yearly average
$\psi^\star$ $\psi - \langle \psi \rangle$, departure from zonal average
ψ_* $\psi - \underline{\psi}$, departure from vertical average

ψ_{**} $\psi - \{\psi\}$, departure from meridional average
ψ_0 $\langle \bar{\psi} \rangle$, zonal mean, climatic average
ψ_1 $\bar{\psi}^*$, standing eddy departure

References

Adem, J. (1962). On the theory of the general circulation of the atmosphere. *Tellus* **14**, 102-115.
Adem, J. (1963). Preliminary computations on the maintenance and prediction of seasonal temperatures in the troposphere. *Mon. Weather Rev.* **91**, 375-386.
Adem, J. (1964a). On the physical basis for the numerical prediction of monthly and seasonal temperatures in the troposphere-ocean-continent system. *Mon. Weather Rev.* **92**, 91-104.
Adem, J. (1964b). On the normal state of the troposphere-ocean-continent system in the Northern Hemisphere. *Geofis. Int. (Mexico)* **4**, 3-32.
Adem, J. (1965). Experiments aiming at monthly and seasonal numerical weather prediction. *Mon. Weather Rev.* **93**, 495-503.
Adem, J. (1970a). On the prediction of mean monthly ocean temperature. *Tellus* **22**, 410-430.
Adem, J. (1970b). Incorporation of advection of heat by mean winds and by ocean currents in a thermodynamic model for long-range weather prediction. *Mon. Weather Rev.* **98**, 776-786.
Adem, J. (1971). Further studies on the thermodynamic prediction of ocean temperatures. *Geofis. Int. (Mexico)* **11**, 7-45.
Adem, J. (1973). Ocean effects on weather and climate. *Geofis. Int. (Mexico)* **13**, 1-71.
Adem, J. (1975). Numerical-thermodynamical prediction of mean monthly ocean temperatures. *Tellus* **27**, 541-551.
Adhémar, J. F. (1842). "Les Revolutions de la Mer Deluges Periodiques." Paris.
Ångström, A. (1926). Energiezufuhr und Temperatur auf verschiedenen Breitengraden. *Gerlands Beitr. Geophys.* **15**, 1-13.
Ashe, S. (1977). On the time average asymmetric circulation of the earth's atmosphere. Ph.D. Thesis, 311 pp. Yale Univ. New Haven, Connecticut.
Atwater, M. A. (1970). Planetary albedo changes due to aerosols. *Science* **170**, 64-66.
Augustsson, T., and Ramanathan, V. (1977). A radiative-convective model study of the CO_2 climate problem. *J. Atmos. Sci.* **34**, 448-451.
Bagnold, R. A. (1954). "The Physics of Blown Sand and Desert Dunes," 265 pp. Methuen, London.
Barrett, E. W. (1971). Depletion of short-wave irradiance at the ground by particles suspended in the atmosphere. *Sol. Energy* **13**, 323-337.
Bates, J. R. (1977). Dynamics of stationary ultra-long waves in middle latitudes. *Q. J. R. Meteorol. Soc.* **103**, 397-430.
Berger, A. L. (1976a). Obliquity and precession of the last 5,000,000 years. *Astron. Astrophys.* **51**, 127-135.
Berger, A. L. (1976b). Long-term variation of daily and monthly insolation during the last ice age. *Eos* **57**, 254. (Abstr.)
Berger, A. L. (1977a). Power and limitation of an energy-balance climate model as applied to the astronomical theory of palaeoclimates. *Palaeogeogr., Paleoclimatol., Palaeoecol.* **21**, 227-235.
Berger, A. L. (1977b). Long-term variation of the earth's orbital elements. *Celestial Mech.* **15**, 53-74.
Berger, A. L. (1977c). Support for the astronomical theory of climatic change. *Nature (London)* **269**, 44-45.
Berliand, T. G. (1960). Methods of climatological computation of total incoming solar radiation. *Meteorol. Gidrol.* No. 6, 9-12.

Birchfield, G. E. (1977). A study of the stability of a model continental ice sheet subject to periodic variations in heat input. *J. Geophys. Res.* **82**, 4909–4913.
Bowling, S. A. (1971). Comments on "The effect of changes in the earth's obliquity on the distribution of mean annual sea level temperatures." *J. Appl. Meteorol.* **9**, 839.
Bray, J. R. (1976). Volcanic triggering of glaciation. *Nature (London)* **260**, 414–415.
Bryson, R. A. (1968). "All other factors being constant. ..." *Weatherwise* **21**, 56–94.
Bryson, R. A., and Dittberner, G. J. (1976). A non-equilibrium model of hemispheric mean surface temperature. *J. Atmos. Sci.* **33**, 2094–2106.
Bryson, R. A., and Dittberner, G. J. (1977). Reply. *J. Atmos. Sci.* **34**, 1821–1824.
Budyko, M. I. (1969). The effect of solar radiation variations on the climate of the earth. *Tellus* **21**, 611–619.
Budyko, M. I. (1970). Comments on a global climatic model based on the energy balance of the earth-atmosphere system. *J. Appl. Meteorol.* **9**, 310.
Budyko, M. I. (1972). The future climate. *Eos* **53**, 868–874.
Budyko, M. I. (1974). "Climate and Life." Academic Press, New York.
Budyko, M. I. (1977). On present-day climatic changes. *Tellus* **29**, 193–204.
Bunker, A. F. (1976). Computations of surface energy flux and annual air-sea interaction cycles of the North Atlantic Ocean. *Mon. Weather Rev.* **104**, 1122–1140.
Cess, R. D. (1974). Radiative transfer due to atmospheric water vapor: Global considerations of the earth's energy balance. *J. Quant. Spectrosc. Radiat. Transfer* **14**, 861–871.
Cess, R. D. (1975). Global climate change: An investigation of atmospheric feedback mechanisms. *Tellus* **27**, 193–198.
Cess, R. D. (1976). Climate change: An appraisal of atmospheric feedback mechanisms employing zonal climatology. *J. Atmos. Sci.* **33**, 1831–1843.
Cess, R. D. (1977). Reply. *J. Atmos. Sci.* **34**, 1826–1827.
Chang, J., ed. (1977). General circulation models of the atmosphere. *Methods Comput. Phys.* **17**, 337 pp.
Charlson, R. J., and Pilat, M. J. (1969). Climate: The influence of aerosols. *J. Appl. Meteorol.* **8**, 1001–1002.
Charlson, R. J., and Pilat, M. J. (1971). Reply (to S. H. Schneider). *J. Appl. Meteorol.* **10**, 841–842.
Charney, J. G. (1947). The dynamics of long waves in a baroclimic westerly current. *J. Meteorol.* **4**, 135–162.
Charney, J. G. (1959). On the general circulation of the atmosphere. "The Atmosphere and the Sea in Motion," Rossby Memorial Volume, pp. 194–211. Rockefeller Univ. Press, New York.
Chorley, R. J., ed. (1969). "Water, Earth, and Man," 588 pp. Methuen, London.
Chýlek, P., and Coakley, J. A., Jr. (1975). Analytical analysis of a Budyko-type climate model. *J. Atmos. Sci.* **32**, 675–679.
Clapp, P. F. (1970). Parameterization of macroscale transient heat transport for use in a mean-motion model of the general circulation. *J. Appl. Meteorol.* **9**, 554–563.
CLIMAP Project Members (1976). The surface of the ice-age earth. *Science* **191**, 1131–1137.
Coakley, J. A. (1977). Feedbacks in vertical-column energy balance models. *J. Atmos. Sci.* **34**, 465–470.
Coakley, J. A., and Grams, G. W. (1976). Relative influence of visible and infrared optical properties of a stratospheric aerosol layer on the global climate. *J. Appl. Meteorol.* **15**, 679–691.
Croll, J. (1875). "Climate and Time in their Geological Relations. A Theory of Secular Changes of the Earth's Climate," 577 pp. Edw. Stanford, London.
Davies, D. R., and Oakes, M. B. (1962). On the problem of formulating a realistic model of the general atmospheric circulation. *J. Geophys. Res.* **67**, 3121–3128.

Defant, A. (1921). Die Zirkulation der Atmosphäre in den gemässigten Breiten der Erde. *Geogr. Ann.* 3, 209–266.
Defant, F. (1950). Das mittlere meridionale Temperaturprofil in der Troposphäre als Effekt von vertikalen und horizontalen Austauschvorgängen und Kondensationsvärme. *Arch. Meteorol., Geophys. Bioklimatol., Ser. A*, 2, 184–206.
deGroot, S. R., and Mazur, P. (1962). "Non-equilibrium Thermodynamics," 510 pp. North-Holland Publ., Amsterdam.
Derome, J., and Wiin-Nielsen, A. (1972). On the maintenance of the axisymmetric part of the flow in the atmosphere. *Pure Appl. Geophys.* 95, 163–185.
Dickinson, R. E. (1971a). Analytic model for zonal winds in the tropics. 1. Details of the model and simmulation of gross features of the zonal mean troposphere. *Mon. Weather Rev.* 99, 501–510.
Dickinson, R. E. (1971b). Analytic model for zonal winds in the tropics. 2. Variation of the tropospheric mean structure with season and differences between hemispheres. *Mon. Weather Rev.* 99, 511–523.
Dietrich, G. (1953). "General Oceanography," 588 pp. Wiley, New York.
Döös, B. R. (1962). The influence of exchange of sensible heat with the earth's surface on the planetary flow. *Tellus* 14, 133–147.
Dolzhanskiy, F. V. (1969). Calculating the zonal atmospheric circulation. *Izv. Acad. Sci. USSR, Atmos. Oceanic Phys.* 5, 659–671.
Dolzhanskiy, F. V. (1971). Numerical experiments on the simulation of the zonal circulation of the earth's atmosphere. *Izv. Acad. Sci. USSR, Atmos. Oceanic Phys.* 7, 3–11.
Donn, W. L., and Shaw, D. M. (1977). Model of climate evolution based on continental drift and polar wandering. *Geol. Soc. Am. Bull.* 88, 390–396.
Drazin, D. G., and Griffel, D. H. (1977). On the branching structure of diffusive climatological models. *J. Atmos. Sci.* 34, 1696–1706.
Dwyer, H. A., and Petersen, T. (1973). Time-dependent global energy modeling. *J. Appl. Meteorol.* 12, 36–42.
Eady, E. T. (1949). Long waves and cyclone waves. *Tellus* 1, 33–52.
Egger, J. (1975). A statistical-dynamical model of the zonally averaged steady-state of the general circulation of the atmosphere. *Tellus* 27, 325–349.
Egger, J. (1976a). The linear response of a hemispheric two-level primitive equation model to forcing by topography. *Mon. Weather Rev.* 104, 351–364.
Egger, J. (1976b). On the theory of the steady perturbations in the troposphere. *Tellus* 28, 381–389.
Egger, J. (1976c). Nonlinear aspects of the theory of standing planetary waves. *Beitr. Phys. Atmos.* 49, 71–80.
Egger, J. (1977). On the linear theory of the atmospheric response to sea surface temperature anomalies. *J. Atmos. Sci.* 34, 603–614.
Eliassen, A. (1951). Slow thermally or frictionally controlled meridional circulations in a circular vortex. *Astrophys. Norv.* 5, 19–60.
Emden, R. (1913). Über Strahlungsgleichgewicht und Atmosphärische Strahlung: Ein Beitrag zur Theorie der Oberen Inversion. *Sitzungsber. Math.-Naturwiss. Kl. Bayer. Akad. Wiss. Muenchen* No.1, 55–142.
Ewing, M., and Donn, W. L. (1956). A theory of ice ages. *Science* 123, 1061–1066.
Faegre, A. (1972). An intransitive model of the earth-atmosphere-ocean system. *J. Appl. Meteorol.* 11, 4–6.
Flint, R. F. (1971). "Glacial and Quaternary Geology," 892 pp. Wiley, New York.
Frankignoul, C. (1977). On the noise level of climate models. *J. Atmos. Sci.* 34, 1827–1831.
Frankignoul, C., and Hasselmann, K. (1977). Stochastic climate models. Part 2. Application to sea-surface temperature anomalies and thermocline variability. *Tellus* 29, 289–305.

Frederiksen, J. S. (1976). Nonlinear albedo-temperature coupling in climate models. *J. Atmos. Sci.* **33**, 2267-2272.
Fritz, S. (1960). The heating distribution in the atmosphere and climatic change. *In* " Dynamics of Climate" (R. L. Pfeffer, ed.), pp. 96-100. Pergamon, New York.
Gal-Chen, T., and Schneider, S. H. (1976). Energy balance climate modeling: Comparison of radiative and dynamic feedback mechanisms. *Tellus* **28**, 108-120.
GARP Joint Organizing Committee (1975). "The Physical Basis of Climate and Climate Modelling," GARP Publ. Ser. No. 16, 265 pp. World Meteorol. Organ., Geneva.
Gates, W. L. (1976). The numerical simulation of ice-age climate with a global general circulation model. *J. Atmos. Sci.* **33**, 1844-1873.
Ghil, M. (1976). Climate stability for a Sellers-type model. *J. Atmos. Sci.* **33**, 3-20.
Golitsyn, G. S. (1970). A similarity approach to the general circulation of planetary atmospheres. *Icarus* **13**, 1-24.
Gordon, H. B., and Davies, D. R. (1974). The effect of changes in solar radiation on climate. *Q. J. R. Meteorol. Soc.* **100**, 123-126.
Green, J. S. A. (1970). Transfer properties of the large-scale eddies and the general circulation of the atmosphere. *Q. J. R. Meteorol. Soc.* **96**, 157-185.
Green, J. S. A. (1977). The weather during July 1976: Some dynamical considerations of the drought. *Weather* **32**, 120-126.
Harshvardhan, and Cess, R. D. (1976). Stratospheric aerosols: Effect upon atmospheric temperature and global climate. *Tellus.* **28**, 1-10.
Hasselmann, K. (1976). Stochastic climate models, Part 1. Theory. *Tellus* **28**, 473-485.
Hasselmann, K. (1977). Application of two-timing methods in statistical geophysics. *J. Geophys.* **43**, 351-358.
Hays, J. D., Imbrie, J., and Shackleton, N. J. (1976). Variations in the earth's orbit: Pacemaker of the ice ages. *Science* **194**, 1121-1132.
Held, I. M., and Suarez, M. J. (1974). Simple albedo feedback models of the icecaps. *Tellus* **26**, 613-628.
Henderson-Sellers, A., and Henderson-Sellers, B. (1975). Comments on "A diffuse thin cloud atmospheric structure as a feedback mechanism in global climatic modeling." *J. Atmos. Sci.* **32**, 2358-2360.
Hess, S. L., and Frank, R. M. (1953). A theory of the temporal and latitudinal distribution of temperature. *J. Meteorol.* **10**, 135-142.
Hoyt, D. V. (1977). Comments on "Climate change: An appraisal of atmospheric feedback mechanisms employing zonal climatology." *J. Atmos. Sci.* **34**, 1824-1826.
Hunt, B. G. (1973). Zonally symmetric global general circulation models with and without the hydrology cycle. *Tellus* **25**, 337-354.
Ingersoll, A. P. (1969). The runaway greenhouse: A history of water on Venus. *J. Atmos. Sci.* **26**, 1191-1198.
Kaplan, W. (1958). "Ordinary Differential Equations," 534 pp. Addison-Wesley, Reading, Massachusetts.
Kessler, E. (1969). On the distribution and continuity of water substance in atmospheric circulations. *Meteorol. Monogr.* No. 32, 84 pp.
Kraus, E. B., and Lorenz, E. N. (1966). Numerical experiments with large-scale seasonal forcing. *J. Atmos. Sci.* **23**, 3-12.
Kubota, I. (1970). Seasonal variation of energy sources in the earth surface layer and in the atmosphere over the northern hemisphere. *J. Meteorol. Soc. Jpn.* **48**, 30-45.
Kubota, I. (1972). Calculation of seasonal variation in the lower tropospheric temperature with heat budget equations. *J. Meteorol. Soc. Jpn.* **50**, 18-34.
Kukla, G. J. (1975). Missing link between Milankovitch and climate. *Nature (London)* **253**, 600-603.

Kuo, H.-L. (1952). Three-dimensional disturbances in a baroclinic zonal current. *J. Meteorol.* **9**, 260–278.

Kuo, H.-L. (1956). Forced and free meridional circulations in the atmosphere. *J. Meteorol.* **13**, 561–568.

Kurihara, Y. (1970). A statistical-dynamical model of the general circulation of the atmosphere. *J. Atmos. Sci.* **27**, 847–870.

Kurihara, Y. (1971). Seasonal variation of temperature in an atmosphere at rest. *J. Meteorol. Soc. Jpn.* **49**, 537–544.

Kurihara, Y. (1973). Experiments on the seasonal variation of the general circulation in a statistical-dynamical model. *J. Atmos. Sci.* **30**, 25–49.

Lacis, A. A., and Hansen, J. E. (1974). A parameterization for the absorption of solar radiation in the earth's atmosphere. *J. Atmos. Sci.* **31**, 118–133.

Lahiff, L. N. (1975). A low-latitude atmosphere-ocean climate model. *J. Atmos. Sci.* **32**, 657–674.

Lee, P. S., and Snell, F. M. (1977). An annual zonally averaged global climatic model with diffuse cloudiness feedback. *J. Atmos. Sci.* **34**, 847–853.

Lemke, P. (1977). Stochastic climate models. Part 3. Application to zonally averaged energy models. *Tellus* **29**, 385–392.

Leovy, C. (1964). Simple models of thermally driven mesopheric circulation. *J. Atmos. Sci.* **21**, 327–341.

Lian, M. S., and Cess, R. D. (1977). Energy balance climate models: A reappraisal of ice-albedo feedback. *J. Atmos. Sci.* **34**, 1058–1062.

Lindzen, R. S., and Farrell, B. (1977). Some realistic modifications of simple climate models. *J. Atmos. Sci.* **34**, 1487–1501.

Lorenz, E. N. (1960). Maximum simplification of the dynamic equations. *Tellus* **12**, 243–254.

Lorenz, E. N. (1964). The problem of deducing the climate from the governing equations. *Tellus* **16**, 1–11.

Lorenz, E. N. (1967). "The Nature and Theory of the General Circulation of the Atmosphere," No. 218, 161 pp. World Meteorol. Organ., Geneva.

Lorenz, E. N. (1968). Climatic determinism. *Meteorol. Monogr.* **5**, 1–3.

Lorenz, E. N. (1970). Climate change as a mathematical problem. *J. Appl. Meteorol.* **9**, 325–329.

Luther, F. M., Wuebbles, D. J., and Chang, J. S. (1977). Temperature feedback in a stratospheric model. *J. Geophys. Res.* **82**, 4935–4942.

MacCracken, M. C. (1973). Zonal atmospheric model ZAM2. *Proc. Conf. Climatic Assess. Program, 2nd, U.S. Dep. Transp.* **DOT-TSC-OST-73-4**, pp. 298–320.

MacCracken, M. C., and Luther, F. M. (1974). Climate studies using a zonal atmospheric model. *Proc. Int. Conf. Struct., Composition Gen. Circ. Upper Lower Atmos. Possible Anthropogenic Peturbations, Melbourne* IAMAP Publ., pp. 1107–1128.

MacDonald, G. J. F. (1964). Tidal friction. *Rev. Geophys.* **2**, 467–541.

Manabe, S. (1971). Estimates of future change of climate due to the increase of carbon dioxide concentration in the air. *In* "Man's Impact on Climate" (W. H. Matthews, W. W. Kellogg, and G. D. Robinson, eds.), p. 256. MIT Press, Cambridge, Massachusetts.

Manabe, S., and Möller, F. (1961). On the radiative equilibrium and heat balance of the atmosphere. *Mon. Weather Rev.* **89**, 503–532.

Manabe, S., and Strickler, R. F. (1964). Thermal equilibrium of the atmosphere with a convective adjustment. *J. Atmos. Sci.* **21**, 361–385.

Manabe, S., and Terpstra, T. B. (1974). The effects of mountains on the general circulation of the atmosphere as identified by numerical experiments. *J. Atmos. Sci.* **31**, 3–42.

Manabe, S., and Wetherald, R. T. (1967). Thermal equilibrium of the atmosphere with a given distribution of relative humidity. *J. Atmos. Sci.* **24**, 241–259.

Maykut, G. A., and Untersteiner, N. (1969). "Numerical Prediction of the Thermodynamic

Response of Arctic Sea Ice to Environmental Changes," RM-6093-PR, 173 pp. Rand Corp., Santa Monica, California.

Maykut, G. A., and Untersteiner, N. (1971). Some results from a time-dependent thermodynamic model of sea ice. *J. Geophys. Res.* **76**, 1550–1575.

Milankovitch, M. (1920). "Théorie Mathématique des Phénomènes Thermiques Produits par le Radiation Solaire," 339 pp. Gauthier-Villars, Paris.

Milankovitch, M. (1930). Mathematische Klimalehre. *In* "Handbuch der Klimatologie" (Köppen-Geiger, ed.), Vol. 1, Part A, 176 pp. Borntraeger, Berlin.

Milankovitch, M. (1941). "Kanon der Erdbestrahlung und seine Anvendung auf des Eiszeitproblem," 633 pp. R. Serb. Acad., Belgrade. (Engl. transl., U.S. Dep. Commer., Clearinghouse Fed. Sci. Tech. Inf., Springfield, Virginia, 1969.)

Minorsky, N. (1962). "Non-linear Oscillations." Van Nostrand, New York.

Mitchell, J. M. (1971). The effect of atmospheric aerosols on climate with special reference to temperature near the earth's surface. *J. Appl. Meteorol.* **10**, 703–714.

Möller, F. (1963). On the influence of changes in CO_2 concentration in air on the radiative balance of the earth's surface and on the climate. *J. Geophys. Res.* **68**, 3877–3886.

Monin, A. S., and Yaglom, A. M. (1971). "Statistical Fluid Mechanics," 769 pp. MIT Press, Cambridge, Massachusetts.

Namias, J. (1974). Longevity of a coupled air-sea-continent system. *Mon. Weather Rev.* **102**, 638–648.

Newell, R. E. (1974). Changes in the poleward energy flux by the atmosphere and ocean as a possible cause for ice ages. *Quat. Res.* **4**, 117–127.

Newell, R. E., Vincent, D. G., Dopplick, T. G., Ferruza, D., and Kidson, J. W. (1969). The energy balance of the global atmosphere. *In* "The Global Circulation of the Atmosphere" (G. A. Corby, ed.), pp. 42–90. R. Meteorol. Soc., London.

North, G. R. (1975a). Analytical solution to a simple climate model with diffusive heat transport. *J. Atmos. Sci.* **32**, 1301–1307.

North, G. R. (1975b). Theory of energy-balance climate models. *J. Atmos. Sci.* **32**, 2033–2043.

Nye, J. F. (1960). The response of glaciers and ice-sheets to seasonal and climatic changes. *Proc. R. Soc., Ser. A* **256**, 559–584.

Nye, J. F. (1963). Theory of glacier variations. *In* "Ice and Snow" (W. D. Kingery, ed.), pp. 151–161. MIT Press, Cambridge, Massachusetts.

Öpik, E. J. (1965). Climatic change in cosmic perspective. *Icarus* **4**, 289–307.

Ohring, G., and Adler, S. (1978). Some experiments with a zonally averaged climate model. *J. Atmos. Sci.* **35**, 186–205.

Ohring, G., and Gyoeri, S. (1977). Some experiments with a zonally averaged climate model. *In* "Radiation in the Atmosphere" (H.-J. Bolle, ed.), pp. 493–495. Sci. Press, Princeton, New Jersey.

Ohring, G., and Mariano, J. (1964). Changes in the amount of cloudiness and the average surface temperature of the earth. *J. Atmos. Sci.* **21**, 448–450.

Oort, A. H., and Vonder Haar, T. H. (1976). On the observed annual cycle in the ocean-atmosphere heat balance over the northern hemisphere. *J. Phys. Oceanogr.* **6**, 781–800.

Paltridge, G. W. (1974). Global cloud cover and earth surface temperature. *J. Atmos. Sci.* **31**, 156–160.

Paltridge, G. W. (1975). Global dynamics and climate—a system of minimum entropy exchange. *Q. J. R. Meteorol. Soc.* **101**, 475–484.

Peixoto, J. P. (1970). Water vapor balance of the atmosphere from five years of hemispheric data. *Nord. Hydrol.* **2**, 120–138.

Penman, H. L. (1948). Natural evaporation from open water, bare soil and grass. *Proc. R. Soc., Ser. A* **193**, 120–145.

Petukhov, V. K. (1974). The long-period process of heat and moisture exchange in the presence of broken clouds. *Izv. Acad. Sci. USSR, Atmos. Oceanic Phys.* **10**, 219–227.

Petukhov, V. K. (1976). A zonally averaged model of heat and moisture exchange in the air-sea system. *Izv. Acad. Sci. USSR, Atmos. Oceanic Phys.* **12**, 1130–1142.

Petukhov, V. K., and Feygel'son, Y. M. (1973). A model of long-period heat and moisture exchange in the atmosphere over the ocean. *Izv. Acad. Sci. USSR, Atmos. Oceanic Phys.* **9**, 352–362.

Pike, A. C. (1971). Intertropical convergence zone studied with an interacting atmosphere and ocean model. *Mon. Weather Rev.* **99**, 469–477.

Pike, A. C. (1972). Response of a tropical atmosphere and ocean model to seasonally variable forcing. *Mon. Weather Rev.* **100**, 424–433.

Plass, G. N. (1956a). The influence of the 15μ carbon-dioxide band on the atmospheric infra-red cooling rate. *Q. J. R. Meteorol. Soc.* **82**, 310–324.

Plass, G. N. (1956b). The carbon dioxide theory of climatic change. *Tellus* **8**, 140–154.

Pollack, J. B., Toon, O. B., Summers, A., Van Camp, W., and Baldwin, B. (1976). Estimates of the climatic impact of aerosols produced by space shuttles, SST's, and other high flying aircraft. *J. Appl. Meteorol.* **15**, 247–258.

Ramanathan, V. (1975). Greenhouse effect due to chlorofluorocarbons: Climatic implications. *Science* **190**, 50–52.

Ramanathan, V. (1976). Radiative transfer within the earth's troposphere and stratosphere: A simplified radiative-convective model. *J. Atmos. Sci.* **33**, 1330–1346.

Ramanathan, V., Callis, L. B., and Boughner, R. E. (1976). Sensitivity of surface temperature and atmospheric temperature to perturbations in the stratospheric concentration of ozone and nitrogen dioxide. *J. Atmos. Sci.* **33**, 1092–1112.

Rasool, S. I., and Schneider, S. H. (1971). Atmospheric carbon dioxide and aerosols: Effects of large increases on global climate. *Science* **173**, 138–141.

Reck, R. (1974). Influence of surface albedo on the change in the atmospheric radiation balance due to aerosols. *Atmos. Environ.* **8**, 823–833.

Reck, R. (1975a). Aerosols and polar temperature changes. *Science* **188**, 728–730.

Reck, R. (1975b). Aerosols and polar temperature. *Science* **190**, 696.

Reck, R. (1976). Stratospheric ozone effects on temperature. *Science* **192**, 557–559.

Richtmyer, R. D. (1957). "Difference Methods for Initial Value Problems," 238 pp. Interscience Publ., New York.

Saltzman, B. (1961). Perturbation equations for the time-average state of the atmosphere including the effects of transient disturbances. *Geofis. Pura Appl.* **48**, 143–150.

Saltzman, B. (1962). Empirical forcing functions for the large-scale mean disturbances in the atmosphere. *Geofis. Pura Appl.* **52**, 173–183.

Saltzman, B. (1963). A generalized solution for the large-scale, time-average, perturbations in the atmosphere. *J. Atmos. Sci.* **20**, 226–235. (Corrigenda, **20**, 465.)

Saltzman, B. (1964). On the theory of the axially-symmetric, time-average, state of the atmosphere. *Pure Appl. Geophys.* **57**, 153–160.

Saltzman, B. (1967). On the theory of the mean temperature of the earth's surface. *Tellus* **19**, 219–229.

Saltzman, B. (1968a). Surface boundary effects on the general circulation and macroclimate: A review of the theory of the quasi-stationary perturbations in the atmosphere. *Meteorol. Monogr.* No. 30, 4–19.

Saltzman, B. (1968b). Steady state solutions for axially-symmetric climatic variables. *Pure Appl. Geophys.* **69**, 237–259.

Saltzman, B. (1977). Global mass and energy requirements for glacial oscillations and their implications for mean ocean temperature oscillations. *Tellus* **29**, 205–212.

Saltzman, B., and Ashe, S. (1976). Parameterization of the monthly mean vertical heat transfer at the earth's surface. *Tellus* **28**, 323–332.

Saltzman, B., and Pollack, J. A. (1977). Sensitivity of the diurnal temperature range to changes in physical parameters. *J. Appl. Meteorol.* **16**, 614–619.

Saltzman, B., and Vernekar, A. D. (1968). A parameterization of the large-scale eddy flux of relative angular momentum. *Mon. Weather Rev.* **96**, 854–857.

Saltzman, B., and Vernekar, A. D. (1971a). An equilibrium solution for the axially-symmetric component of the earth's macroclimate. *J. Geophys. Res.* **76**, 1498–1524.

Saltzman, B., and Vernekar, A. D. (1971b). Note on the effect of earth orbital radiation variations on climate. *J. Geophys. Res.* **76**, 4195–4197.

Saltzman, B., and Vernekar, A. D. (1972). Global equilibrium solutions for the zonally-averaged macroclimate. *J. Geophys. Res.* **77**, 3936–3945.

Saltzman, B., and Vernekar, A. D. (1975). A solution for the northern hemisphere climatic zonation during a glacial maximum. *Quat. Res.* **5**, 307–320.

Sankar-Rao, M., and Saltzman, B. (1969). On a steady state theory of global monsoons. *Tellus* **21**, 308–330.

Sasamori, T. (1968). The radiative cooling calculation for application to general circulation experiments. *J. Appl. Meteorol.* **7**, 721–729.

Sasamori, T. (1970). Simplification of radiative cooling calculation for application to atmospheric dynamics. *WMO Tech. Note* No. 104, 479–488.

Sasamori, T. (1975). A statistical model for stationary atmospheric cloudiness, liquid water content, and rate of precipitation. *Mon. Weather Rev.* **103**, 1037–1049.

Schneider, E. D., and Lindzen, R. S. (1977). Axially symmetric steady-state models of the basic state for instability and climate studies. Part I. Linearized calculations. *J. Atmos. Sci.* **34**, 263–279.

Schneider, S. H. (1971). A comment on "Climate: The influence of aerosols." *J. Appl. Meteorol.* **10**, 840–841.

Schneider, S. H. (1972). Cloudiness as a global feedback mechanism: The effects on the radiation balance and surface temperature of variations in cloudiness. *J. Atmos. Sci.* **29**, 1413–1422.

Schneider, S. H. (1975). On the carbon dioxide-climate confusion. *J. Atmos. Sci.* **32**, 2060–2066.

Schneider, S. H., and Dickinson, R. E. (1974). Climate modeling. *Rev. Geophys. Space Phys.* **12**, 447–493.

Schneider, S. H., and Dickinson, R. E. (1976). Parameterization of fractional cloud amounts in climatic models: The importance of modeling multiple reflections. *J. Appl. Meteorol.* **15**, 1050–1056.

Schneider, S. H., and Gal-Chen, T. (1973). Numerical experiments in climate stability. *J. Geophys. Res.* **78**, 6182–6194.

Schneider, S. H., and Mass, C. (1975). Volcanic dust, sunspots, and temperature trends. *Science* **190**, 741–746.

Schneider, S. H., and Mesirow, L. E. (1976). "The Genesis Strategy: Climate and Global Survival," 419 pp. Plenum, New York.

Sela, J., and Wiin-Nielsen, A. (1971). Simulation of the atmospheric annual energy cycle. *Mon. Weather Rev.* **99**, 460–468.

Sellers, A., and Meadows, A. J. (1975). Long term variations in the albedo and surface temperature of the earth. *Nature (London)* **254**, 44.

Sellers, W. D. (1969). A global climatic model based on the energy balance of the earth-atmosphere system. *J. Appl. Meteorol.* **8**, 392–400.

Sellers, W. D. (1970a). The effect of changes in the earth's obliquity on the distribution of mean annual sea-level temperatures. *J. Appl. Meteorol.* **9**, 960–961.

Sellers, W. D. (1970b). Reply. *J. Appl. Meteorol.* **9**, 311.

Sellers, W. D. (1973). A new global climatic model. *J. Appl. Meteorol.* **12**, 241–254.
Sellers, W. D. (1974). A reassessment of the effect of CO_2 variations on a simple global climatic model. *J. Appl. Meteorol.* **13**, 831–833.
Sellers, W. D. (1976). A two-dimensional global climatic model. *Mon. Weather Rev.* **104**, 233–248.
Sergin, V. Y., and Sergin, S. Y. (1976). Paper 1 *In* "The Simulation of the 'Glaciers-Ocean-Atmosphere' Planetary System" (S. Y. Sergin, ed.), pp. 5–51. Far East Sci. Cent., USSR Acad. Sci., Vladivostok. (In Russ.)
Shaw, D. M., and Donn, W. L. (1968). Milankovitch radiation variations: A quantitative evaluation. *Science* **162**, 1270–1272.
Shaw, D. M., and Donn, W. L. (1971). A thermodynamic study of Arctic paleoclimatology. *Quat. Res.* **1**, 175–187.
Shumskiy, P. A., Krenke, A. N., and Zotikov, I. A. (1964). Ice and its changes. *In* "Solid Earth and Interface Phenomena" (H. Odishaw, ed.), Research in Geophysics, Vol. 2, pp. 425–460. MIT Press, Cambridge, Massachusetts.
Smagorinsky, J. (1963). General circulation experiments with the primitive equations. *Mon. Weather Rev.* **91**, 99–164.
Smagorinsky, J. (1964). Some aspects of the general circulation. *Q. J. R. Meteorol. Soc.* **90**, 1–14.
Smagorinsky, J. (1974). Global atmospheric modeling and the numerical simulation of climate. *In* "Weather and Climate Modification" (W. N. Hess, ed.), pp. 633–686. Wiley, New York.
Starr, V. P. (1968). "Physics of Negative Viscosity Phenomena," 256 pp. McGraw-Hill, New York.
Stone, P. H. (1972). A simplified radiative-dynamical model for the static stability of rotating atmospheres. *J. Atmos. Sci.* **29**, 405–418.
Stone, P. H. (1973). The effect of large-scale eddies on climatic change. *J. Atmos. Sci.* **30**, 521–529.
Stone, P. H. (1974). The meridional variation of the eddy heat fluxes by baroclinic waves and their parameterization. *J. Atmos. Sci.* **31**, 444–456.
Su, C. H., and Hsieh, D. Y. (1976). Stability of the Budyko climate model. *J. Atmos. Sci.* **33**, 2273–2275.
Suarez, M. J., and Held, I. M. (1976). Modeling climatic response to orbital parameter variations. *Nature (London)* **263**, 46–47.
Talbot, R. J., Butler, D. M., and Newman, M. J. (1976). Climatic effects during passage of the solar system through interstellar clouds. *Nature (London)* **262**, 561–563.
Taylor, K. (1976). The influence of subsurface energy storage on seasonal temperature variations. *J. Appl. Meteorol.* **15**, 1129–1138.
Taylor, K. (1978). The role of mean meridional motions in climate modeling. Ph.D. Thesis, Yale Univ., New Haven, Connecticut.
Temkin, R. L., and Snell, F. M. (1976). An annual zonally averaged hemispherical climatic model with diffuse cloudiness feedback. *J. Atmos. Sci.* **33**, 1671–1685.
Temkin, R. L., Weare, B. C., and Snell, F. M. (1975). Feedback coupling of absorbed solar radiation by three model atmospheres with clouds. *J. Atmos. Sci.* **32**, 873–880.
Van Mieghem, J. (1973). "Atmospheric Energetics," 306 pp. Oxford Univ. Press, London and New York.
Vernekar, A. D. (1967). On mean meridional circulations in the atmosphere. *Mon. Weather Rev.* **95**, 705–721.
Vernekar, A. D. (1972). Long-period global variations of incoming solar radiation. *Meteorol. Monogr.* **12**, No. 34, 21 pp.
Vernekar, A. D. (1975). A calculation of normal temperature at the earth's surface. *J. Atmos. Sci.* **32**, 2067–2081.

Vernekar, A. D., and Chang, H. D. (1978). A statistical-dynamical model for stationary perturbations in the atmosphere. *J. Atmos. Sci.* **35**, 433–444.
Wang, W., and Domoto, G. A. (1974). The radiative effect of aerosols in the earth's atmosphere. *J. Appl. Meteorol.* **13**, 521–534.
Weare, B. C., and Snell, F. M. (1974). A diffuse thin cloud atmospheric structure as a feedback mechanism in global climatic modeling. *J. Atmos. Sci.* **31**, 1725–1734.
Weare, B. C., and Snell, F. M. (1975). Reply. *J. Atmos. Sci.* **32**, 2361–2363.
Webster, P. J. (1972). Response of the tropical atmosphere to local, steady forcing. *Mon. Weather Rev.* **100**, 518–541.
Webster, P. J. (1973a). Remote forcing of the time-independent tropical atmosphere. *Mon. Weather Rev.* **101**, 58–68.
Webster, P. J. (1973b). Temporal variation of low-latitude zonal circulations. *Mon. Weather Rev.* **101**, 803–816.
Webster, P. J., and Lau, K. M. W. (1977). A simple ocean-atmosphere climate model: Basic model and a simple experiment. *J. Atmos. Sci.* **34**, 1063–1084.
Weertman, J. (1964). Rate of growth or shrinkage of nonequilibrium ice sheets. *J. Glaciol.* **6**, 145–158.
Weertman, J. (1976). Milankovitch solar radiation variations and ice age sheet sizes. *Nature (London)* **261**, 17–20.
Wells, J. W. (1963). Coral growth and geochronometry. *Nature (London)* **197**, 948–950.
Wigley, T. M. L. (1976). Spectral analysis and the astronomical theory of climatic change. *Nature (London)* **264**, 629–631.
Wiin-Nielsen, A. (1970). A theoretical study of the annual variation of atmospheric energy. *Tellus* **22**, 1–16.
Wiin-Nielsen, A. (1972). Simulations of the annual variation of the zonally averaged state of the atmosphere. *Geofys. Publ.* **28**, 1–45.
Wiin-Nielsen, A. C., and Fuenzalida, H. (1975). On the simulation of the axisymmetric circulation of the atmosphere. *Tellus* **27**, 199–214.
Wiin-Nielsen, A., and Sela, J. (1971). On the transport of quasi-geostrophic potential vorticity. *Mon. Weather Rev.* **99**, 447–459.
Williams, G. P., and Davies, D. R. (1965). A mean motion model of the general circulation. *Q. J. R. Meteorol. Soc.* **91**, 471–489.
Williams, J., Barry, R. G., and Washington, W. M. (1974). Simulation of the atmospheric circulation using the NCAR global circulation model with ice age boundary conditions. *J. Appl. Meteorol.* **13**, 305–317.
Wilson, A. T. (1964). Origin of ice ages: An ice shelf theory for Pleistocene glaciation. *Nature (London)* **201**, 147–149.
Woronko, S. F. (1977). Comments on "A non-equilibrium model of hemispheric mean surface temperature." *J. Atmos. Sci.* **34**, 1820–1821.
Yamamoto, G., and Tanaka, M. (1972). Increase of global albedo due to air pollution. *J. Atmos. Sci.* **29**, 1405–1412.

SUBJECT INDEX

A

Atmospheres, of planets, see Planetary atmospheres
Atmospheric motion systems, 87–130
  basic scale equations in, 117–128
    for high latitude belt, 126
    for low latitude belt, 126–128
    for middle latitude belt, 118–119
    for moderately large-scale motion, 119–127
    for very large-scale systems, 122–126
  characterization of, 97–106
  horizontal wind in, 109–113
  scale analysis of, 87–130
    of continuity equation, 100–102
    of equation of adiabatic transformations, 106–109
    of equations of horizontal motion, 103–106
    of kinetic energy equation of balance, 128–130
    of vorticity equation, 113–117
  scale parameters of, 91–97
  numerical values, 100
Atmospheric radiation, upwelling type, 69–79
Azimuthal disturbances, on planets, 163–170

B

Band absorption, 23–59
  band emissivity, 57–59
  comparison of models for, 50
  correlations of, 51–54
  large path length limit of, 28
  large pressure limit of, 27–28
  limiting forms of, 24–28
  linear approximation of, 25–26
  narrow band models of, 28–42
  quasi-random band model in, 41–42
  square-root limit of, 27
  statistical model of, 33–41
    exponential wide band absorptance from, 48–51
  strong line approximation, 26–27
  weak line approximation, 26
  wide band models of, 43–51
    box type, 43
    exponential type, 43–51
    result comparison, 54–57
Box model, of wide band absorption, 43–51
Budyko–Sellers ice-albedo feedback models, for terrestrial climate, 238–248

C

Climate, terrestrial statistical dynamic models of, 183–304
Coffin model, of wide band absorption, 43–51

E

Earth, atmospheric convection on, 161–163
Elsasser band model
  of band absorption, 28–33
  exponential wide band absorptance from, 44–51

H

Hadley circulation, model of, 133

I

Infrared atmospheric radiation, 1–85
  band absorption, 23–59
  integrated absorptance of, 59–69
  models for, 1–85
  spectral line absorption, 3–23
  transmittance of, 59–69
    computational procedures, 60–62
  upwelling atmospheric radiation, 69–79
    computation, 71–72
    model calculations, 72–77

J

Jupiter
  circulation on, 134
  of atmosphere, 157–161
  bands, 175

K

Kinetic energy equation, scale analysis of, 128–130

L

Lorentz–Doppler line profile, transmittance of, 9
Lorentz line profile, transmittance of, 9

M

Mayer–Goody model, of band absorption, 33
Momentum models for terrestrial climate, 250–266

P

Planetary atmospheres, 131–181
  azimuthal disturbances in, 163–170
    finite, 170
    rotating annulus experiment, 164–170
  convection on heated sphere, 134–163
    on earth, 161–163
    Jupiter atmospheric circulation, 157–161
    nonuniform heating in, 143–149
    nonuniform heating with rotation, 151–156
    symmetric baroclinic instability, 156–157
    uniform heating with rotation, 149–151
  nonlinearity in, 170–175
    numerical studies, 172–175
  symmetric circulations of, 131–181

S

Saltzman–Vernekar model, for terrestrial climate, 253–263
Sea ice–ocean temperature oscillator model, of terrestrial climate, 281–290

Spectral lines
  absorption by, 3–23
    in finite spectral interval, 17–20
    in infinite spectral interval, 11–17
    overlapping lines, 21–23
  radiative transmittance by, 8–10

T

Terrestrial climate, 183–304
  asymmetric models, 264–265
  biolithosphere role in, 231–232
  climatic averaging, 193–205
    climatological-mean equations, 194–202
    ensemble monthly average, 193–194
    spatial, 202–205
  climatic zonation in, 248–249
  energy balance and, 229–230
  fundamental equations governing, 185–193
    constitutive equations, 193
    continuity equations, 186–188
    equations of motion, 188–191
    thermodynamical energy equation, 191–193
  hydrologic cycle in, 249–250
  ice-albedo feedback models in, 238–247
  modeling the evolution of, 267–280
    earth orbital variations, 274–276
    equilibrium solutions, 267–271
    free changes, 277–279
    forced changes, 271–272
    nonastronomical forcing, 276–277
    seasonal cycles, 272–274
  momentum models of, 221–223, 250–266
    Saltzman–Vernekar model, 253–263
    symmetric models, 252–264
  nonzonal circulations in, 249
  ocean role in, 230–231
  prediction of, 279–280
  sea ice–ocean temperature oscillator model of, 281–290
    equilibrium conditions and constants, 286–288
    ice limit equation, 282–283
    ocean temperature equation, 285–286
    surface heat balance, 283–285
  statistical–dynamical models for, 183–304
    index system, 211–215
    overview, 206–225
    survey, 216–223

subsurface factors in, 232–233
thermodynamic models for, 216–220
    vertically integrated, 225–280
time average state studies on, 265–266
water mass balance and, 226–229

V

Voight profile, derivation of, 6–7
Vorticity equation, scale analysis of, 113–117